Laboratory Manual for General, Organic, and Biological Chemistry

Cindy Applegate & Mary Bethé Neely
University of Colorado, Colorado Springs

Michael Sakuta
Perimeter College at Georgia State University

LABORATORY MANUAL FOR GENERAL, ORGANIC, AND BIOLOGICAL CHEMISTRY.

Published by McGraw-Hill Education, 2 Penn Plaza, New York, NY 10121.

This book is printed on acid-free paper.

5 6 7 8 9 0 CCD 27 26 25 24 23

ISBN 978-0-07-351125-2
MHID 0-07-351125-0

mheducation.com/highered

TABLE OF CONTENTS

This lab manual contains experiments that illustrate the concepts covered in chemistry courses designed for health science majors. Experiments are designed to be completed in a 2:40 hour lab period. Two experiments are virtual labs available by accessing a website.

Some experiments may be combined into one lab period if substantial work is done ahead of time. Suggested combinations of labs are:

- "Density and Specific Gravity" and "Atomic Structure and Periodic Properties"
- "Solutions" and "Titration of the Acid Content in Vinegar"
- "Carbohydrate Structures and Models" and "Carbohydrate Tests"

Each experiment is presented in the following format:

Goals: Specific goals covered in the experiment are listed. Students should be prepared to work alone or with others to accomplish the goals in the time allotted.

Materials: All materials and equipment for the experiment are listed. Students should be familiar with the techniques required for the experiment or ask the lab instructor for clarification before beginning the experiment.

Discussion: The topics covered in the experiment are explained and examples may be given. Additional information found in the textbook and in lecture may be helpful for further clarification.

Experimental Procedures: Directions are given for each step of the experiment. Read through the entire procedure before beginning the experiment.

Pre-lab Questions: This sheet must be completed **before** entering the lab and submitted to the instructor **before** beginning the experiment. Information generally is found in the "Discussion" or the "Experimental Procedures" sections of the lab. Additional information may be found in the textbook or in lecture.

Report Sheet: This section provides space for recording data from the experiment and contains questions and problems applicable to the concepts of the experiment.

SAFETY AND SPECIAL DIRECTIONS: Bold and boxed areas contain special information. *ITALICIZED* words usually indicate a definition.

Safety in the laboratory is of particular importance. Safety procedures are covered at the beginning of the semester and a passing grade on a safety quiz is required. SAFETY GOGGLES ARE REQUIRED AT ALL TIMES IN THE LABORATORY.

The authors wish to acknowledge Dr. Dave Anderson for providing the virtual lab experiments.

Mary Bethé Neely, Cindy Applegate — University of Colorado at Colorado Springs

Dr. Michael Sakuta — Perimeter College at Georgia State University

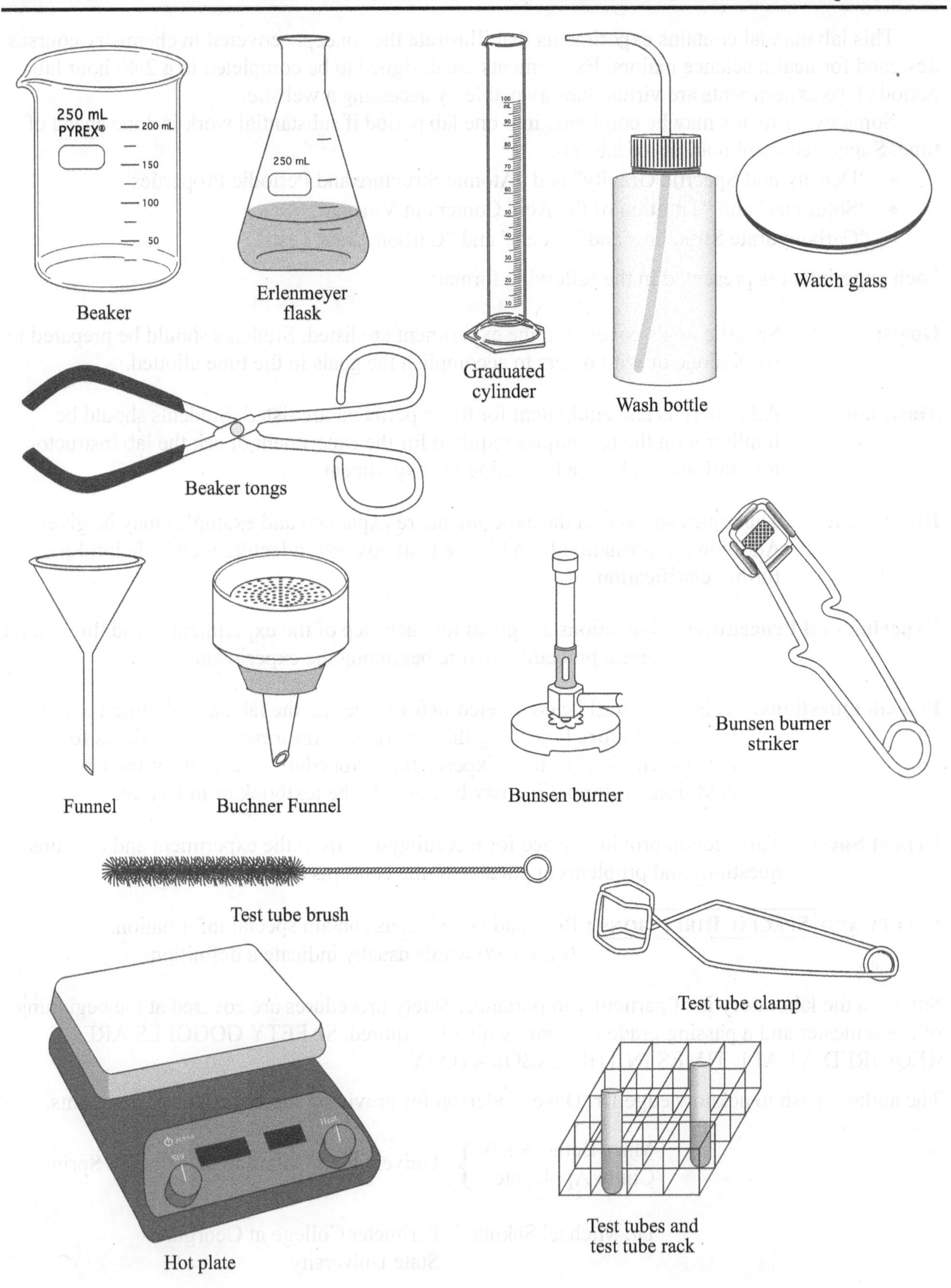
250 mL
PYREX®
200 mL
150
100
50
Beaker
250 mL
Erlenmeyer flask
Graduated cylinder
Wash bottle
Watch glass
Beaker tongs
Funnel
Buchner Funnel
Bunsen burner
Bunsen burner striker
Test tube brush
Test tube clamp
Hot plate
Test tubes and test tube rack

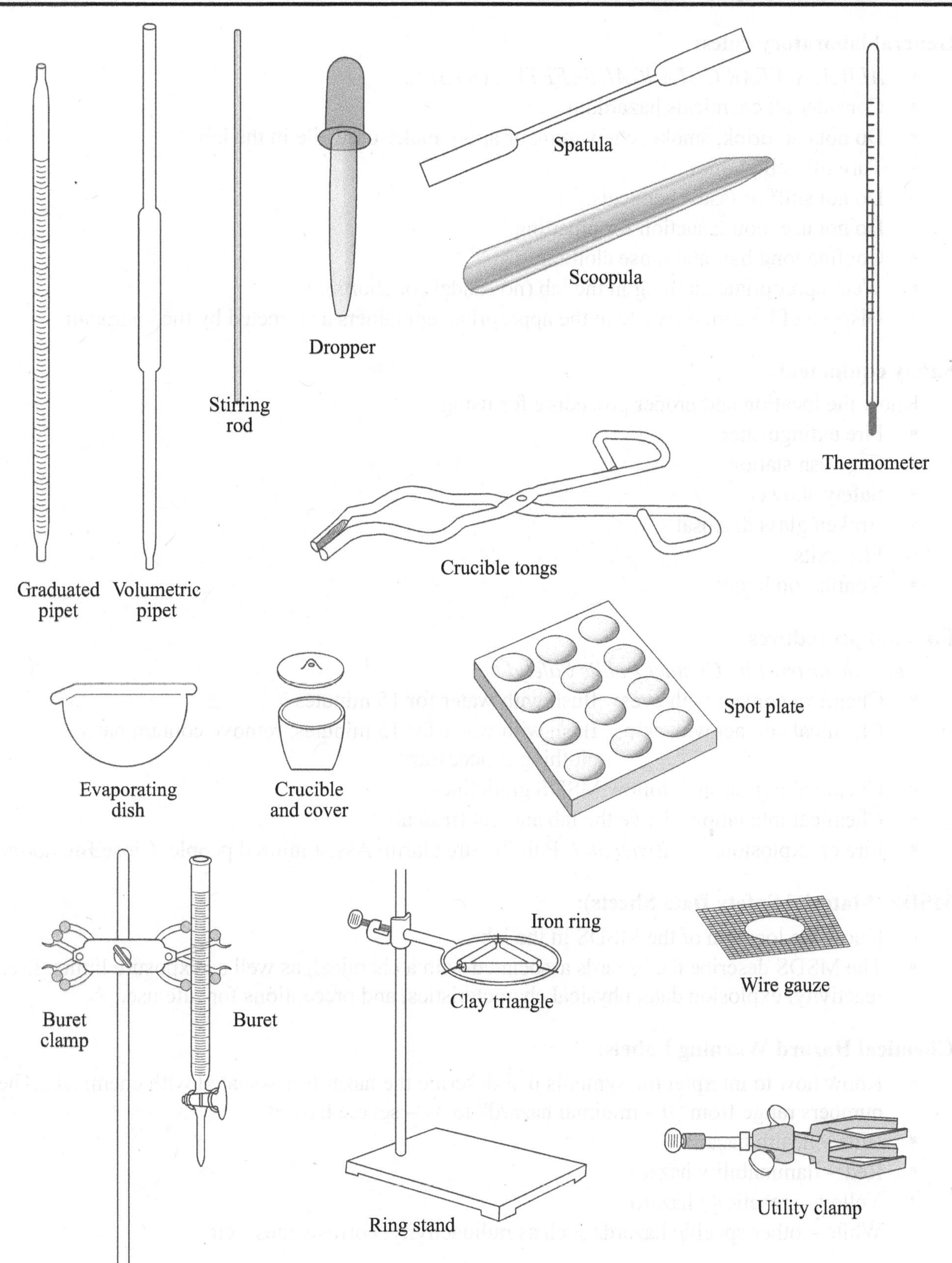
Graduated pipet
Volumetric pipet
Stirring rod
Dropper
Spatula
Scoopula
Thermometer
Crucible tongs
Spot plate
Evaporating dish
Crucible and cover
Iron ring
Wire gauze
Clay triangle
Buret clamp
Buret
Ring stand
Utility clamp

General laboratory rules:

- *ALWAYS WEAR CHEMICAL SAFETY GOGGLES.*
- Consider all chemicals hazardous.
- Do not eat, drink, smoke, chew gum, or apply make-up while in the lab.
- Turn off cell phones.
- Do not sniff or taste chemicals.
- Do not use mouth suction for pipetting.
- Confine long hair and loose clothing.
- Wear appropriate clothing in the lab (no sandals or shorts).
- Dispose of hazardous waste in the appropriate containers as directed by the instructor.

Safety equipment:

Know the location and proper procedure for using

- Fire extinguisher
- Eyewash station
- Safety shower
- Broken glass disposal
- Fire exits
- Ventilation hoods

First aid procedures:

Alert the instructor! Contact Public Safety!

- Chemical contact with eyes: flush with water for 15 minutes
- Chemical contact with skin: flush with water for 15 minutes; remove contaminated clothing if necessary
- Chemical ingestion: follow MSDS guidelines
- Chemical inhalation: leave the lab and get fresh air
- Fire or explosion: *Evacuate!* Pull the fire alarm. Assist injured people. Close fire doors.

MSDS (Material Safety Data Sheets):

- Know the location of the MSDS in the lab.
- The MSDS describe the hazards associated with a chemical, as well as exposure limits, fire, reactivity, explosion data, physical characteristics, and precautions for safe use.

Chemical Hazard Warning Labels:

- Know how to interpret the symbols that describe the hazards associated with chemicals. The numbers range from "0 – minimal hazard" to "4 – severe hazard."
- Blue – health hazard
- Red – flammability hazard
- Yellow – reactivity hazard
- White – other specific hazards such as radioactivity, corrosiveness, etc.

Materials

- Bunsen burner
- Striker or matches

In the chemistry laboratory, substances are often heated with a hotplate or a Bunsen burner. The Bunsen burner produces a single open gas flame, much like the burners of a gas stove. Natural gas and air mix in the Bunsen burner tube, and this mixture is ignited at the top of the tube. Heat (and light) is produced in the ensuing combustion reaction. The gas flow can be controlled by adjusting the gas lever at the bench, or by adjusting the needle valve at the base of the Bunsen burner. The gas lever is closed (off) when the lever is perpendicular to the body of the valve. When the lever is turned 90°, the gas valve is open.

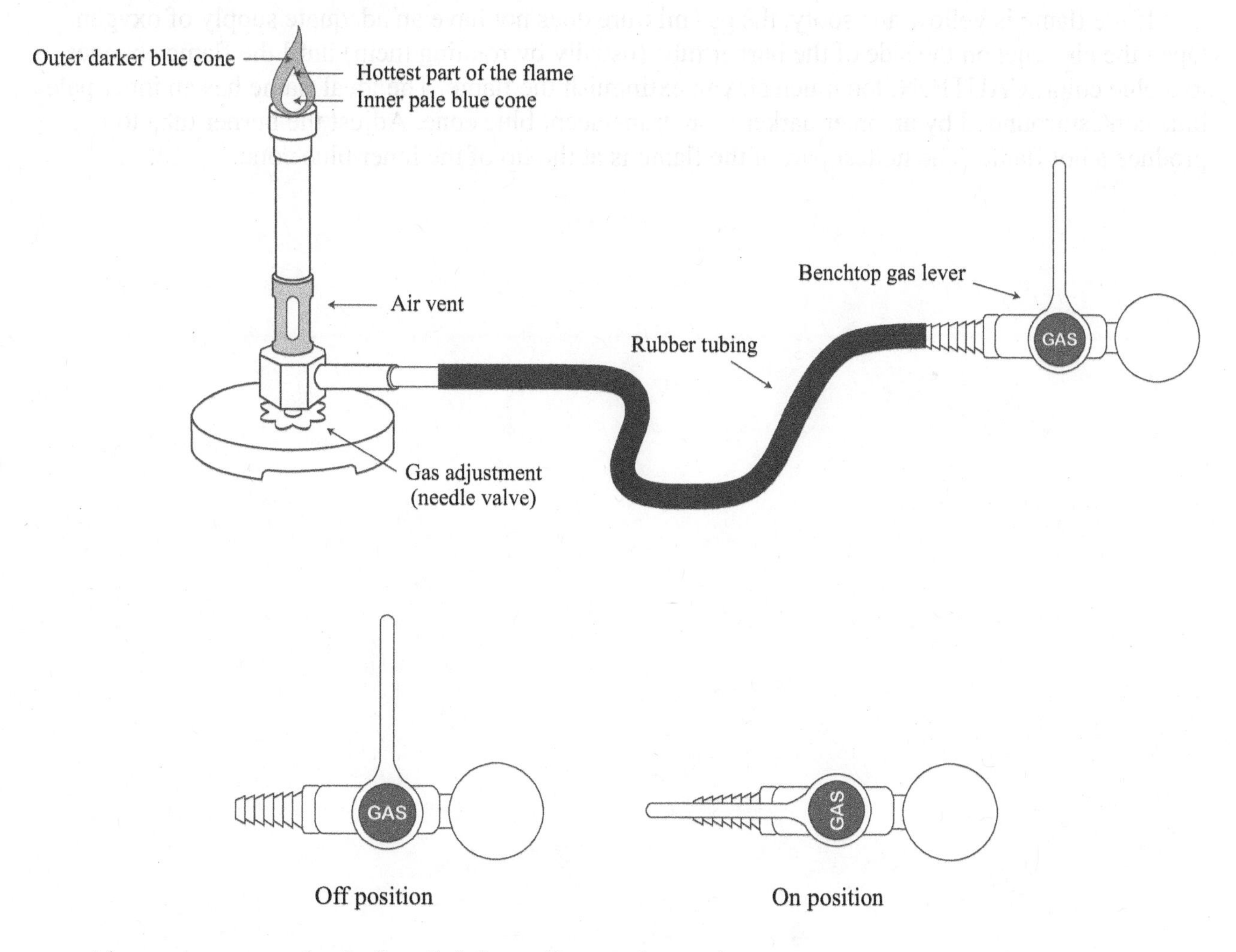

Always wear goggles before lighting a Bunsen burner!

Before lighting the Bunsen burner, inspect the rubber tubing for cracks or tears. Get a new piece of tubing if it is defective. If using a striker to light the burner, check that it produces sparks readily. If it doesn't, check that a flint is installed and that it is not worn out. If the striker doesn't work properly, inform the instructor and request a new one.

The amount of gas entering the Bunsen burner tube is regulated by turning the needle valve at the base of the burner. If you were to look at the Bunsen burner from the bottom, turn clockwise to close and counterclockwise to open. However, since you will be looking at the burner from above, the directions are reversed. Turn counterclockwise to close and turn clockwise to open.

To light the burner, turn on the gas at the source to the full on position. Open the needle valve approximately one-quarter to one-half turn. Bring the striker to the side and just above the top of the barrel. Hold it at a 45° angle and strike. Repeat striking, if necessary. If the burner doesn't light after several strikes, turn the gas off at the source and consult the instructor.

If using a match to light the Bunsen burner, strike the match first. Then turn the gas on as above. Bring the match to the side of the Bunsen burner so it does not get blown out by the gas. Once lit, the flame may be a little small. Open the needle valve on the base of the Bunsen burner to let more gas in for combustion. CAUTION: too much gas flow can extinguish the flame.

If the flame is yellow and sooty, the gas mixture does not have an adequate supply of oxygen. Open the air vents on the side of the burner tube (usually by rotating them) until the flame changes to a blue color. CAUTION: too much air can extinguish the flame. The ideal flame has an inner pale blue cone surrounded by an outer darker more translucent blue cone. Adjust the burner tube to produce a hot flame. The hottest part of the flame is at the tip of the inner blue cone.

After organizing data in tables, the data may be displayed in a graph. A graph is a pictorial representation of the data that shows the relationship of one variable to another. A graph makes interpretation and analysis of data easier. With a graph, it is possible to:

- show how a variable changes over time
- determine one variable given the other
- show trends in the data
- make predictions about data that is not measured

Data table

Assume that you measured the pressure of a fixed amount of gas at various temperatures. Note that each column in the table is clearly labeled with the quantity measured and its unit (in parentheses).

Temperature (Kelvin)	**Pressure (atm)**
125	0.46
158	0.66
200	0.73
235	0.86
315	0.95
363	0.35
380	1.39

Laying out the graph

First, draw a vertical and horizontal axis on the paper. Leave room in the margin for numbers and labels.

Next, determine which quantity depends on the other. One of the variables in the above table is the *independent variable* and the other is the *dependent variable*. In this example, the temperature of the gas was changed and then the pressure was measured. Thus, we can say that the pressure depends on the temperature. Pressure is the dependent variable while temperature is the independent variable.

Labeling the axes

The dependent variable goes on the *y* axis (vertical) while the independent variable goes on the *x* axis (horizontal). Label each axis with the quantity being measured and the units of measurement (in parentheses). Select a scale for each axis that allows the data to fill out most of the graph area. The intervals on the axes must be **equally spaced**. The intervals on one axis need not be the same as the interval on the other axis.

Add a title to the graph. The title should explain what is shown in the graph. At the very least, it should include the dependent and independent variables.

Plot the data points

Each data point should be plotted in its proper position. If a data point does not fall on a grid line, you should estimate how far it is between two adjacent grid lines and place a dot there. The dot can be black circle, an open circle, a square, etc., just be consistent. DO NOT, under any circumstances, connect successive data points with a series of straight lines ("connect the dots").

A smooth line or curve is then drawn that best fits the data points. However, sometimes it is impossible to get all the points to fit on the line or curve you drew. Some data points are a long way from the general trend line or curve. These anomalies should be ignored. Draw a line that comes as close as possible to all the remaining data points. If some data points are not on the line or curve, try to have as many above the line as below it. Try to minimize the distance from the line to each of the data points. There are computer programs that can determine the best fit line and its resulting equation. The above guidelines are for eyeballing the data and drawing the line by hand.

To calculate the slope of the best fit line, two points are needed. Mark two points on the line that are **not** points that were plotted using the experimental data. These points will be called (x_1,y_1) and (x_2,y_2). Mathematically, the slope (usually given the symbol m) is given by the following equation:

$$m = \frac{y_2 - y_1}{x_2 - x_1}$$

Summary—Characteristics of a good graph

1. Both axes are labeled with the quantity being measured and its units. They are not labeled x and y.
2. The independent variable is plotted on the x axis (horizontal).
3. The dependent variable is plotted on the y axis (vertical).
4. A best fit line is drawn showing the trend of the data. The anomalous data point has been ignored.
5. The graph is clearly titled.
6. The axes are properly scaled so that the graph fits the space, the grids are consistently scaled, and all of the data fits on the graph.

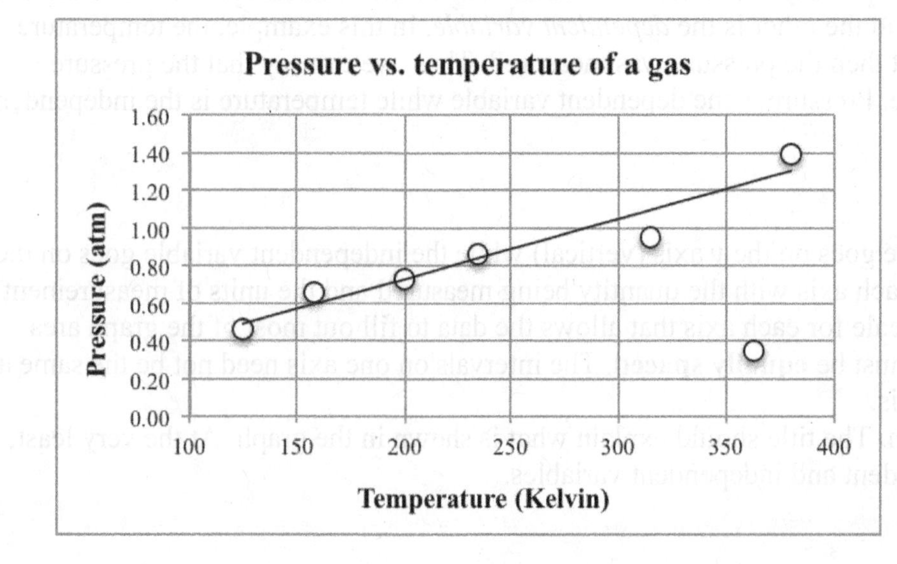

Example of an unsatisfactory graph

1. The axes are labeled, but without units.
2. Only a small portion of graph area is used. This bunches everything in one corner.
3. The data points were connected in a "connect the dots" manner.
4. There is no title.
5. There is no best fit line. It is difficult to tell that this data has a linear relationship.

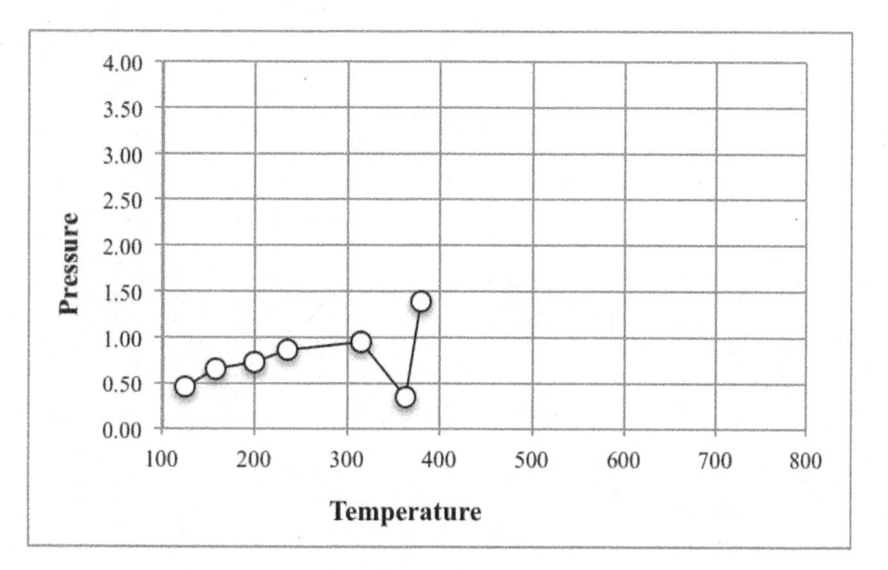

Name ______________________ Date ______________________

Partner(s) ______________________ Section ______________________

______________________ Instructor ______________________

Math is a key component to success in chemistry. Basic math and algebraic skills will be reviewed in this Math Module. Not all math required for the course will be covered, but an early assessment might be helpful to determine whether supplemental instruction will be needed.

Significant figures are not addressed in this review, but will be covered in the next lab. Additional skill reviews may also be provided by the instructor.

Goals

- Use scientific notation for large and small numbers.
- Use scientific notation in calculations.
- Become familiar with the functions of a scientific calculator.
- Round off numbers correctly.
- Review simple algebraic functions.
- Convert temperatures between degrees Fahrenheit and degrees Celsius.

Materials

- Scientific calculator

Scientific Notation

In science, work often involves small numbers such as 0.000000025 m and large numbers such as 4,000,000 g. It is sometimes convenient to express such large and small numbers in terms of a power of 10.

$0.00001 = 10^{-5}$	$10{,}000 = 10^{4}$
$0.001 = 10^{-3}$	$1{,}000 = 10^{3}$
$0.01 = 10^{-2}$	$100 = 10^{2}$
$0.1 = 10^{-1}$	$10 = 10^{1}$

To write a number in scientific notation, use the following rules:

For numbers larger than 10:

- Move the decimal point to the left until it follows the first non-zero digit in the number.
- State a positive power of ten that is equal to the number of places the decimal point moved to the left.

For numbers smaller than 1:

- Move the decimal point to the right until it is placed just after the first non-zero digit in the number.
- State a negative power of ten that is equal to the number of places the decimal point moved to the right.

Examples:

Write the following numbers in scientific notation:

1. 35,000 ⇒ 35,000. ⇒ 3.5×10^4
2. 0.0042 ⇒ 0.0042 ⇒ 4.2×10^{-3}
3. 0.0000815 ⇒ 0.0000815 ⇒ 8.15×10^{-5}

Scientific notation format may be summarized with these points:

- The method is used to represent very large or very small numbers.
- The general form is $M \times 10^n$ where: M = mantissa (a number between 1 and 10)
  n = exponent (a positive or negative integer)
- Positive exponents (+) indicate large numbers (e.g., 5.6×10^4).
- Negative exponents (–) indicate small numbers (e.g., 7.3×10^{-1}).

Practice Problems

Convert the following numbers from regular numbers to scientific notation:

1. 0.000 000 5893 ______________
2. 299,793,000 ______________
3. 3410 ______________
4. 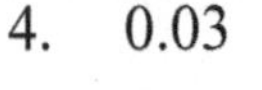 0.03 ______________
5. 15.045 ______________
6. 710,000,000 ______________
7. 2370 ______________
8. 0.000 045 ______________
9. 6003 ______________

Convert the following numbers from scientific notation to regular numbers:

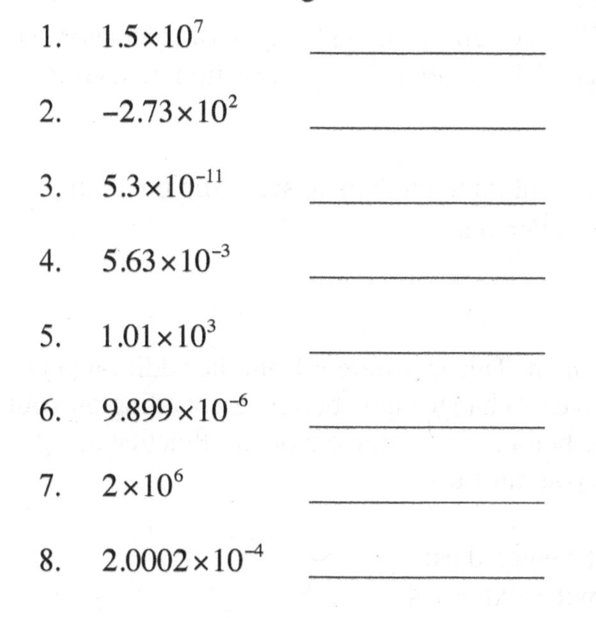

1. 1.5×10^{7} ____________
2. -2.73×10^{2} ____________
3. 5.3×10^{-11} ____________
4. 5.63×10^{-3} ____________
5. 1.01×10^{3} ____________
6. 9.899×10^{-6} ____________
7. 2×10^{6} ____________
8. 2.0002×10^{-4} ____________

To multiply or divide numbers in scientific notation without using a scientific calculator, multiply or divide the nonexponential numbers (M) and then add (if multiplying) or subtract (if dividing) the exponents (n). Be sure that the final answer is in correct scientific notation (one and only one non-zero digit to the left of the decimal point).

Examples:

1. $(3.5 \times 10^{4})(2.0 \times 10^{2}) \Rightarrow (3.5)(2.0) \times 10^{4+2} \Rightarrow 7.0 \times 10^{6}$
2. $\dfrac{3.8 \times 10^{5}}{1.9 \times 10^{2}} \Rightarrow \left(\dfrac{3.8}{1.9}\right) \times 10^{5-2} \Rightarrow 2.0 \times 10^{3}$

To add or subtract numbers in scientific notation without using a scientific calculator, the exponents must be the same. First, rewrite one of the numbers so the exponents are the same. Then add or subtract the M values. Be sure that the final answer is in correct scientific notation (one and only one non-zero digit to the left of the decimal point).

Examples:

3. $3.2 \times 10^{3} + 1.8 \times 10^{3}$ $\Rightarrow$ exponents are the same $\Rightarrow$ 5.0×10^{3}
4. $1.53 \times 10^{2} + 2.5 \times 10^{3}$ $\Rightarrow$ exponents must agree so change to
   $0.153 \times 10^{3} + 2.5 \times 10^{3}$ $\Rightarrow$ 2.653×10^{3}
5. $8.15 \times 10^{-2} - 5.3 \times 10^{-3}$ $\Rightarrow$ exponents must agree so change to
   $8.15 \times 10^{-2} - 0.53 \times 10^{-2}$ $\Rightarrow$ 7.62×10^{-2}

Using a Scientific Calculator

Instructions for using a scientific calculator will vary depending on the type of calculator. The following guidelines apply to the most common types. Please refer to the instruction manual for your particular calculator.

- Most calculators have an **EE** or **EXP** button for entering numbers in scientific notation. For example, the number 5.8×10^3 would be entered as:
  5 decimal 8 EE 3 or 5 decimal 8 EXP 3.
- EE or EXP takes the place of "$\times 10$"
- The +/– button changes the sign on the exponent. This is different from the addition (+) or subtraction (–) buttons. Some calculators have a "change sign" button. Depending on your particular calculator, the +/– may be entered before or after the exponent. Practice using your calculator to determine which method you must use.

  For example, the number 7.3×10^{-4} would be entered as:
  7 decimal 3 EE +/– 4 or 7 decimal 3 EXP +/– 4

  OR, depending on your calculator, it might be entered as:
  7 decimal 3 EE 4 +/– or 7 decimal 3 EXP 4 +/–

Practice Problems

Perform the indicated operations and write your answer in scientific notation:

1. $10^{-2} \times 10^5 =$ ____________

2. $\dfrac{10^7}{10^3} =$ ____________

3. $\dfrac{10^2}{10^{-5}} =$ ____________

4. $(5.4 \times 10^2)(2.5 \times 10^9) =$ ____________

5. $(1.2 \times 10^{-5})(5.4 \times 10^6)(1.5 \times 10^{-3}) =$ ____________

6. $\dfrac{3.3 \times 10^{-7}}{6.6 \times 10^{-7}} =$ ____________

7. $\dfrac{2.56 \times 10^3}{1.00 \times 10^{-1}} =$ ____________

8. $\dfrac{(3.11\times10^{3})(8.92\times10^{7})}{6.27\times10^{4}} =$ ______________

9. $\dfrac{2.7\times10^{-2}}{(7.9\times10^{2})(5.3\times10^{3})} =$ ______________

10. $3.5\times10^{4}+6.23\times10^{5} =$ ______________

11. $6.33\times10^{6}-5.8\times10^{5} =$ ______________

12. $6.44\times10^{-1}+8\times10^{-3} =$ ______________

13. $1.6\times10^{-4}-5.07\times10^{-3} =$ ______________

Rounding

Rounding is important when trying to write a number with the correct number of significant figures after performing a mathematical operation.

- If the first number that must be dropped is less than 5, drop it and all the following numbers. For example, rounding 32.431914 to the hundredths place is 32.43.
- If the first number to be dropped is 5 or greater, round up the last number that will be retained and drop all following numbers. For example, rounding 103.770 to the tenths place is 103.8.

Numbers greater than 1 that do not have a decimal must retain zeros as placeholders when rounding. For example, rounding 745,678 to the hundreds place is 745,700 (no decimal is used).

Practice Problems

Round the following numbers correctly:

1. 225.757 to the hundredths place ______________

2. 43.13 to the ones place ______________

3. 1.8947368 to the hundredths place ______________

4. 1.8947368 to the thousandths place ______________

5. 1.8947368 to the tenths place ______________

6. 87.025 to the tenths place ______________

7. 604,108 to the tens place ______________

8. 5,635 to the thousands place ______________

9. 52,049 to the hundreds place ______________

Algebraic Functions

Many problems in science are most readily solved by means of algebraic functions. Sometimes all that is required is substitution of values in a fundamental equation. Other times the equation must be rearranged and manipulated into a more usable format.

The simplest type of algebraic equation is known as a "first order" equation with only one unknown. This type of equation is solved by applying a fundamental principle of algebra that states, "An equation remains valid if the same operation is performed on both sides." Specifically, that means that the following operations can be done without changing the equation:

- Add or subtract the same quantity to both sides.
- Multiply or divide both sides by the same quantity.

These rules may be used to isolate the unknown quantity "x" on one side of an equation, with all of the other known quantities on the other side of the equation.

Examples:

1. Solve the equation: $8x = 4$
   Step 1 Divide each side of the equation by 8:

$$\frac{8x}{8} = \frac{4}{8}$$

   Step 2: Solve for x:

$$x = \frac{4}{8} = 0.5$$

2. Solve the equation: $\frac{2.0}{x} = 3.25$

Because x is in the denominator, there are two common methods to solve this problem: using inversion and cross multiplication. Both methods are shown below.

Using inversion	Using cross multiplication
Step 1 Invert both sides of the equation: $\frac{x}{2.0} = \frac{1}{3.25}$	Step 1 Multiply both sides of the equation by x: $\frac{2.0x}{x} = 3.25x$
Step 2 Multiply both sides of the equation by 2.0: $\frac{2.0x}{2.0} = \frac{2.0}{3.25}$	Step 2 Divide both sides of the equation by 3.25: $\frac{2.0}{3.25} = \frac{3.25x}{3.25}$
Step 3 Solve for x: $x = \frac{2.0}{3.25} = 0.62$	Step 3 Solve for x: $x = \frac{2.0}{3.25} = 0.62$

Practice Problems

Complete the following algebraic equations. Show your work. Some problems will have numerical answers and others will require rearrangement of a formula to isolate a letter value (variable). Do not leave numerical answers as fractions; use the decimal equivalent.

1. $\text{density} = \dfrac{\text{mass}}{\text{volume}}$ in short, $d = \dfrac{m}{v}$ Solve for volume.

2. $15 = \dfrac{5}{x}$ Solve for x.

3. $PV = nRT$ Solve for V.

4. $16x = 4$ Solve for x.

5. $\dfrac{P_1V_1}{T_1} = \dfrac{P_2V_2}{T_2}$ Solve for T_2.

6. $2x + 8 = -4$ Solve for x.

Temperature Conversion Problems

The relationship between the temperature expressed in degrees Fahrenheit (°F) and degrees Celsius (°C) is given by the equations:

$$°F = (1.8 \times °C) + 32 \quad \text{or} \quad °F = \left(\frac{9}{5}\right)°C + 32$$

Using the algebraic skills from the previous section, it is easy to rearrange the equation $°F = (1.8 \times °C) + 32$ to arrive at the equation $°C = \frac{(°F - 32)}{1.8}$.

Step 1 Subtract 32 from both sides of the equation:

$$°F - 32 = (1.8 \times °C) + 32 - 32$$

$$°F - 32 = 1.8 \times °C$$

Step 2 Divide both sides of the equation by 1.8:

$$\frac{°F - 32}{1.8} = \frac{1.8 \times °C}{1.8}$$

$$\frac{°F - 32}{1.8} = °C$$

Example: Normal body temperature is 98.6°F. What is this temperature in degrees Celsius? Now that the equation has been rearranged in Step 2, only one more step is needed.

Step 3 Substitute in the known value for degrees Fahrenheit and solve:

$$\frac{98.6 - 32}{1.8} = °C$$

$$\frac{66.6}{1.8} = °C$$

$$37.0 = °C$$

Practice Problems

Complete the following temperature conversions. Show your work.

1. Silver melts at 960.8°C. At what Fahrenheit temperature does silver melt?

2. The boiling point of water is 212°F. At what Celsius temperature does water boil?

3. An oven has a temperature of 325°F. What Celsius temperature does this correspond to?

4. A normal body temperature for a dog is about is 38°C. What is this temperature in degrees Fahrenheit?

Goals

- Determine the correct device to use when measuring mass, volume, length and temperature.
- Read a measurement device to the correct number of significant figures.
- Use the correct units for measurements.
- Use the correct number of significant figures when performing calculations.
- Round off calculated answers correctly.
- Identify and use metric-metric and metric-U.S. conversion factors in calculations.

Materials

- Metric ruler
- 100 mL graduated cylinder
- 10 mL graduated cylinder
- 50 mL beaker
- 3 pennies
- Electronic balances
- Thermometer
- Ice
- Hot plate

Discussion

Laboratory experiments often require taking measurements. Each measurement is a quantitative reading consisting of a numerical value and a unit. Any measurement has some degree of uncertainty and while selecting the correct measurement device and taking care when making the reading can minimize the uncertainty, it can never be completely eliminated. The terms *precision* and *accuracy* are used to describe a measurement.

Precision and accuracy

Precision refers to the reproducibility of a measurement. It compares several measured numerical values obtained in the same way under the same conditions. For example, a student may measure the length of a line three times using the same metric ruler, obtaining the values 2.58 cm, 2.56 cm, and 2.57 cm. These measurements are said to have high precision since there is only a 0.02 difference between the highest and lowest numerical values. Multiple measurements are frequently taken and averaged so that small random errors are minimized.

Accuracy is a comparison of the measured value to the accepted (true) value. In the example above, if the true length of the line was 3.63 cm, the accuracy of the measurements would be low since the difference between the true value (3.63 cm) and the average of the measured values (2.57 cm) is relatively large. Lack of accuracy can be attributed to improper use of the measurement device or other errors such as an incorrect calibration that is made each time the measurement is taken. Instruments that have inscribed or painted lines, use gauges or meters, or have other types of measurement markings, are "calibrated" at the factory to ensure that the markings are correct. Some electronic instruments such as pH meters also require periodic manual calibrations to match the instrumental readout to a known standardized solution.

Accuracy and precision can be visualized by using a dartboard analogy. Imagine throwing three darts at a dartboard. The example on the left in Figure 1 is the worst case. The darts are not near each other and not near the bull's-eye (low precision and low accuracy). In the center example, the darts are clustered together (high precision), but they are not near the bull's-eye (low accuracy). Science students should always strive for high precision and high accuracy, as shown in the example on the right in Figure 1.

Figure 1 ***Dartboard analogy***

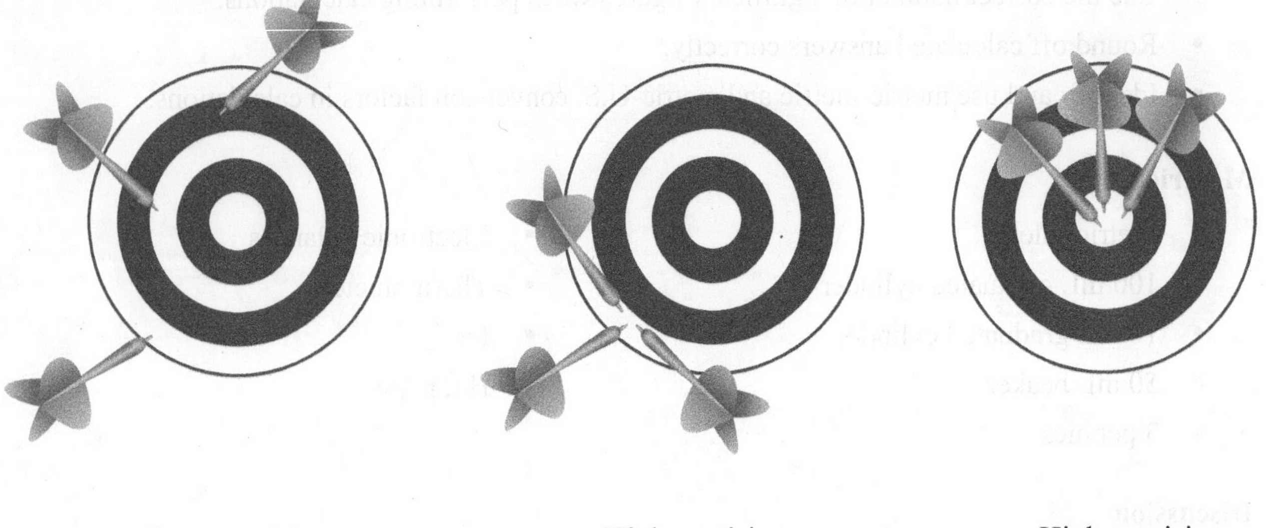

Low precision Low accuracy	High precision Low accuracy	High precision High accuracy

Exact and measured numbers

Exact numbers are obtained by counting, not measuring, objects. Counting five books gives "5" as an exact number. A dozen eggs gives "12" as an exact number. "Dozen" can be counted, but it is also a definition for the number twelve. Exact numbers also include *defined numbers* such as 1 meter = 100 cm, 1 foot = 12 inches, and 1 hour = 60 minutes. Defined numbers are typically found in reference tables such as Tables 2, 3, 4, and 6 in this lab and define a relationship between units. Exact, counted, or defined numbers are **excluded** from consideration when determining the number of significant figures in numerical values.

Measured numbers are the numerical values obtained when using a measurement device. A calibrated scale divides the measuring tool into segments. Since the distance between two measurement increments can be estimated, there is a degree of uncertainty in the measured value. The degree of uncertainty is reflected in the number of significant figures shown when the value is recorded. It is common practice to estimate the value of the digit beyond the closest scaled subdivision. Thus, the recorded value should include the significant digits that are known with certainty plus the estimated digit.

Significant figures

In measured numbers, all reported figures are called *significant figures* (sig figs). When reading a numerical value, the first significant figure is the FIRST NON-ZERO DIGIT and the last

significant figure is the ESTIMATED digit. All zero and non-zero numbers in-between those two numbers are also significant. "Leading" zeros are not significant; they are placeholders. Table 1 summarizes the significant figure rules.

Table 1 ***Significant Figure Rules***

Rule	Measured numbers	# of sig figs
A number is a significant figure if it is:		
a nonzero digit	982 mg 3.34 mL	3 3
a zero between nonzero digits	304 m 56.02 L	3 4
a zero at the end of a decimal number (a.k.a. trailing zeros)	250. g 25.0 cm	3 3
A number is NOT significant if it is:		
a zero at the beginning of a decimal number (a.k.a. leading zeros)	0.002 kg 0.0831 mm	1 3
a placeholder zero in a non-decimal number	200,000 cg 1,500,000 km	1 2

Measurement instruments with calibrated scales such as graduated cylinders and metric rulers contain marks and printed numerical values at regular intervals. When a measurement is read, the value should be estimated one digit further than the smallest calibrated mark. If the measurement is exactly on the calibrated mark, then a "0" must be used for the estimated digit. For example, in the diagrams below, ruler A has calibrated marks representing 1 cm and ruler B has calibrated marks representing 0.1 cm. When measuring the length of a metal rod, ruler A would have a measurement including estimation between the 1 cm markings and thus would have the answer recorded to the 0.1 cm digit. Similarly, ruler B would be estimated to the 0.01 cm digit.

Consider the two rulers shown below and determine the lengths of the metal rods in Figure 2:

Figure 2 ***Rulers A and B***

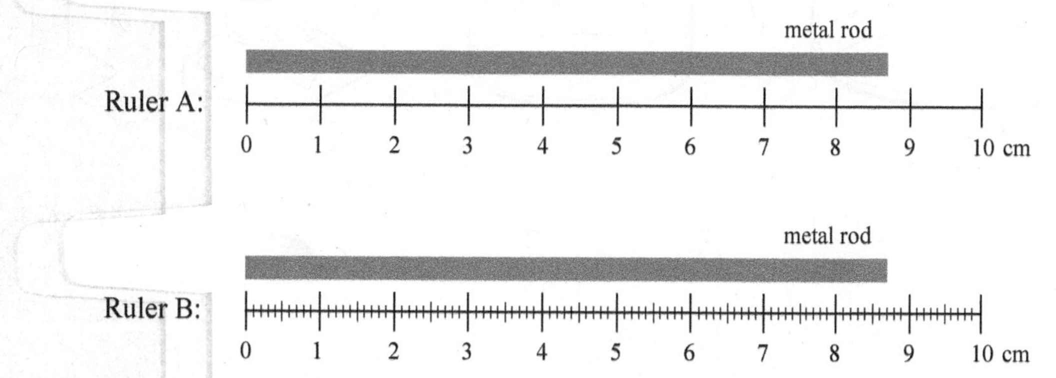

The recorded length of the metal rod measured with Ruler A would be 8.6 cm and using Ruler B it would be 8.69 cm.

When using a measurement tool with a calibrated scale, the measurement must be reported as precisely as possible. The number of significant figures that can be reported depends upon the lines marked on the tool. Just as Ruler A and Ruler B in Figure 2 have different calibration scales, volume measuring devices also may have a variety of calibration scales. The smaller the incremental markings, also known as gradations, the more precisely the measurement may be made. Beakers may have volume markings that may be more than adequate for an experiment that asks for "approximately 100 mL" or "about 30 mL," but the markings on beakers are approximations (see Figure 3). They are not calibrated to measure volumes accurately. Graduated cylinders, on the other hand, are calibrated to measure volumes accurately. When any measurement device is used, it is important to choose the best device for the precision required. A 100 mL graduated cylinder may have markings at every 1 mL and that could be used if the experiment asks for 32.5 mL. But if the experiment asks for "exactly 8.50 mL," then a 10 mL graduated cylinder with markings at every 0.1 mL would be more appropriate.

Figure 3 ***Markings on Various Laboratory Glassware***

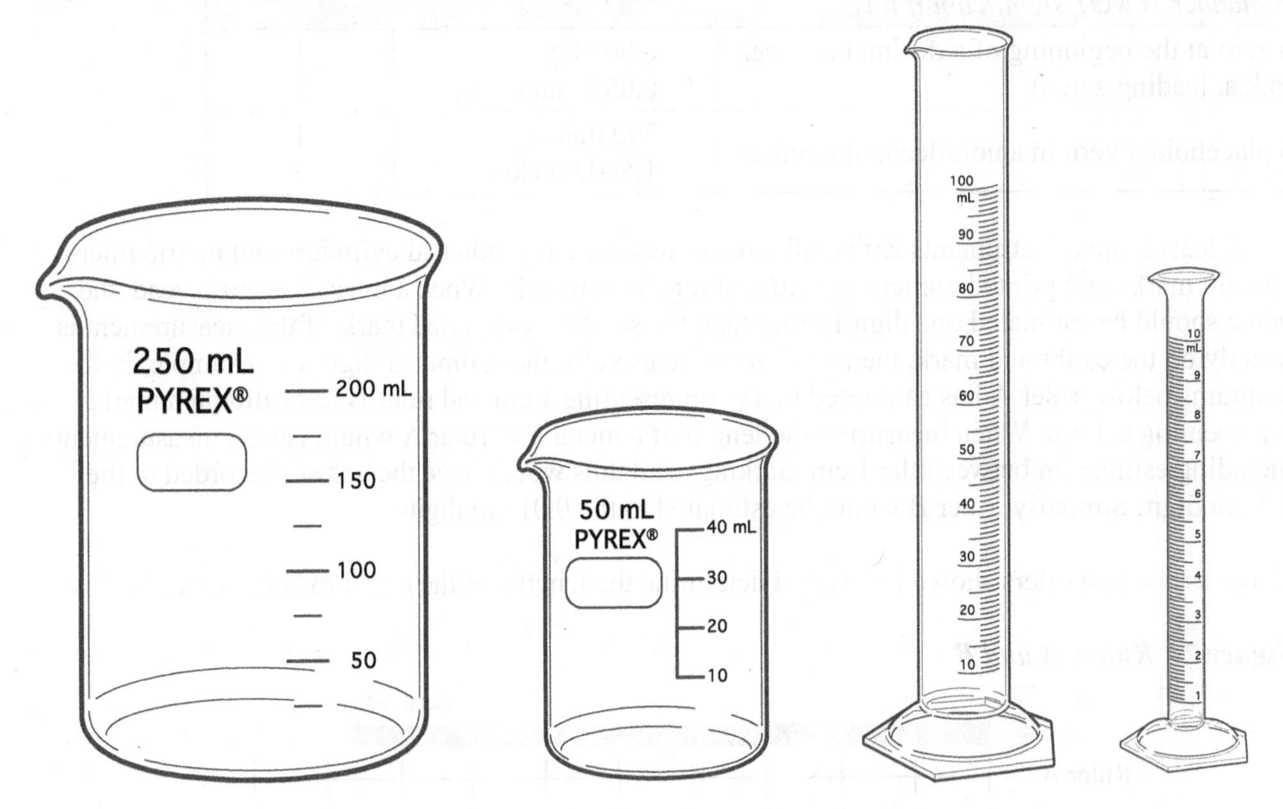

Measuring length

The standard unit of length in the metric system is the *meter* (*m*). Metric prefixes are used to indicate lengths that are larger or smaller than the meter. The same prefixes also may be used for other units in the metric system such as grams and liters. Commonly used metric units are shown in Table 2.

Table 2 ***Metric Length Units***

Length unit	Symbol	Conversion factor
meter	m	
kilometer	km	1 km = 1000 m 0.001 km = 1 m
centimeter	cm	1 cm = 0.01 m 100 cm = 1 m
millimeter	mm	1 mm = 0.001 m 1000 mm = 1 m

A meter stick is divided into 100 cm and may be further divided into mm increments. When reading a meter stick, observe the numbers printed on the device as well as the size of the increments. Centimeters are approximately the width of your little finger and millimeters are the size of the thickness of a few sheets of paper. Determine the last digit of a measurement by estimating the distance between the smallest interval markings.

Measuring volume

In the metric system, the unit for measuring volume is the *liter* (*L*). This is a measurement of the space that a substance occupies. Some of the commonly used prefixes for volumes are shown in Table 3. Note that one cubic centimeter (cm^3 or cc) is equal to one milliliter (mL). These units are used interchangeably.

Table 3 ***Metric Volume Units***

Volume unit	Symbol	Conversion factor
liter	L	
milliliter	mL	1 mL = 0.001 L 1000 mL = 1 L 1 mL = 1 cm^3 = 1 cc

Note that when measuring volumes, the liquid surface is curved. This curved surface is called a *meniscus* (from the Greek root meaning moon). The smaller the diameter of a liquid measurement device, the easier it is to see the meniscus. This phenomenon usually contributes to more precise measurements of volume if the calibration lines are marked in smaller increments. To correctly read the volume measurement you should always have the bottom of the meniscus at eye level and report the reading at the bottom of the meniscus. A few liquids, such as mercury in a barometer, curve the opposite direction and should be read at the top of the curve. Remember to estimate the final digit one place further than the calibrated mark. The volume of the liquid in the graduated cylinder in Figure 4 should be recorded as 23.2 mL.

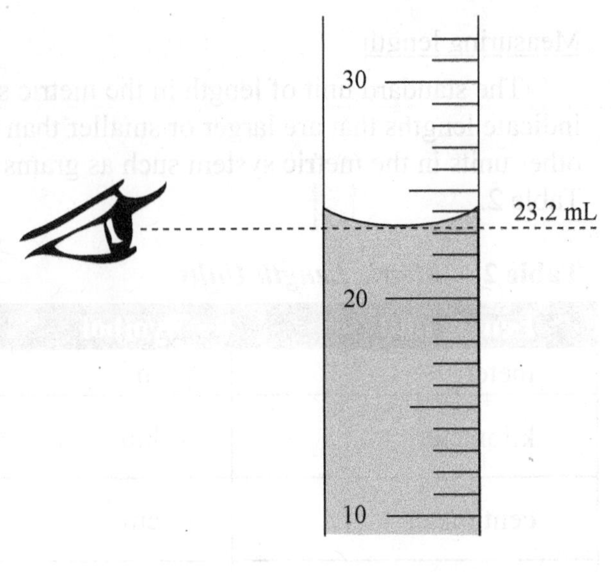

Figure 4 ***Graduated Cylinder***

Measuring mass

The *mass* of an object is a measurement of the amount of matter. The *weight* of an object is a measurement of the attraction of gravity for that object. The terms mass and weight are often used interchangeably since they are proportional, but technically, they are not. You have the same mass here on Earth as you do on the moon. However, you *weigh* differently on the moon because the gravitational attraction is less. The metric unit for mass is the *gram* (g). Commonly used metric units to measure mass are shown in Table 4.

Table 4 ***Metric Mass Units***

Mass unit	Symbol	Conversion factor
gram	g	
kilogram	kg	1 kg = 1000 g 0.001 kg = 1 g
centigram	cg	1 cg = 0.01 g 100 cg = 1 g
milligram	mg	1 mg = 0.001 g 1000 mg = 1 g

Measuring temperature

Temperature is the measurement of the average kinetic energy of the particles of a substance. Substances with lower temperature values feel "cold" while substances with higher temperature values feel "hot." But "cold" and "hot" are simply relative terms. Using a calibrated thermometer allows for numerical values to be obtained, and the precision of those values depends upon the calibration marks on the thermometer. When reading the temperature, remember to estimate the final digit one place further than the calibrated mark.

The metric system of temperature measurement uses the Celsius and Kelvin scales. Their relationships to the Fahrenheit scale are summarized in Table 5. Note that the Kelvin unit (K) does not have a degree symbol (°).

Table 5 ***Metric Temperature Units***

Celsius to Fahrenheit	Fahrenheit to Celsius	Celsius/Kelvin conversions
$°F = \left(\frac{9}{5}\right)°C + 32$ or	$°C = \left(\frac{5}{9}\right)(°F - 32)$ or	$K = °C + 273$
$°F = (1.8 \times °C) + 32$	$°C = \frac{(°F - 32)}{1.8}$	$°C = K - 273$

Calculations

When using measured numbers in calculations, the reported answers must reflect the precision of the original measurements. This requires identifying the correct number of significant figures in each measurement, performing the necessary calculations using any necessary conversion factors, determining the correct number of significant figures for the answer, and rounding off the answer correctly.

Conversion factors are mathematical equalities written as fractions. For example, from Table 4 you can determine that:

$$1 \text{ kg} = 1000 \text{ g}$$

This can be expressed as two different conversion factors:

$$\frac{1 \text{ kg}}{1000 \text{ g}} \quad \text{or} \quad \frac{1000 \text{ g}}{1 \text{ kg}}$$

Either of these conversion factors can be used to solve a mathematical problem, depending on which arrangement provides the correct placement of the units.

Conversion factors are also necessary when converting between the metric system and U.S. units. Some of the equalities that might be needed to set up these conversion factors are shown in Table 6.

Table 6 ***Metric – U.S. Unit Conversion Factors***

Property	Conversion factors
Length	2.54 cm = 1 in 1.61 km = 1 mile
Volume	0.946 L = 1 qt 1 L = 1.06 qt
Mass	454 g = 1 lb 1 kg = 2.20 lb

Significant figures used in calculations

The correct number of significant figures in a calculation depends upon the number of significant figures in the measurements made. Two separate rules govern the determination.

Multiplication/division: when numbers are multiplied or divided, the answer will have the same number of *significant figures* as the measured number with the "fewest" significant figures.

Sample problems:

$4.73 \times 2.0 = 9.46 \quad \Rightarrow \quad 9.5$ (2 significant figures allowed)

$\frac{3000}{1.5} = 2000 \quad \Rightarrow \quad 2000$ (1 significant figure allowed)

Addition/subtraction: when numbers are added or subtracted, the answer will have the same number of *decimal places* as the measured number with the "fewest" decimal places.

Sample problems:

$$\begin{array}{r} 1.008 \\ +\ 0.2 \\ \hline 1.208 \end{array} \quad \Rightarrow \quad 1.2 \quad \text{(1 number after the decimal)}$$

$$\begin{array}{r} 13.5 \\ -\ 2.5 \\ \hline 11 \end{array} \quad \Rightarrow \quad 11.0 \quad \text{(1 number after the decimal)}$$

To obtain the correct number of significant figures in a calculated answer, it may be necessary to round off the answer. If the first of the nonsignificant numbers to be dropped is *less than 5*, all the nonsignificant numbers are dropped. If the first of the nonsignificant numbers to be dropped is *equal to or greater than 5*, the value of the last significant digit to be retained is increased by one and all the nonsignificant numbers are dropped. Occasionally it is necessary to add significant zeros at the end of a number to obtain the correct number of significant figures.

Sample problems:

7706 rounded to 3 significant figures = 7710

0.00373 rounded to 2 significant figures = 0.0037

Experimental Procedures

Eye protection and appropriate clothing must be worn at all times.

Record all measurements with the correct number of significant figures and units.

Discard all wastes properly as directed by the instructor.

A. Measuring length

1. Using a metric ruler, measure the length (the longer measurement) and the width (the shorter measurement) of this lab manual. Record the measurements in centimeters, remembering to estimate one digit past the smallest calibrated mark on the metric ruler.
2. Convert the measurements to inches.
3. Convert the measurements to meters.
4. To calculate the area of a rectangle use the formula: area = length × width. Calculate the area of the lab manual in cm^2, in^2, and m^2. Record your answers using the correct number of significant figures and units.

B. Measuring volume

1. Using a 50 mL beaker, measure approximately 30 mL of water. Record the volume in milliliters, remembering to estimate between the smallest calibrated marks.
2. Pour the water from the beaker into a 100 mL graduated cylinder. Record the volume again.
3. Convert the volume from B.2 to liters and quarts and record your answers.
4. Empty and dry the 100 mL graduated cylinder.
5. Measure approximately 9 mL of water in the 100 mL cylinder. Record the volume.
6. Pour the water from the 100 mL cylinder into a 10 mL graduated cylinder. Record the volume again.
7. Convert the volume from B.6 to liters and quarts and record your answers.

C. Measuring mass

1. Follow the instructions given by your lab instructor for proper use of the balance.
   i. Electronic balances do not have calibration marks and no "estimated" digit is used. Always record the entire mass shown on the digital display.
   ii. Do not place any chemicals directly on the balance. Use an appropriate container as directed by the instructor.
   iii. Keep area around balance clean.
   iv. Turn off the balance or reset to zero.
2. Since balances may have slight variations in their accuracy, it is important to use the same balance for all measurements in an experiment. This procedure will use two different balances to compare their accuracies when measuring mass.

3. Obtain 3 pennies. Measure and record the mass in grams of each penny.
4. Use a different balance to measure and record the mass of the same three pennies.
5. Complete the calculations on the report sheet to obtain the total mass of the pennies. Convert the total mass to kilograms, ounces, and pounds.

D. Measuring temperature

1. NEVER SHAKE THE THERMOMETER. Take care not to hit the walls or bottom of the container. Hold the thermometer in the center of the liquid when reading the temperature.
2. Obtain a 250 mL beaker. Fill the beaker half full with ice. Add enough water to cover the ice.
3. Place a thermometer in the ice bath, gently stirring the mixture.
4. After a minute, read the thermometer and record the temperature in degrees Celsius with the correct number of significant figures and units.
5. Empty the beaker and refill it half full with tap water. Measure and record the temperature of the tap water.
6. Place the beaker containing the tap water on the hot plate. Turn the hot plate to high and heat until the water boils.
7. Place the thermometer in the boiling water.
8. After a minute, read the thermometer and record the temperature with the correct number of significant figures and units.
9. Convert the temperatures to Fahrenheit and Kelvin.

PRE-LAB QUESTIONS: MEASUREMENTS, SIGNIFICANT FIGURES, CALCULATIONS

Name ______________________ Date ______________________

Partner(s) ______________________ Section ______________________

______________________ Instructor ______________________

1. Describe the proper procedure for weighing chemicals on a balance.

2. How many significant figures are in each of the following numbers? Circle the estimated digit in the number.

   a) 175.0 g ________

   b) 0.0034 cm ________

   c) 739 mL ________

   d) 750,000 km ________

   e) 43,014 m ________

3. How do you determine the last number recorded when using a measuring device?

4. Complete the following problems and record the answers to the correct number of significant figures.

   a) $(4.73\ \text{cm})(2.1\ \text{cm}) =$ ________

   b) $1.250 + 2.2 =$ ________

   c) $9.74 - 1.362 =$ ________

   d) $\dfrac{3.841}{1.23} =$ ________

REPORT SHEET: MEASUREMENTS, SIGNIFICANT FIGURES, CALCULATIONS

Name ______________________ Date ______________________

Partner(s) ______________________ Section ______________________

______________________ Instructor ______________________

Read the procedures before completing the following tables. Show the correct number of significant figures and units for all measurements and calculations.

A. Measuring length

Length of the lab manual in centimeters (cm): ____________

(circle the estimated digit)

Number of sig figs in the measurement: ____________

Calculate the length of the lab manual in inches (in) and show calculations: ____________

Calculate the length of the lab manual in meters (m) and show calculations: ____________

Width of the lab manual in centimeters (cm): ____________

(circle the estimated digit)

Number of sig figs in the measurement: ____________

Calculate the width of the lab manual in inches (in) and show calculations: ____________

Calculate the width of the lab manual in meters (m) and show calculations: ____________

Area calculations:

Area of the lab manual in cm^2 and show calculations: ____________

Area of the lab manual in in^2 and show calculations: ____________

Area of the lab manual in m^2 and show calculations: ____________

B. Measuring volume

1. Volume of water in 50 mL beaker (mL): ______________ (circle the estimated digit)

   Number of sig figs in the measurement: ______________

2. Volume of water poured into 100 mL graduated cylinder from *Procedure #2*: ______________ (circle the estimated digit)

   Number of sig figs in the measurement: ______________

3. Convert the volume from line B.2 to liters and show calculations: ______________

4. Convert the volume from line B.2 to quarts and show calculations: ______________

5. Volume of water in 100 mL graduated cylinder from *Procedure #5*: ______________ (circle the estimated digit)

   Number of sig figs in the measurement: ______________

6. Volume of water poured into 10 mL graduated cylinder from *Procedure #6*: ______________ (circle the estimated digit)

   Number of sig figs in the measurement: ______________

7. Convert the volume from line B.6 to liters and show calculations: ______________

8. Convert the volume from line B.6 to quarts and show calculations: ______________

C. Measuring mass

	First balance	Second balance
Mass of penny #1	________	________
Mass of penny #2	________	________
Mass of penny #3	________	________
Total mass of three pennies (calculated)	________	________

Show calculations for the conversion of the total mass of the three pennies from the second balance to kilograms, pounds, and ounces:

________ kg

________ lb

________ oz

D. Measuring temperature

Temperature of the ice bath (°C): ______________

(circle the estimated digit)

Number of sig figs in the measurement: ______________

Calculate the temperature in degrees Fahrenheit and show calculations: ______________

Calculate the temperature in Kelvin and show calculations: ______________

Temperature of the tap water bath (°C): ______________

(circle the estimated digit)

Number of sig figs in the measurement: ______________

Calculate the temperature in degrees Fahrenheit and show calculations: ______________

Calculate the temperature in Kelvin and show calculations: ______________

Temperature of the boiling water bath (°C): ______________

(circle the estimated digit)

Number of sig figs in the measurement: ______________

Calculate the temperature in degrees Fahrenheit and show calculations: ______________

Calculate the temperature in Kelvin and show calculations: ______________

Problems and questions (show all calculations, correct significant figures and units)

1. A sample obtained from the National Bureau of Standards had a calibrated weight of 20.000 g. When a student measured the weight three times with a balance the results were: 19.603 g, 19.599 g, and 19.600 g. What conclusions may be drawn regarding the accuracy and precision of the balance used by the student?

2. Convert a temperature of 25°C to °F and Kelvin.

3. Why do scientists prefer to use the metric system?

4. List two circumstances in which you might use the metric system. (Other than chemistry lab!)

5. Using conversion factors, complete the following problems:

   a. A pen has a length of 7.6 cm. What is its length in millimeters? In meters?

   b. A patient weighs 155 lbs. What is the patient's weight in grams?

   c. A doctor has ordered 225 mg of morphine. It is only available as a 0.40 g/mL solution. How many mL of solution are required?

6. When measuring the mass of the pennies in part C, did you obtain the same values for balances 1 and 2? Explain your answer.

Goals

- Read and record measurements with the correct number of significant figures.
- Determine the density of a regular solid.
- Determine the density of an irregular solid using volume by displacement.
- Identify an unknown by comparing its density with known densities of solids.
- Determine the density of a liquid.
- Graph the relationship between mass and volume of a liquid.
- Identify an unknown by comparing its density with known densities of liquids.
- Calculate the specific gravity of a liquid.
- Measure the specific gravity of a liquid using a hydrometer.

Materials

- Metric ruler
- Solid object with regular block shape
- Balance
- Solid object with irregular shape
- Five unknown irregularly-shaped solids labeled #1 – #5 (solids from Table 1)
- String
- 100 mL graduated cylinder
- 10 mL graduated cylinder
- Five unknown liquids labeled #1 – #5 (liquids from Table 2)
- Hydrometer

Discussion

This experiment will introduce the use of measurement devices used in the lab. Any measurement device has a level of uncertainty and the value recorded reflects this. The *accuracy* of a measured value is a comparison of the measurement to the accepted or true value. *Precision* refers to the reproducibility of the measurement. Measurements may be very precise but not very accurate if an error is made each time the measurement is taken. Three measurements such as 3.56 g, 3.54 g, and 3.55 g are "precise" since they are in very close agreement with each other. But if the true value is 3.20 g, none of the measurements are "accurate."

A measurement device with many smaller gradations allows for greater precision than a device with fewer larger gradations. For example, when measurements of a liquid volume are required, a 10 mL graduated cylinder with increments marked in 0.1 mL would allow for more precise readings than a 100 mL graduated cylinder that only has markings in 1 mL increments.

The volume of an object can be measured by two methods. If the object has a regular shape such as a rectangular solid or a cylinder, measurements may be made using a metric ruler and the volume then calculated using mathematical formulas such as: $V = L \times W \times H$ for a rectangular solid or $V = \pi \times r^2 \times H$ for a cylinder. If the object has an irregular shape, the volume can be measured by "displacement." A known volume of liquid, usually water, is measured in a container large enough to also hold the unknown object. After the initial volume is recorded, the object is carefully lowered into the container until it is completely covered by the liquid. The final volume of the liquid, which is higher than the initial level, is recorded. The difference between the initial and final volumes of the liquid is equal to the volume of the submerged object.

Thus, appropriate measurement devices should be selected when determining the density of solid objects and liquids, since the smaller the gradations on the devices, the more precise the measurements will be. Also, smaller gradations allow for more significant figures in each measurement and since density is a calculated value of the ratio of mass and volume measurement, the more significant figures in the measurements, the more significant figures can be included in the density value.

Density is defined as mass per unit volume and can be calculated by using the equation:

$$\text{density} = \frac{\text{mass of sample}}{\text{volume of sample}} \qquad \text{or} \quad D = \frac{M}{V}$$

If the mass is measured in grams and the volume in milliliters, the density will have units of g/mL. If the volume is measured in cm^3 then the density will have units of g/cm^3.

Specific gravity of a liquid is a comparison of the density of that liquid with the density of water. The density of water at 4°C is 1.00 g/mL. The formula to calculate specific gravity is:

$$\text{Specific gravity (sp gr)} = \frac{\text{density of liquid } (\text{g}/\text{mL})}{\text{density of water } (\text{g}/\text{mL})}$$

Since the units of density in the numerator and denominator cancel out, specific gravity has no units. Specific gravity may also be determined by using a hydrometer and this technique is frequently used to determine the specific gravity of urine for medical tests. A hydrometer is also used to measure the specific gravity of liquid in a car battery, or to measure the alcohol content in the wine-making industry.

When using a hydrometer, place it in the liquid to be tested and spin it slowly to keep it from sticking to the sides of the container. The specific gravity is read at the bottom of the meniscus of the liquid. Some hydrometers use the European decimal point (a comma) instead of a period. For example: 0,953 (European system) = 0.953 (U.S. system).

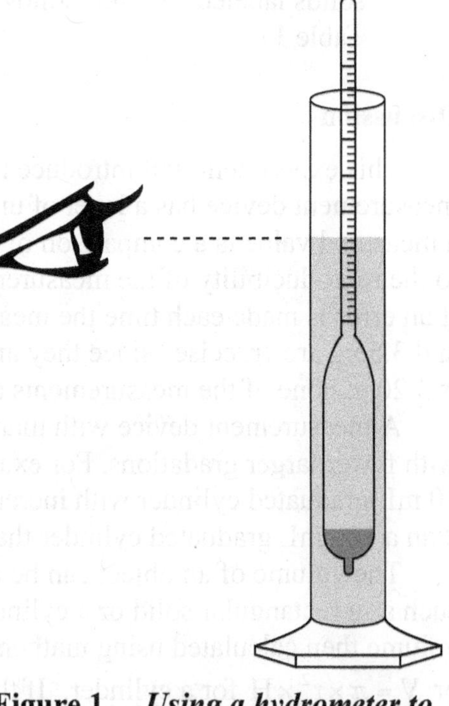

Figure 1 ***Using a hydrometer to measure specific gravity***

Experimental Procedures

Eye protection and appropriate clothing must be worn at all times.

Record all measurements with the correct number of significant figures and units.

Since balances may have slight variations in their accuracy, it is important to use the same balance for all measurements in an experiment.

Wash any contacted skin area immediately with running water and alert the instructor. Discard all wastes properly as directed by the instructor.

A. Determining the density of a regular solid

1. Mass of the solid: Obtain a regularly-shaped solid block. Determine its mass and record.
2. Volume of the solid: Use a metric ruler to measure the dimensions of the solid block in centimeters. Record all three dimensions (length, width, and height).
3. Calculate the volume of the solid and record the answer.
4. Calculate the density of the solid and record the answer.

B. Determining the density of an irregularly-shaped solid and identification of an unknown solid

1. Mass of the solid: Obtain an irregularly-shaped solid object. Determine its mass and record.
2. Volume of the solid using displacement: Use a graduated cylinder that is large enough to hold the solid object. Fill the cylinder about half full with water. Carefully read and record the water level, remembering to estimate the final digit in between the smallest gradations. Attach a string to the solid object and gently lower the object into the water, taking care not to drop it to the bottom. Make sure the object is totally submerged under the water. Once again, carefully read and record the volume of the water. Calculate the volume of the solid using the equation:

   volume of solid = final water level – initial water level
3. Remove the object and empty out the water.
4. Calculate the density of the solid.
5. Repeat the procedure using one of the unknown solids. Record the number of the unknown on the report sheet.
6. Using the calculated density of the unknown, identify the unknown by comparing its density with the densities listed in Table 1.

C. Determining the density of a liquid and identification of an unknown liquid

1. Weigh and record the mass of an empty 10 mL graduated cylinder.
2. Fill the graduated cylinder with water so that the meniscus remains below the 10 mL marking.
3. Weigh and record the mass of the graduated cylinder and water.
4. Carefully read and record the volume of water.
5. Calculate the density of the water.
6. Empty and dry the graduated cylinder.
7. Repeat the procedure using one of the unknown liquids. Record the number of the unknown and calculate the density on the report sheet.
8. Using the calculated density of the unknown, identify the unknown by comparing its density with the densities listed in Table 2.

Table 1 ***Densities of Some Solids***

Solid	Density @ 25°C (g/mL)
Copper	8.92
Lead	11.3
Aluminum	2.70
Zinc	6.14
Iron	7.86

Table 2 ***Densities of Some Liquids***

Liquid	Density @ 25°C (g/mL)
Gasoline	0.675
Ethyl alcohol	0.791
Olive oil	0.918
Glycerin	1.260
Carbon tetrachloride	1.595

D. Graphing the relationship between mass and volume

1. Weigh and record the mass of an empty 100 mL graduated cylinder.
2. Pour in approximately 10 mL of water.
3. Weigh and record the mass of the graduated cylinder and water.
4. Carefully read and record the volume of water. **DO NOT EMPTY ANY OF THE WATER FOR THE REST OF THE PROCEDURE.**
5. Add another approximately 10 mL of water.
6. Weigh and record the new mass of the graduated cylinder and water.

7. Carefully read and record the new volume of water.
8. Repeat steps 5 – 7 until you have a total of **five** volumes of water.
9. Using the graph provided on the report sheet, graph the relationship between mass and volume of the water. The mass of the water is calculated by subtracting the mass of the graduated cylinder from the total mass of the graduated cylinder and water samples.
10. Label the vertical axis as "Mass" and the horizontal axis as "Volume." Mark appropriate values along each axis, using as much of the graph as possible. Each axis should start with the value of "0." Label each axis with the correct units.
11. Use a ruler to draw a **straight** line (best fit line) through the data points, showing the relationship of mass and volume. Do NOT "connect the dots."

    Sometimes it may not be possible to have all of the data points lying on a straight line. In that case, try to have as many points above the line as below it. Try to minimize the distance from the line to each of the data points.

    Continue drawing a "dotted" line until it intersects with one of the axes.
12. The slope of the line represents the density. Mark two points on the line that are **not** points that were plotted using the experimental data. Divide the difference between the two mass values by the difference between the two volume values to determine the density.

$$\text{density} = \frac{\text{mass}(2) - \text{mass}(1)}{\text{volume}(2) - \text{volume}(1)}$$

E. Determining specific gravity

1. Calculate the specific gravity of the unknown liquid used in Procedure C using the measurements from part C. Record the result.
2. Use a hydrometer to measure the specific gravity of the unknown liquid used in Procedure C. Record the result.

Name ______________________ Date ______________________

Partner(s) ______________________ Section ______________________

______________________ Instructor ______________________

1. What is the formula for calculating density?

2. What is meant by “volume by displacement?”

3. Units of volume can be milliliters or cubic centimeters. Which unit would normally apply to liquids? Regularly-shaped solids? Irregularly-shaped solids?

4. If the smallest marking of a graduated cylinder is in the tenths place, to what place should the measurement be made?

5. What is the proper method for disposal of broken glassware?

Name ______________________ Date ______________________

Partner(s) ______________________ Section ______________________

______________________ Instructor ______________________

Read the procedures before completing the following tables. Show the correct number of significant figures and units for all measurements and calculations.

A. Density of a regular solid

Mass of the solid ______________

Length of the solid ______________

Width of the solid ______________

Height of the solid ______________

Volume of the solid ______________

Density of solid ______________
(Show calculations)

B. Density of an irregular solid

Mass of the solid ______________

Initial water level ______________

Final water level ______________

Volume of the solid ______________

Density of solid ______________
(Show calculations)

Unknown solid # ________

Mass of unknown solid	________
Initial water level	________
Final water level	________
Volume of unknown solid	________
Density of unknown solid (Show calculations)	________

Identity of unknown solid ________

C. Density of a liquid

	Water	Unknown liquid # ________
Mass of empty graduated cylinder	________	________
Mass of cylinder and liquid	________	________
Mass of liquid	________	________
Volume of liquid	________	________
Density of liquid (Show calculations)	________	________

Identity of unknown liquid ________

D. Graphing the relationship between mass and volume of water

Sample	Total mass	Mass of water	Volume of water
Empty cylinder			
#1			
#1 + #2			
#1 + #2 + #3			
#1 + #2 + #3 + #4			
#1 + #2 + #3 + #4 + #5			

Mass of Water vs. Volume graph

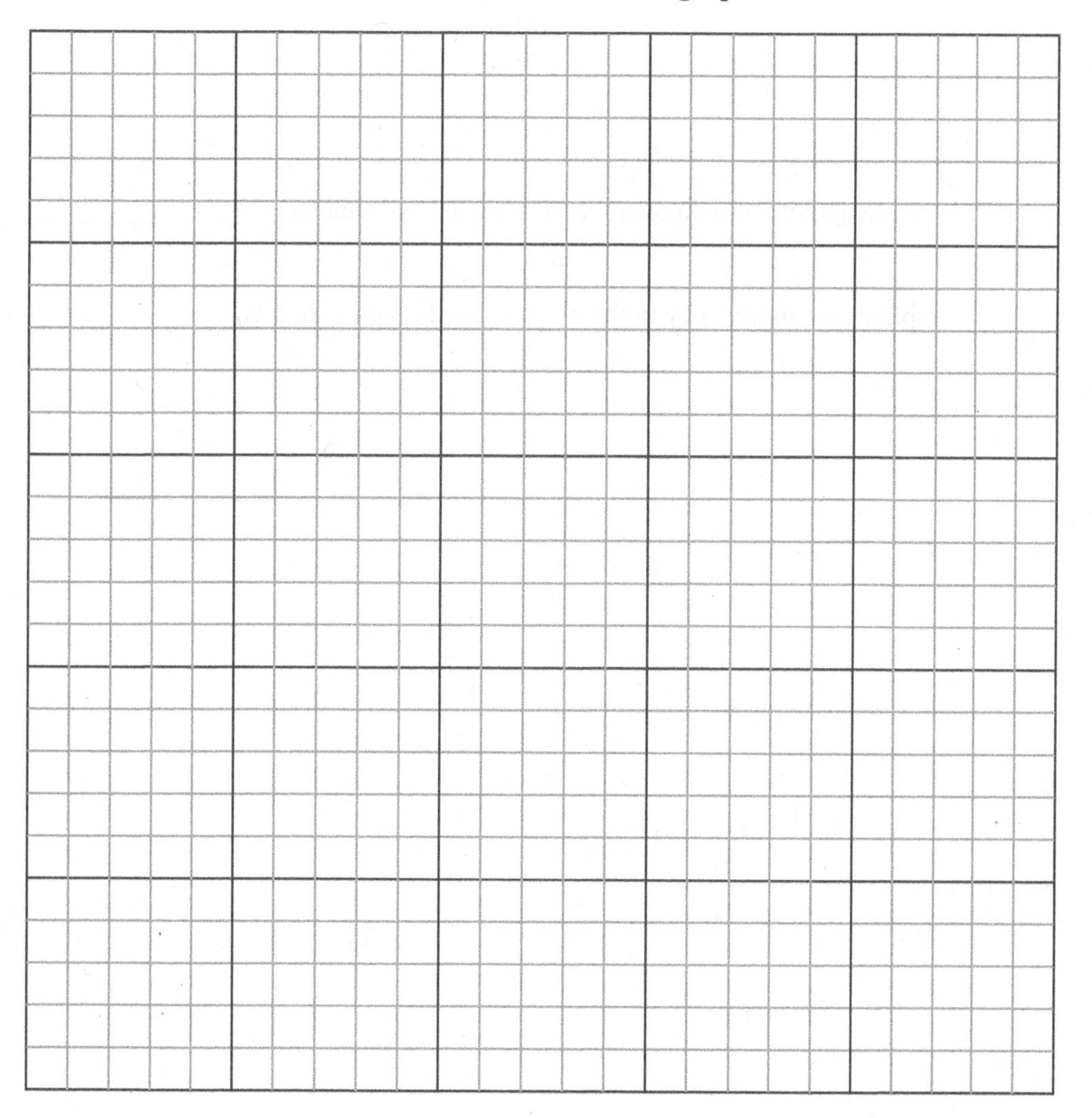

Show calculations to determine the density
using the slope of the line from the graph ______________

E. Determining Specific Gravity

1. Show calculations to determine the specific gravity of the unknown used in Procedure C ______________

2. Specific gravity of unknown liquid using the hydrometer ______________

3. Explain any discrepancy in the two values of specific gravity.

Problems and questions (show all calculations, correct significant figures and units)

1. What is the mass of an object that has a density of 11.3 g/cm^3 and a volume of 6.45 cm^3?

2. Using the density of ethyl alcohol from Table 2, what volume of alcohol would be needed to obtain a mass of 12.5 g?

3. Circle the type of glassware that is the most precise.

   a) 10 mL graduated cylinder

   b) 100 mL graduated cylinder

   c) 100 mL beaker

   Explain your answer.

4. Using the graph from Procedure D, what is the value your graph shows for the mass of the liquid when the volume is 0 mL? What *should* the value be? Explain your results and why they might not have given the expected value.

5. What are the units for specific gravity?

Some of the procedures for this lab (parts B–F) may be completed before the lab period begins as directed by the instructor.

Goals

- Write the correct symbol and name of an element.
- Observe and describe the physical properties of some elements.
- Classify an element as a metal or nonmetal from its physical properties.
- Determine the atomic number, mass number, and number of protons, neutrons and electrons for an element using the periodic table.
- Describe the atomic structure of isotopes of an element.
- Write the electron configuration of an element.
- Graph the relationship between ionization energy and atomic number.
- Interpret the trends of ionization energy within a family and within a period from a graph.
- Describe and differentiate among the colors produced by flame tests of various elements.
- Identify unknown elements using flame tests.

Materials

- Display of elements
- Colored pencils
- Bunsen burner
- Spot plate
- Flame test wire
- 0.1 M solutions of NaCl, KCl, $CaCl_2$, $BaCl_2$, $SrCl_2$, $CuCl_2$
- 1.0 M HCl
- 50 mL beaker
- Three unknown solutions

Discussion

Elemental properties

This experiment will provide an opportunity to observe a number of elements. By observing their physical properties and their placement on the periodic table, elements may be classified as metals or nonmetals. Metals are elements that are usually shiny, ductile (can be drawn into a wire), malleable (can be formed into a flat sheet or other shape), and conduct heat and electricity. Some metals, particularly those in Groups IA and IIA, react with air to form an oxide coating on the surface. Cutting through the coating exposes the shiny metallic surface underneath. Nonmetals are elements that are usually dull, brittle, and are not good conductors of heat and electricity. A number of nonmetals are gases at room temperature.

Periodic table

The periodic table provides information about each element. The horizontal rows are called *periods*. The vertical columns are called *groups* or *families*. Each *group* consists of elements that exhibit similar physical and chemical properties. The groups are numbered at the top of each column. Two numbering systems are frequently used: the newer system using Arabic numerals from 1 – 18 and the older system using Arabic or Roman numerals and the letters A and B. Some groups have common names. For example, elements in Group 1 or IA are referred to as alkali metals, Group 2 or IIA as alkaline earth metals, and Group 17 or VIIA as halogens. Elements in Group 18 or VIIIA are known as the noble gases or the inert gases and are generally unreactive. A zigzag line separates the periodic table into *metals* (to the left) and *nonmetals* (to the right). Most of the elements in contact with the line are referred to as *metalloids*. Metalloids exhibit properties of both metals and nonmetals.

Subatomic particles

Each element consists of a number of different subatomic particles. While there are more than 100 different types of subatomic particles, chemistry focuses on three: protons, neutrons and electrons. *Protons* (p^+) are positively charged particles, *neutrons* (n^0) are neutral (no charge), and *electrons* (e^-) are negatively charged. The protons and neutrons are found in the center of the atom called the *nucleus*. The electrons are found moving in the empty space surrounding the nucleus. Protons and neutrons have approximately the same mass, but electrons are so small that their mass is comparatively negligible. The *atomic number* is the whole number found on the periodic table with the elemental symbol and is equal to the number of protons. In a neutral atom, the number of protons must equal the number of electrons. The *mass number* of an element indicates the number of protons plus neutrons.

atomic number = number of protons (p^+) = number of electrons (e^-)
mass number = number of protons + number of neutrons ($p^+ + n^0$)

Atomic masses on the periodic table are generally not whole numbers because they are weighted averages of all the isotopes of that element.

Isotopes

Isotopes are atoms of the same element that differ only in the number of neutrons. Since isotopes are atoms of the same element, the number of protons must be the same and thus the atomic number remains the same. But since the mass number equals the sum of the protons and neutrons, it will differ with each isotope. Carbon has several different isotopes, three of which are shown in Table 1.

Table 1 *Principal Isotopes of Carbon*

Isotope	# Protons	# Neutrons	# Electrons	Atomic Number	Mass Number
carbon-12	6	6	6	6	12
carbon-13	6	7	6	6	13
carbon-14	6	8	6	6	14

These isotopes also may be represented using their isotope symbols:

carbon-12 ⇒ mass number / atomic number $^{12}_{6}C$ carbon-13 ⇒ $^{13}_{6}C$ carbon-14 ⇒ $^{14}_{6}C$

Carbon-12 is a stable isotope and is the most abundant naturally occurring isotope of carbon. In the medical field, carbon-13, also a stable isotope, is being used to determine metabolic changes in the brain using magnetic resonance imaging (MRI) spectroscopy. One goal is the early diagnosis of neuro-psychiatric disorders. Finally, carbon-14 is a radioactive isotope and is often used to determine the age of plant and animal items that were alive at one time. Although the three isotopes react the same chemically, they can be separated and identified by various means.

Electron configuration

The position of electrons is described by a notation system known as *electron configuration.* Electrons with similar energies are grouped into energy levels, also called shells. These energy levels (shells) are then divided into subshells. In the electron configuration notation the energy level is indicated with a numerical coefficient, the subshell is indicated with an *s, p, d,* or *f*, and the number of electrons in each subshell is indicated with a numerical superscript. The periodic table is arranged according to the subshell blocks (see Figure 1). Groups IA and IIA are the *s block.* The *p block* consists of Groups IIIA – VIIIA. The *d block* is found in the center of the periodic table with the elements known as the transition elements, Groups IB – VIIIB. The *f block* elements are the bottom two rows of the periodic table, the lanthanides and the actinides.

Figure 1 ***Subshell Blocks in the Periodic Table***

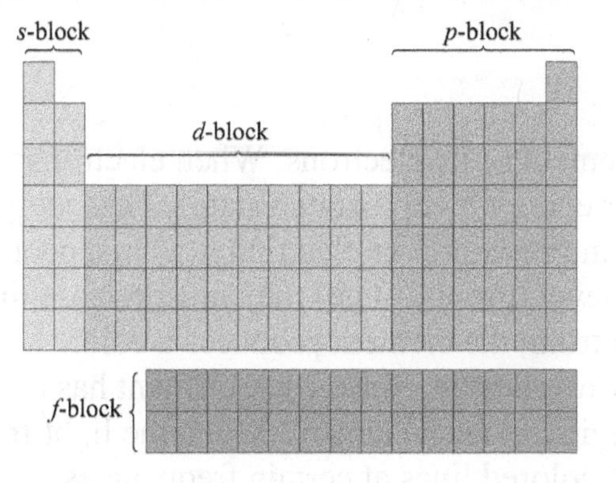

The last electron filled in an element's *electron configuration* can be determined by identifying which block the element is in and which period it is in. The following examples show the proper method of writing electron configuration:

Period 2:	Li	$1s^22s^1$	O	$1s^22s^22p^4$
Period 3:	Na	$1s^22s^22p^63s^1$	S	$1s^22s^22p^63s^23p^4$

Period 4 starts by filling the 4*s* block. After the 4*s* block has been filled, the next ten elements, scandium through zinc, fill the 3*d* block before filling the 4*p* block:

Sc $1s^22s^22p^63s^23p^64s^23d^1$ Zn $1s^22s^22p^63s^23p^64s^23d^{10}$

The energy level of each *d* block is one less than its period number. The *f* block follows the 6*s* block and the energy level of each *f* block is two less than the corresponding period number. For example, the electronic configuration of praseodymium (element 59) is $1s^22s^22p^63s^23p^64s^23d^{10}4p^65s^24d^{10}5p^66s^24f^3$.

Valence electrons are the electrons found in the outer shell of an atom. The *group number* of a representative element (an element in the *s*- or *p*-block) corresponds to the number of valence electrons. For example, in the older system, oxygen is in group VIA. The Roman numeral VI indicates that oxygen has six valence electrons. The group number can NOT be used to determine the number of valence electrons for the transition elements.

Periodic tendencies

The periodic table reflects trends of elemental properties that show *periodic* (cyclic) tendencies. Values for electronegativity (attraction of an atom for an electron), ionization energy (energy required to remove an electron from an atom), and atomic size will show a repetitive cycle when graphed versus the atomic numbers of the elements. In the 1800's Dmitri Mendeleev used these tendencies when arranging the elements to form the periodic table. He also recognized that elements were missing in the repetitive cycles and predicted the characteristics of elements that were discovered later.

Flame tests

The chemistry of an element is related to the arrangement of its electrons. When electrons absorb a specific amount of energy they jump to a higher energy level (excited state). This energy may result from the element being "excited" by heat (as in fireworks) or electricity (as in a neon sign). When the electrons return to their original energy level (ground state) they must release the absorbed energy. If this energy is in the visible light spectrum, the element produces a color. Elements in Group IA and IIA produce colorful flames when heated. Since each element has a unique arrangement of electrons, the color emitted is specific to each element. When the light from one of these flames is passed through a prism, a series of colored lines at certain frequencies appears, separated by dark spaces. This is known as a *line* or *emission spectrum* and is somewhat like a fingerprint that may be used to identify elements in the sun and other stars as well as in drugs, foods, and water. Helium, for instance, was discovered on the sun before it was discovered on Earth.

Experimental Procedures

Eye protection and appropriate clothing must be worn at all times.

Hydrochloric acid (HCl) is highly corrosive. Wash any contacted skin area immediately with running water and alert the instructor. Be sure to secure long hair and loose clothing when using a Bunsen burner.

Discard all wastes properly as directed by the instructor.

A. Physical properties of elements

1. Observe the elements provided for the experiment.
2. Record the symbol and atomic number for each element **in order of atomic number** (from low to high).
3. Record some of the physical properties observed for each element.
4. Using the observed properties, identify each element as a metal, nonmetal, or metalloid.

B. Periodic table

1. Write the elemental symbols and atomic numbers from Procedure A in the proper boxes on the blank periodic table in the report sheet. Write the group numbers on the top of the *representative* element vertical columns using **both** numbering systems. Write the period numbers for each of the horizontal rows. Draw a heavy line to separate the metals and nonmetals.
2. Six of the fourteen actinides were named for scientists. Write the elemental symbols and atomic numbers in the proper boxes on the blank periodic table in the report sheet for the elements named for the following famous scientists: Marie Curie (known for her pioneering work on radioactivity), Albert Einstein (developed the theory of general relativity), Dmitri Mendeleev (created the periodic table), and Alfred Nobel (inventor of dynamite and sponsor of the Nobel prizes).
3. Use colored pencils to label and shade the columns that contain the alkali metals, alkaline earth metals, halogens, and noble gases.

C. Subatomic particles

Complete the information for each element given: numbers of protons, neutrons, electrons, atomic number and mass number.

D. Isotopes

Complete the information for each of the isotopes of sulfur given: name, nuclear symbol, number of protons, neutrons and electrons.

E. Electron configuration

Complete the information in the table: electron configuration, group number and number of valence electrons.

F. Periodic tendencies

Use the information provided in Table 2 to graph the ionization energy of each element versus the atomic number of the element. Plot the atomic numbers on the horizontal axis and the ionization energy on the vertical axis. Be sure to connect the points on the graph, label each axis, and scale the graph to use as much of the graph paper as possible.

Table 2 ***Ionization Energy for Elements 1 to 20***

Element	Ionization Energy (kJ/mole)	Element	Ionization Energy (kJ/mole)
H	1311	Na	496
He	2370	Mg	737
Li	521	Al	576
Be	899	Si	786
B	799	P	1052
C	1087	S	1000
N	1404	Cl	1245
O	1314	Ar	1521
F	1682	K	419
Ne	2080	Ca	590

G. Flame tests

1. The instructor will demonstrate the proper procedures for cleaning the flame test wire with HCl and testing the solutions using a Bunsen burner.
2. Pour approximately 20 mL of HCl in a 50 mL beaker. Dip the loop end of the flame test wire into the solution and then hold the loop in the Bunsen burner flame, as shown below. The hottest part of the flame is at the tip of the inner cone. Repeat the cleaning process until the flame does not change color.

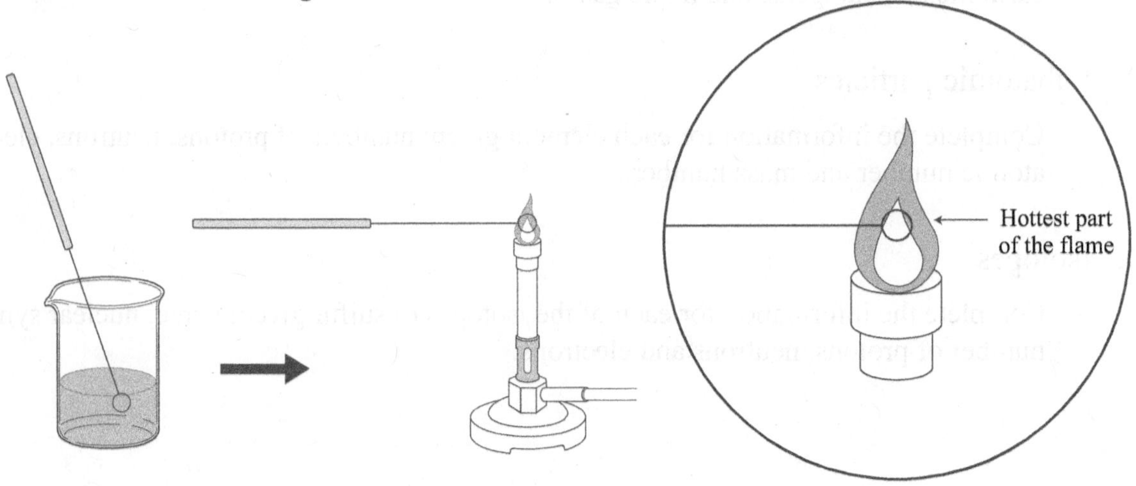

3. Place several drops of each solution to be tested into separate wells of a spot plate. Draw a sketch of the spot plate and label the positions of the solutions.
4. Dip the cleaned wire into one of the solutions on the spot plate. Hold the loop in the hottest part of the flame. The color change is often short-lived. Repeat the test until a good color change is observed. The color change is due to the metallic ion in each compound (i.e., Na^+ or Ba^{2+}). Record the color change.
5. Clean the wire with HCl after testing each solution.
6. Test the remaining solutions and record the color change for each.
7. Obtain one or more unknown solutions as directed by the instructor. Use the flame test procedure to identify the solutions. Be sure to record the unknown number(s) on the report sheet.
8. It may be necessary to retest some of the known solutions to verify the colors in order to match an unknown to a known.

Name ______________________ Date ______________________

Partner(s) ______________________ Section ______________________

______________________ Instructor ______________________

1. For the following elements, underline the transition elements, circle the noble gases, and draw a rectangle around the halogens.

   Cl Ne Fe Cu I Rn

2. Write the name and symbol for each of the elements described below:

   a) halogen in period 4 ______________ ______

   b) alkali metal in period 2 ______________ ______

   c) element in group IIIA, period 3 ______________ ______

   d) noble gas in period 1 ______________ ______

   e) alkaline earth metal in period 3 ______________ ______

3. Describe the procedure that should be carried out before performing each individual flame test.

4. While cooking pasta, some of the salted water spills onto the gas burner. The gas flame momentarily changes to a bright orange color. Explain what has happened.

5. How long should eyes be rinsed in the eyewash after they are exposed to chemical contact?

Name ______________________ Date ______________________

Partner(s) ______________________ Section ______________________

______________________ Instructor ______________________

A. Physical properties of elements

Element	Symbol	Atomic number	Color	Luster	Metal, Nonmetal, or Metalloid?

B. Periodic table

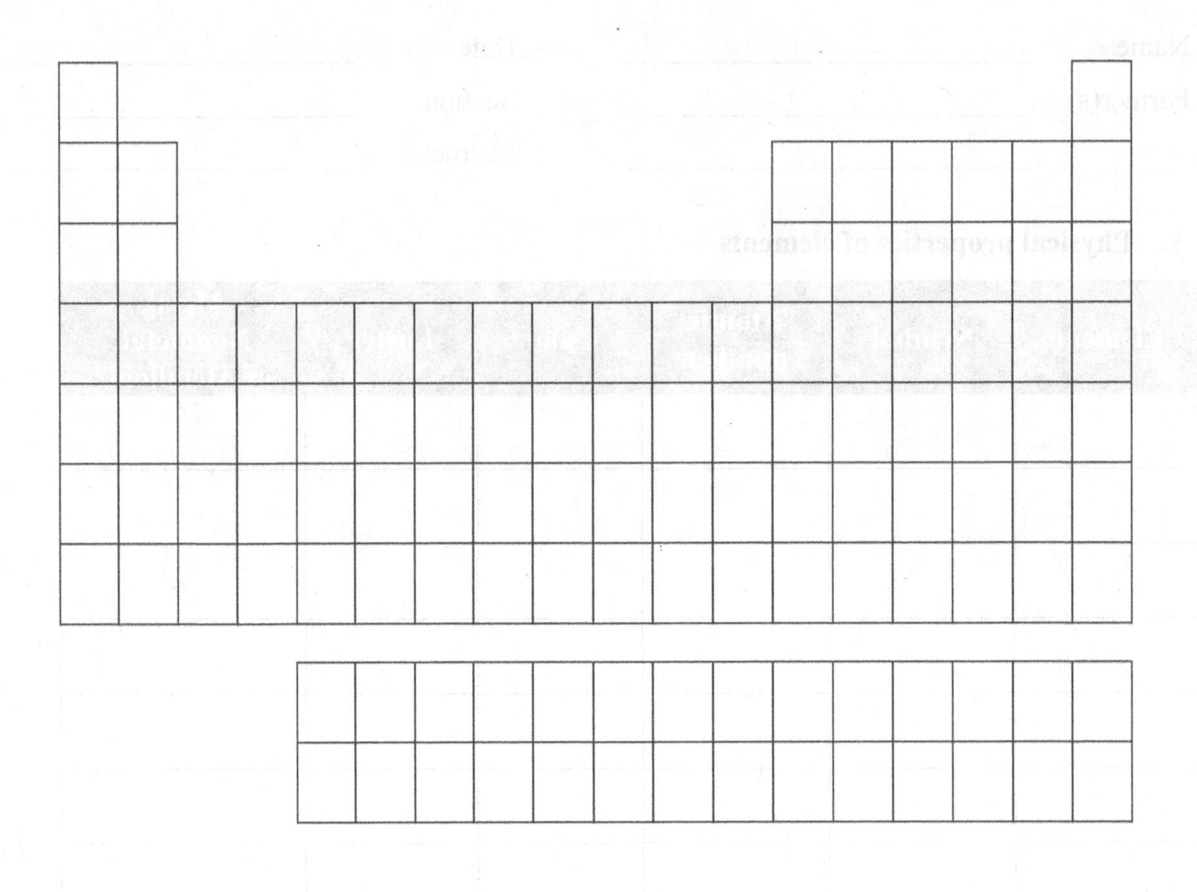

C. Subatomic particles

Element	Atomic number	Mass number	Protons	Neutrons	Electrons
sodium				12	
	6	12			
				8	8
copper		64			
	1			0	
chlorine		35			

D. Isotopes

Isotope	Isotope Symbol	Protons	Neutrons	Electrons
sulfur-31				
	$^{34}_{16}S$			
		16	21	

E. Electron configuration

Element	Electron configuration	Group Number	Number of valence electrons
H			
Be			
N			
Ne			
Na			
Mg			
Si			
Cl			
Ca			
V			
Ge			
Kr			

F. Graphing periodic properties

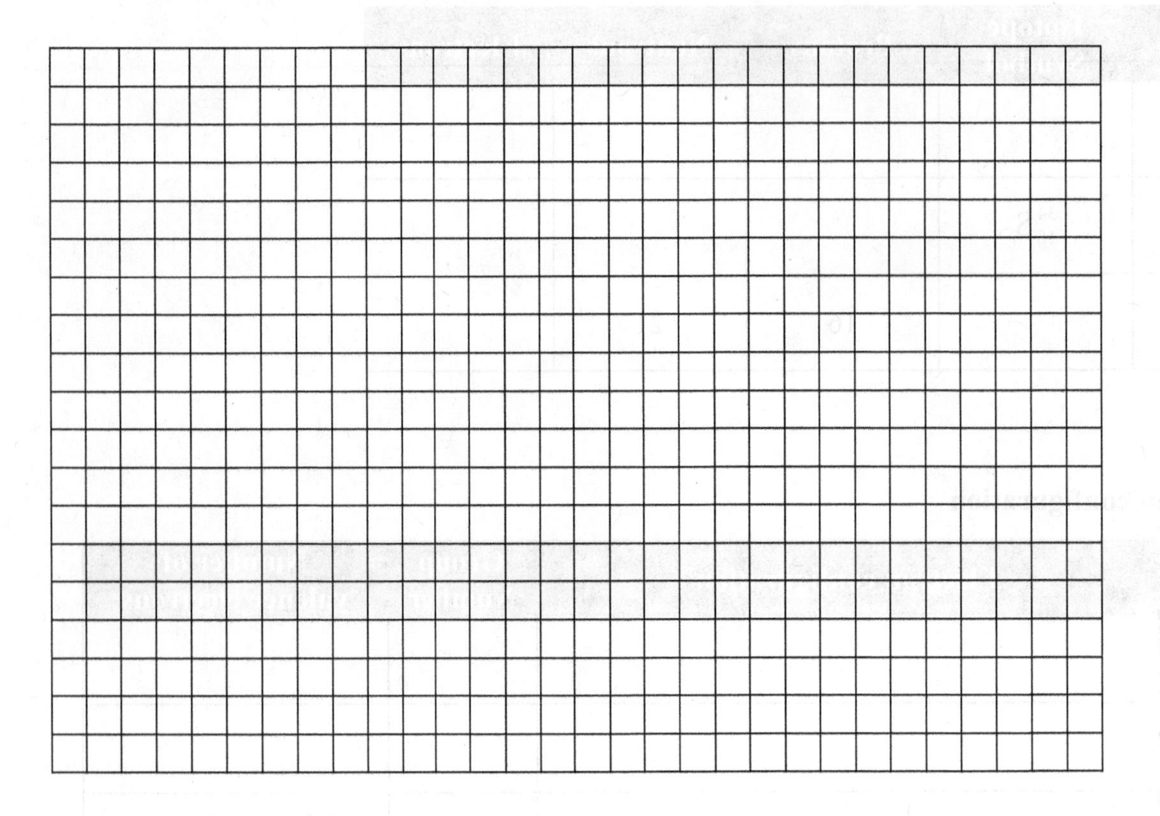

G. Flame tests

Solution	Metal Ion	Color
NaCl		
KCl		
$BaCl_2$		
$CaCl_2$		
$SrCl_2$		
$CuCl_2$		
Unknown #______		
Unknown #______		
Unknown #______		

Problems and questions

1. Write the symbol for the element that meets the following conditions:

    ________has 4 electrons in shell 2

    ________has five 3*p* electrons

    ________has 1 electron in the 3*d* subshell

    ________has 2 electrons in the 5*p* subshell

    ________completes the 6*s* subshell

2. Explain why elements in group VIIIA don't normally form compounds.

3. Several different elements were identified using the flame tests. Which of these elements would be a good choice for a fireworks display? Why?

4. Name two elements that should exhibit properties of metalloids.

Goals

- Investigate how three forms of radiation pass through different types of materials.
- Identify the type of radiation produced by various nuclides by measuring the amount of radiation penetration through shielding materials.
- Examine the effectiveness of different thicknesses of shielding against three forms of radiation.
- Examine the relationship between the intensity of radiation from a radioactive source and the distance from the source.
- Estimate the age of various samples using radiocarbon dating.

Materials

The website address for the experiment is
http://highered.mheducation.com/sites/0073511250/student_view0/nuclear_chemistry_online_lab.html.

Discussion

In nuclear chemistry discussions, specific atoms are usually called *nuclides* rather than isotopes. Nuclides may be divided into two types based on nuclear stability: stable nuclides and unstable nuclides. A *stable nuclide* is an atom with a stable nucleus that does not readily undergo change. Conversely, an *unstable nuclide* contains an unstable nucleus that spontaneously undergoes change. Unstable nuclides emit radiation from the nucleus (nuclear decay) to form a more stable nucleus. These radioactive isotopes (nuclides) are usually called *radioisotopes* or *radionuclides*.

Different forms of radiation are emitted when a radionuclide is converted to a more stable nuclide, including alpha particles, beta particles, and gamma rays. There are other forms of radiation, but this experiment focuses on the three above.

Types of radiation

An alpha particle is a high-energy particle that contains two protons and two neutrons. Alpha particles are identical to the nuclei of helium-4 atoms. Textbooks commonly use one of three symbols for an alpha particle:

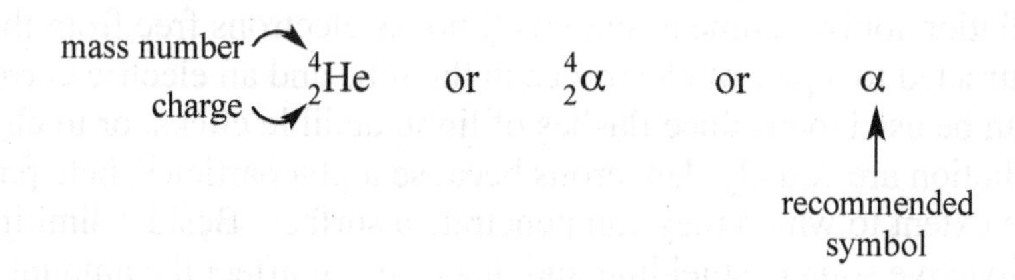

The subscript indicates that the particle has a +2 charge (from the two protons), while the superscript indicates that that the particle has a mass of 4 u (an older abbreviation for atomic masses, *amu*, may be used in some textbooks).

A beta particle is a high-energy particle whose mass and charge are identical to an electron. Textbooks commonly use one of three symbols for a beta particle:

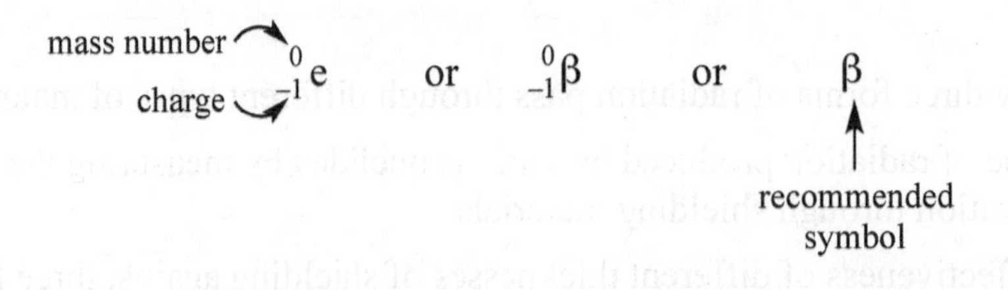

The subscript indicates that that the particle has a –1 charge. Beta particles have a mass, but the number is so small (0.00055 u) it is usually rounded to zero.

A gamma ray is a form of high-energy radiation without mass or charge that is emitted from a radionuclide. In one of the two symbols for a gamma ray, the zeros are shown explicitly:

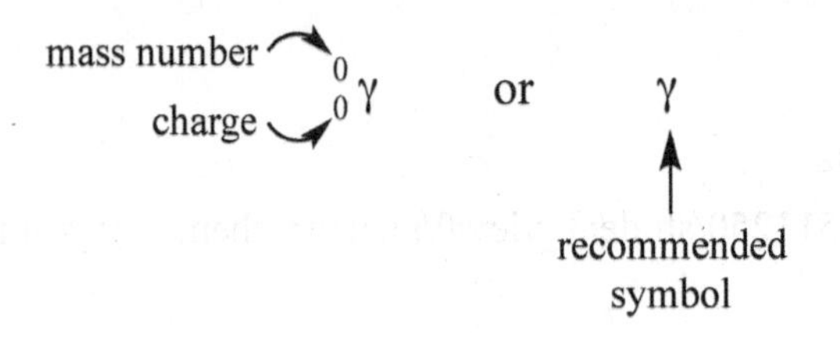

Effects of Radioactivity

Radiation of various types can be classified as *nonionizing radiation* or *ionizing radiation*. Nonionizing radiation is radiation with insufficient energy to remove an electron from an atom or molecule. Examples of nonionizing radiation include radio waves, microwaves, infrared light, and visible light.

Ionizing radiation is radiation with sufficient energy to remove an electron from an atom or molecule. Radiation associated with radioactive decay is ionizing radiation. Additionally, cosmic rays, X-rays, and ultraviolet light are also forms of ionizing radiation.

When ionizing radiation passes through the cells of the body, electrons are knocked away from atoms or molecules, forming ions and free radicals. The ions and free radicals are very reactive and interact with many different biological molecules, in turn damaging or killing the cell. The more time spent near a radioactive source, the greater the amount of radiation received and the more damage that can occur.

In order to detect radiation, a device called a Geiger-Müller tube (also known as a Geiger counter) is used. Radiation passes through a tube containing an inert gas (usually helium or argon) at low pressure. The radiation ionizes some gas atoms (knocks electrons free from the atoms). The ions and electrons are attracted to a pair of electrodes in the tube and an electric current is produced. The current produced can be used to produce flashes of light, audible clicks, or to change a counter.

Not all forms of radiation are equally dangerous because alpha particles, beta particles, and gamma rays differ in the extent to which they can penetrate a surface. Besides limiting the amount of time spent near a radioactive source, shielding and distance can affect the amount of radiation received.

Shielding

Alpha particles are the most massive and also the slowest particles (on the order of one-tenth the speed of light) of the three principal forms of radiation. Because they are slow, alpha particles have low penetrating power and cannot penetrate the body's outer (dead) layers of skin. A piece of paper will stop alpha particles. If inhaled or ingested, however, alpha particles can cause serious damage to internal organs.

Beta particles are much faster (on the order of nine-tenths the speed of light) and can penetrate much deeper than alpha particles. While beta particles are absorbed before they can reach internal organs, they can be damaging to the skin. Thicker shielding such as heavy clothing and gloves, or metals such as aluminum are needed to stop these particles.

Finally, gamma rays travel at the speed of light and they readily penetrate deeply into the body. Thick lead or concrete is needed to stop gamma rays. X-rays and gamma rays are similar except that X-rays originate from electron fields surrounding the nucleus or are machine-produced while gamma rays originate from inside the nucleus. Also, X-rays are of lower energy. The X-rays used in diagnostic medicine have about 10% of the energy of gamma rays. However, X-rays are still a form of ionizing radiation and can damage the body. Modern diagnostic medicine strives to use the lowest possible X-ray energy level to avoid unnecessary tissue damage.

Distance

The farther away people are from a radiation source, the less their exposure. Distance is a prime concern when dealing with gamma rays because they can travel long distances. Alpha particles and beta particles don't have enough energy to travel very far. A typical alpha particle travels about 6 cm in air while a typical beta particle travels about 1000 cm in air.

The intensity of the radiation follows an inverse-square law. In mathematical terms:

$$\text{Intensity} \propto \frac{1}{\text{distance}^2}$$

If the distance doubles, the intensity of the radiation is one-fourth the original value. Halving the distance increases the intensity by a factor of 4. If the intensity of the radiation and distance are measured at two different distances, then the following equation describes the relationship:

$$\frac{I_i}{I_f} = \frac{D_f^2}{D_i^2}$$ where "i" indicates initial conditions and "f" indicates final conditions.

Half-life

Radionuclides decay at different rates. Some decay very rapidly, while others decay slowly. The greater the decay rate, the lower the stability of the nuclide. The *half-life* ($t_{1/2}$) of a radionuclide is a number that conveys nuclear stability. The half-life of a radionuclide is the time required for one-half of a given quantity of a radionuclide to decay. For example, iodine-131 has a half-life of 8 days. When a 4.00 mg sample decays, only 2.00 mg of the sample (one-half of the original amount) will remain undecayed after one half-life; the other half will have decayed into some other substance. After two half-lives, one-fourth of the material remains ($\frac{1}{2} \times \frac{1}{2}$). After three half-lives, one-eighth of the material remains ($\frac{1}{2} \times \frac{1}{2} \times \frac{1}{2}$), etc. This is seen graphically below.

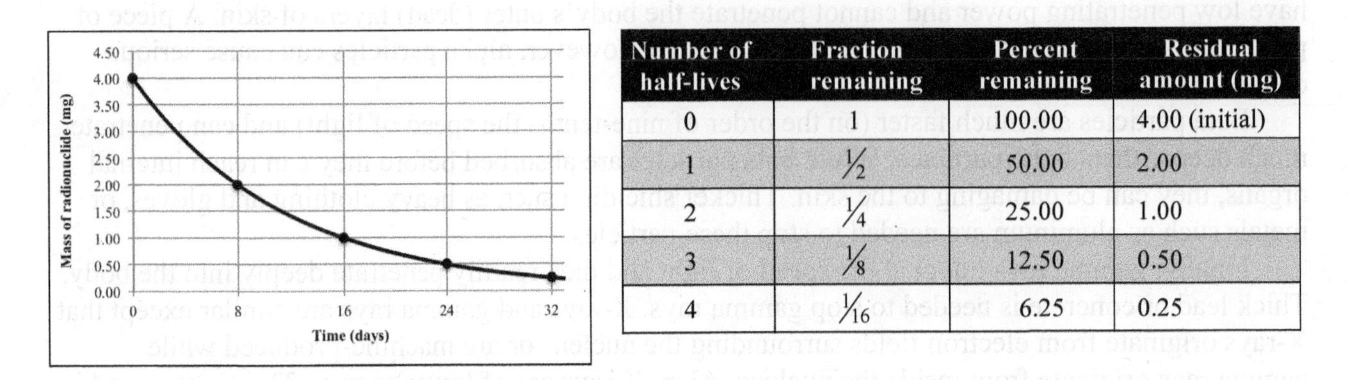

Number of half-lives	Fraction remaining	Percent remaining	Residual amount (mg)
0	1	100.00	4.00 (initial)
1	$\frac{1}{2}$	50.00	2.00
2	$\frac{1}{4}$	25.00	1.00
3	$\frac{1}{8}$	12.50	0.50
4	$\frac{1}{16}$	6.25	0.25

The half-life of a radionuclide is constant. It is independent of physical conditions such as temperature, pressure, and quantity. If a nuclide is radioactive, nothing will stop it from decaying and nothing will increase or decrease its decay rate.

There are equations for determining half-life or time elapsed if measurements of concentration and time are taken, but they are usually reserved for more advanced chemistry classes. In this lab, half-life or time elapsed can be approximated by reading and interpreting a graph.

There is a wide range of half-lives. Some radionuclides have long half-lives measured in billions of years. They can be used to determine the age of rocks or of the Earth. Other radionuclides have short half-lives measured in fractions of a second.

Radiocarbon dating

When archaeologists report that a scroll was written in Roman times, or a mummy has been dated to 1000 B.C., or a piece of bone from a human is 10,000 years old, how can they be certain? Archaeologists use the half-life of the radionuclide carbon-14 to determine the age of carbon-containing material derived from plants or animals (such as wood, fiber, natural pigments, bone, and cotton and woolen clothing).

Carbon-14 is created in the upper atmosphere by the bombardment of ${}^{14}_{7}N$ by high-energy neutrons from cosmic rays.

$$ {}^{14}_{7}N + {}^{1}_{0}n \longrightarrow {}^{14}_{6}C + {}^{1}_{1}H $$

Nitrogen in atmosphere	Neutron from cosmic rays	Radioactive carbon-14	Proton

The carbon-14 reacts with oxygen in the atmosphere to form radioactive carbon dioxide, which eventually makes its way into the biosphere where it is absorbed by living plants during photosynthesis. Animals eat the plants, ultimately exhaling the ${}^{14}_{6}C$ again as CO_2. In this way, a steady concentration of ${}^{14}_{6}C$ is maintained in living tissue. Although ${}^{14}_{6}C$ emits ionizing radiation and can therefore damage cells, the natural abundance of ${}^{14}_{6}C$ is very small: there is only one atom of ${}^{14}_{6}C$ in every 10^{12} atoms of carbon.

Once an organism dies, the ${}^{14}_{6}C$ is no longer replenished and the amount of ${}^{14}_{6}C$ steadily decreases as it undergoes β decay:

$$ {}^{14}_{6}C \longrightarrow {}^{14}_{7}N + \beta $$

In a process called radiocarbon dating, scientists use the half-life of carbon-14 (5730 years) to calculate the length of time since the plant or animal died. For example, a wooden box found in an ancient hut might have one-fourth the carbon-14 found in a living tree, indicating that two half-lives have passed. Therefore, it can be concluded that the tree from which the box was made was cut down about 11,460 years ago ($2 \times 5730 = 11460$).

If the sample undergoing radiocarbon dating is either too young or too old, this method will not work. If a piece of bone is only a couple of years old, there will have been too little decay of ${}^{14}_{6}C$ to be able to measure a decrease. Likewise, if it is more than about 40,000 years old, there will be too little ${}^{14}_{6}C$ left to measure accurately.

Experimental Procedures

1. Go to the website provided by the instructor.
2. Each of the five experiments online has a tab for "Lab Manual" and a tab for "Worksheets." Printed versions of the "Lab Manual" tabs are provided on the next five pages. Using the online "Lab Manual" tab is optional.
3. For the first four experiments, follow these directions:
   a. Open the "Worksheet" tab and enter all data into the Excel worksheet. The graphs (if applicable) will be automatically generated.
   b. Transfer all data and graphs to the printed report sheets.
   c. Do not print out the computer-generated worksheets.
4. For experiment 5, the graphs are provided and will be used to determine the age of the objects.
5. Turn in your:
   a. Written pre-lab
   b. "Report Sheets" including all data and graphs for the five experiments
   c. "Problems and questions" including each Analysis for the five experiments

Experiment 1: Radiation and Matter

1. Open the worksheet for Experiment 1. Enter the data within the Excel spreadsheet. Follow the directions from your instructor regarding how to submit the worksheet.
2. Move the cursor over the apparatus until the "lighted area" appears. Click on this to start the experiment. The experimental setup includes three radioactive sources (one each of an α, β, and γ emitter), different types of shielding, and the apparatus to measure the radioactivity. On the right is a Geiger counter that will measure radiation in counts per second. On the left is a holder for the radioactive source, and in the middle is a holder for the shielding material.
3. Click on the drop-down list of radioactive sources and choose radon-222, an α emitter. Drag the source into the sample holder. (There is information on each nuclide in the box that pops up. After reading it, you can close the box, or it will close on its own when you click on something else.)
4. Click on the Geiger counter switch to turn it on. Read the activity from the gauge. The needle on the gauge may move around a bit. You should try to get an average reading. (Note that the scale on the gauge is not linear.) Record the activity of the sample in counts/sec in the worksheet.
5. Click the switch again to turn it off.
6. Click on the drop-down list of shielding materials and choose one. Drag the shielding into its holder in the apparatus.
7. Again click on the Geiger counter switch, and again record the activity in the appropriate column in the worksheet.
8. Repeat steps 6 and 7 using the other types of shielding material. Record the activity values in the appropriate cells of the worksheet.
9. Repeat steps 3–8 for the other radioactive sources: iron-59 (β) and strontium-85 (γ).

Experiment 2: Types of Radiation

1. Open the worksheet for Experiment 2. Enter the data within the Excel spreadsheet. Follow the directions from your instructor regarding how to submit the worksheet.
2. Move the cursor over the apparatus until the "lighted area" appears. Click on this to start the experiment. The experimental setup includes a number of radioactive sources, different types of shielding, and the apparatus to measure the radioactivity. On the right is a Geiger counter that will measure radiation in counts per second. On the left is a holder for the radioactive source, and in the middle is a holder for the shielding material.
3. Click on the drop-down list of radioactive sources and choose a nuclide. Drag the source into the sample holder. Record the name (e.g., iron-59) or the symbol (e.g., ^{59}Fe) of the source in the worksheet. (There is information on each nuclide in the box that pops up. After reading it, you can close the box, or it will close on its own when you click on something else.)
4. Click on the Geiger counter switch to turn it on. Record the activity of the sample in counts/sec. The needle on the gauge may move around a bit. You should try to get an average reading. (Note that the scale on the gauge is not linear.)
5. Click the switch again to turn it off.
6. Click on the drop-down list of shielding materials and choose one. Drag the shielding into its holder in the apparatus.
7. Again click on the Geiger counter switch, and again record the activity in the appropriate column in the worksheet.
8. Repeat steps 6 and 7 for the other types of shielding material.
9. Repeat steps 3–8 until you have identified the type of radiation given off by several nuclides. Try to find at least one for each type of radiation: α, β, and γ.

Experiment 3: Shielding

1. Open the worksheet for Experiment 3. Enter the data within the Excel spreadsheet so that the graph can be created. After completing the experiment, copy all data and graphs to the printed report sheet.
2. Move the cursor over the apparatus until the "lighted area" appears. Click on this to start the experiment. The experimental setup includes three radioactive sources (one each of an α, β, and γ emitter), several 1-mm thick sheets of the three types of shielding, and the apparatus to measure the radioactivity. On the right is a Geiger counter that will measure radiation in counts per second. On the left is a holder for the radioactive source, and in the middle is a holder for the shielding material.
3. Click on the drop-down list of radioactive sources and choose radon-222, an α emitter. Drag the source into the sample holder. (There is information on each nuclide in the box that pops up. After reading it, you can close the box, or it will close on its own when you click on something else.)
4. Click on the Geiger counter switch to turn it on. Read the activity from the gauge. The needle on the gauge may move around a bit. You should try to get an average reading. (Note that the scale on the gauge is not linear.) Record the activity of the sample in counts/sec in the worksheet.
5. Click the switch again to turn it off.
6. Click on the drop-down list of shielding materials and choose a single sheet (1 mm) of one of them. Drag the shielding into its holder in the apparatus.
7. Again click on the Geiger counter switch, and again record the activity in the appropriate column in the worksheet.
8. Repeat steps 6 and 7 for the other types of shielding material.
9. Click on the drop-down list of radioactive sources and choose iron-59, a β emitter. Drag the source into the sample holder and measure its radioactivity.
10. Choose one of the shielding materials and measure the radioactivity using 1, 2, 3, and 4 mm thicknesses. Record the activity values in the appropriate cells of the worksheet. Note the graph developing as you enter data.
11. Repeat step 10 for the other shielding materials.
12. Repeat steps 9–11 for strontium-85, a γ emitter.

Experiment 4: Radiation Intensity vs Distance

1. Open the worksheet for Experiment 4. Enter the data within the Excel spreadsheet so that the graph can be created. After completing the experiment, copy all data and graphs to the printed report sheet.
2. Move the cursor over the apparatus until the "lighted area" appears. Click on this to start the experiment. The experimental setup includes the radioactive source (gallium-67, a γ emitter), a ruler to measure the distance from the source to the detector, and a Geiger counter to measure the radiation in counts per second.
3. Click on the drop-down list to retrieve the source. Drag the source into the holder.
4. Click on the 1-cm mark on the ruler to position the source holder at 1 cm distance from the detector.
5. Click on the Geiger counter switch to turn it on. Record the activity of the sample in the appropriate cell in the worksheet. The needle on the gauge may move around a bit. You should try to get an average reading. (Note that the scale on the gauge is not linear.)
6. Click the switch again to turn it off.
7. Click at a point on the ruler to move the source a distance away from the detector.
8. Again click on the Geiger counter switch, and record the activity in the appropriate cell in the worksheet.
9. Repeat steps 6 and 7 for the other distances in the worksheet.

Experiment 5: Radiocarbon Dating

1. Open the worksheet for Experiment 5. Enter the data within the Excel spreadsheet. After completing the experiment, transfer the data to the report sheet.
2. Move the cursor over the apparatus until the "lighted area" appears. Click on this to start the experiment. The experimental setup includes a number of samples of ^{14}C isolated from objects discovered in archaeological excavations, and the apparatus to measure the radioactivity. On the right is a Geiger counter that will measure radiation in counts per second. On the left is a holder for the sample.

Determining the graphs for the "present-day" materials

3. Using the table for the "present-day" materials, select the first sample and drag that sample into the sample holder until it "clicks." (There is information on each sample in the box that pops up. After reading it, you can close the box, or it will close on its own when you click on something else.)
4. Click on the Geiger counter switch to turn it on. Record the activity of the sample in counts/sec. The needle on the gauge may move around a bit. You should try to get an average reading. (Note that the scale on the gauge is not linear.)
5. Type the sample's activity into the "Initial Activity" cell above the graph and Enter. This will generate a decay curve for ^{14}C based on that activity. Notice the values on each axis of the graph that is generated and select the matching pre-printed graph on the report sheet in your lab manual. Enter the name of the material into the title box. (Note that the shape of each graph's curve is the same but the values on the axis differ.)
6. Repeat steps 3–5 for each of the other "present-day" materials. You should now have each of the three graphs labeled correctly as paper, bone, or wood as determined by the values on the graph axis.

Determining the age of the "archeological" samples

7. Select the first "archeological" sample and place it in the holder. Record the activity of the sample on the table.
8. Repeat step 7 for each of the "archeological" samples.
9. Identify all of the "archeological" samples that are wood. Use the wood graph to determine the age of each sample and record your answers in the table in the lab manual.
10. Repeat step 9 for all of the paper "archeological" samples and all of the bone "archeological" samples. Be sure to use the correct corresponding graph.

Name	__________	Date	__________
Partner(s)	__________	Section	__________
	__________	Instructor	__________

Become familiar with a radioisotope used in the heath field. The instructor will assign one isotope to each student the week prior to the lab. CIRCLE YOUR ASSIGNED ISOTOPE. Information may be found in textbooks, online, or other reference materials. Write a paragraph describing the isotope. Include radiation type, half-life, uses in the medical field, uses in other fields, special precautions needed when handling, source, cost, and any other relevant details. Turn in the written information and prepare a brief (1 – 2 minute) oral summary to present in class.

Isotope assignment list:

C-11	Ca-47	Ce-141	Co-57	Co-60	Cr-51
Cs-137	Fe-59	Ga-67	Ga-68	Gd-153	H-3
I-123	I-125	I-131	Ir-192	Na-24	P-32
Sr-85	Sr-89	Sr-90	Tc-99*	Tl-201	Xe-133

Name ____________________ Date ____________________

Partner(s) ____________________ Section ____________________

____________________ Instructor ____________________

Experiment 1: Radiation and Matter

Source	Type of radiation	Activity (counts/sec)				
		No shielding	Paper	1 mm Cardboard	1 mm Aluminum	1 mm Lead
Radon-222	alpha					
Iron-59	beta					
Strontium-85	gamma					

Experiment 2: Types of Radiation

Source	Activity (counts/sec)					Type of radiation
	No shielding	Paper	1 mm Cardboard	1 mm Aluminum	1 mm Lead	

Experiment 3: Shielding

Type of radiation:	alpha particles
Source:	radon-222

Shielding	Activity (counts/sec)
No shielding	
1 mm cardboard	
1 mm aluminum	
1 mm lead	

Type of radiation:	beta particles
Source:	iron-59

Shield thickness (mm)	Activity (counts/sec)		
	Cardboard	Aluminum	Lead
0			
1			
2			
3			
4			

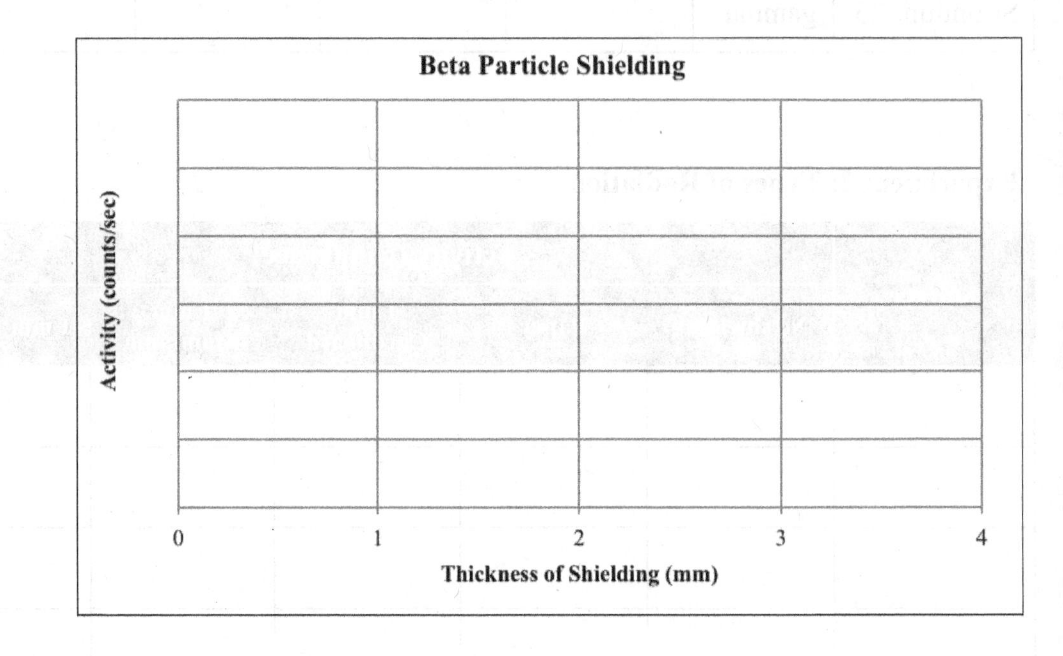

Type of radiation: gamma rays
Source: strontium-85

Shield thickness (mm)	Activity (counts/sec)		
	Cardboard	Aluminum	Lead
0			
1			
2			
3			
4			

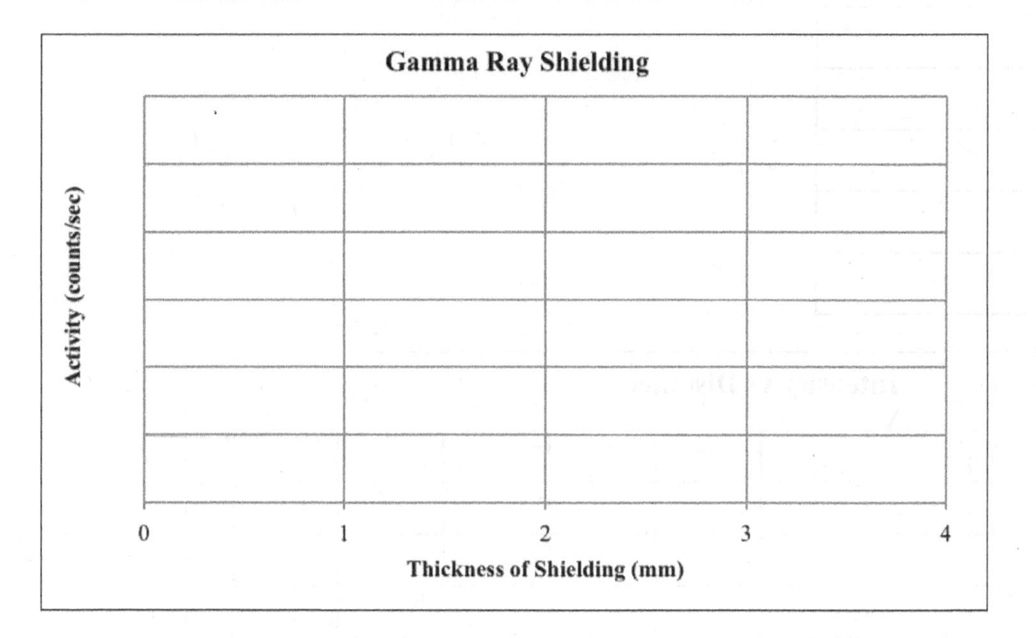

Experiment 4: Radiation Intensity vs Distance

Type of radiation:	gamma rays
Source:	gallium-67

Distance (cm)	Activity (counts/sec)
1	
2	
3	
4	
5	
6	
7	
8	
9	
10	

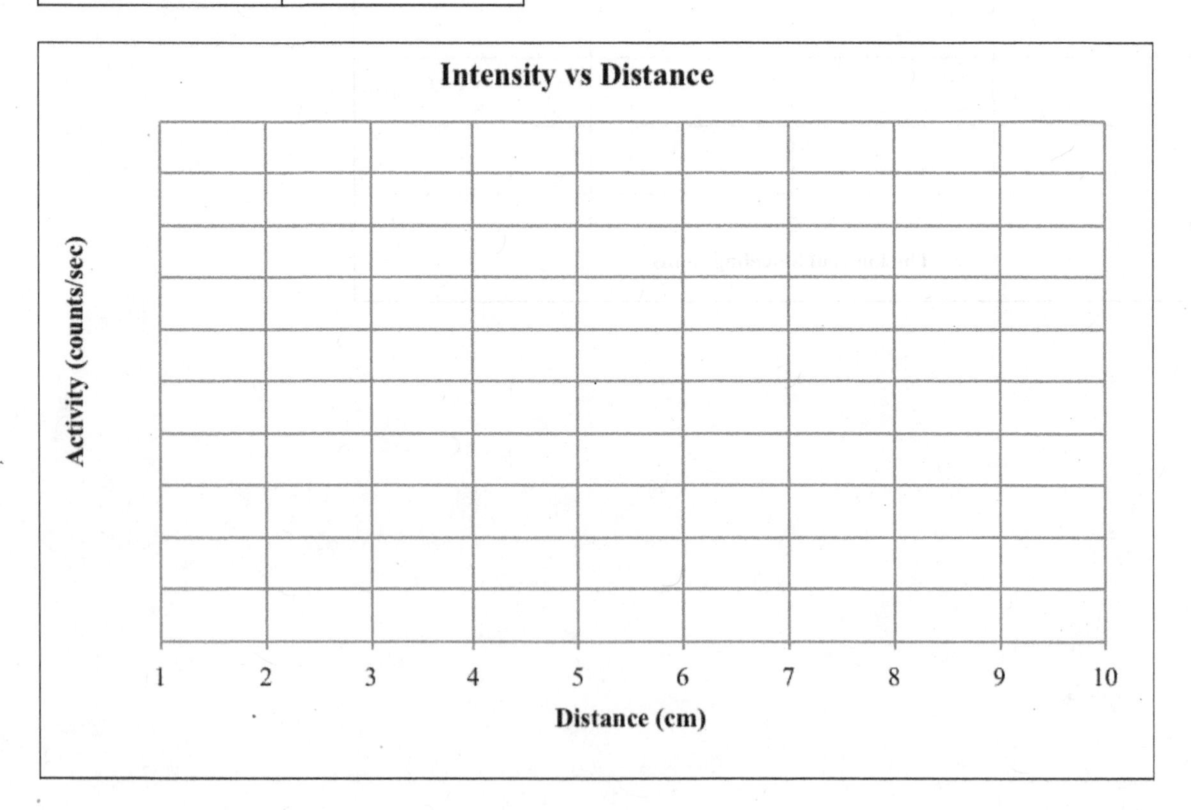

Experiment 5: Radiocarbon Dating

Present day materials	
Sample	**Activity (counts/sec)**
wood	
paper	
bone	

Archaelogical objects		
Sample	**Activity (counts/sec)**	**Age (years)**
bone, Bering land bridge		
wood, tomb of pharaoh Zoser		
bone, La Brea tar pits		
paper, Dead Sea scrolls		
skull, Laguna Beach, CA		
wooden timber, Stonehenge		
bone, Pedra Furada, Brazil		
wooden beam, India		

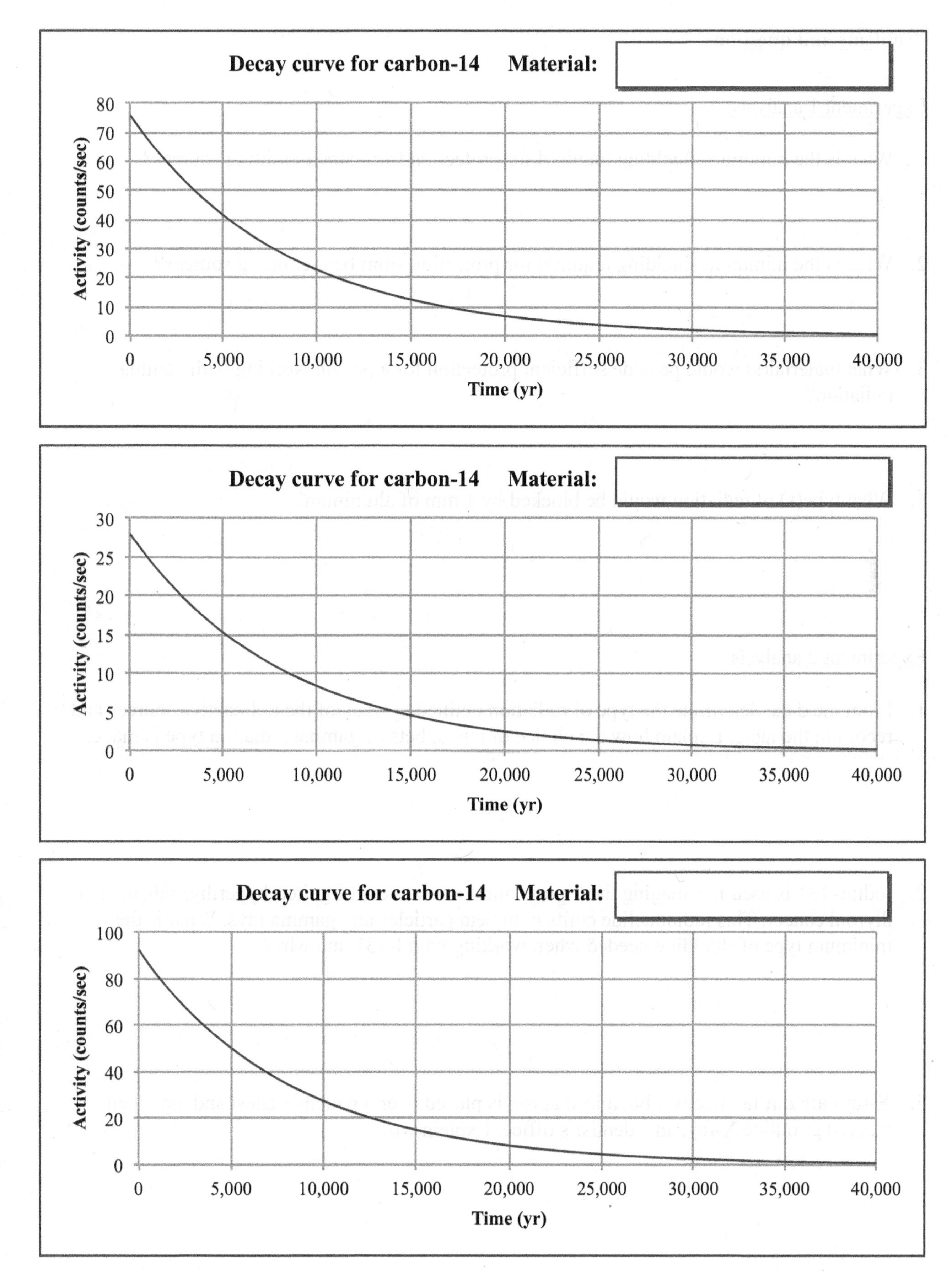

Decay curve for carbon-14
Material:
Activity (counts/sec)
80
70
60
50
40
30
20
10
0
0
5,000
10,000
15,000
20,000
25,000
30,000
35,000
40,000
Time (yr)
Decay curve for carbon-14
Material:
Activity (counts/sec)
30
25
20
15
10
5
0
0
5,000
10,000
15,000
20,000
25,000
30,000
35,000
40,000
Time (yr)
Decay curve for carbon-14
Material:
Activity (counts/sec)
100
80
60
40
20
0
0
5,000
10,000
15,000
20,000
25,000
30,000
35,000
40,000
Time (yr)

Problems and questions

Experiment 1 analysis:

1. What is the minimum shielding required for protection from alpha-emitting sources?

2. What is the minimum shielding required for protection from beta-emitting sources?

3. What material(s) would provide sufficient protection for a person working with gamma radiation?

4. What type(s) of radiation would be blocked by 1 mm of aluminum?

Experiment 2 analysis:

1. From the data, determine the type of radiation emitted by each of the radioactive sources and record in the table. Explain how the choice of alpha, beta or gamma radiation type is made.

2. Iodine-131 is used for imaging the thyroid and for treatment of goiter, hyperthyroidism, and thyroid cancer. This radionuclide emits both beta particles and gamma rays. What is the minimum type of shielding needed when working with I-131 and why?

3. X-rays are **not** radioactive, but a lead apron is placed over a patient's chest and lap when receiving routine X-rays in a dentist's office. Explain why.

Experiment 3 analysis:

1. What was the most effective type of shielding against alpha particles?

2. Use the graph to estimate the percent of beta particle radiation that passed through:

   a) One sheet of cardboard __________

   b) Two sheets of cardboard __________

3. How many sheets of cardboard would be required to block 99% of beta particles?

4. Use the graph to estimate the percent of beta particle radiation that passed through:

   a) One sheet of aluminum __________

   b) Two sheets of aluminum __________

5. How many sheets of aluminum would be required to block 99% of beta particles?

6. What was the most effective type of shielding against gamma rays?

7. What percentage of the gamma radiation passed through four sheets of lead?

   Does this provide adequate shielding? Explain.

Experiment 4 analysis:

1. Describe in words the relationship between radiation intensity and distance from the source.

2. Rearrange the mathematical formula describing the relationship between intensity and distance to solve for the final intensity.

3. Phosphorus-32 is a beta-emitter used to treat leukemia and pancreatic cancer. When a sample was placed 1.00 cm from the detector, the intensity of the radiation was measured at 491 counts per second. Calculate its intensity at:

   a) 2.00 cm from the detector __________

   b) 5.00 cm from the detector __________

   Explain how your answers reflect your statement in question #1.

4. X-rays are frequently performed in medical and dental offices. Although X-rays are not radioactive, what precaution must technicians (not the patients) take to protect themselves and why?

Experiment 5 analysis:

Use the appropriate carbon-14 decay curve from the report sheet to help answer the following questions.

1. On September 19, 1991 two hikers in the Ötztal Alps discovered the body of a man sticking half-way out of the ice. Originally, it was thought that the man, nicknamed "Ötzi," was about 500 years old, but the artifacts near the body hinted at a much older age. Radiocarbon-14 dating of milligram amounts of bone revealed that Ötzi died about 5,250 years ago. What was the carbon-14 activity (in counts/sec) of the bone fragments?

2. A collector put up for auction a papyrus (a paper-like material) supposedly written by a scribe in Egyptian pharoah Tutankhamun's court. Tutankhamun ruled about 1332 B.C. – 1323 B.C. A cautious buyer had a small piece of the papyrus dated. The activity of carbon-14 measured was 71 counts/sec. Is the papyrus genuine, or is it a forgery? Explain how you arrived at your conclusion.

3. A wooden object from the site of an ancient temple has a carbon-14 activity of 50 counts/sec. In what **year** (including A.D. or B.C) did the tree die that was used to make the object? (Hint: first determine how old the object is.)

Some of the procedures for this lab may be completed before the lab period begins as directed by the instructor.

Goals

- Compare the physical properties of compounds and their elements.
- Identify a compound as ionic or covalent.
- Write Lewis dot structures for compounds.
- Write correct names and formulas for compounds.

Materials

- Display of elements and compounds
- Fe metal filings
- S powder
- Fe and S mixture
- FeS solid
- Magnet
- *Merck Index* or *CRC Handbook of Chemistry and Physics*
- 0.1 M solutions of $FeSO_4$, $AgNO_3$, and $Al(NO_3)_3$ labeled for students as:
  Cation #1—Fe^{2+}
  Cation #2—Ag^{+}
  Cation #3—Al^{3+}
- 0.1 M solutions of Na_2S, NaOH, and Na_2CO_3 labeled for students as:
  Anion #1—S^{2-}
  Anion #2—OH^{-}
  Anion #3—CO_3^{2-}
- Spot plate

Discussion

An *element* is a pure substance that cannot be broken down by ordinary chemical methods to form simpler substances. *Compounds* are composed of two or more elements chemically bonded in a definite ratio and may be broken down into simpler substances by chemical methods. *Mixtures* are composed of two or more substances (elements or compounds) in variable ratios and may be separated by physical methods.

This laboratory experiment will provide an opportunity to compare the properties of elements, mixtures and compounds. The structure of compounds depends upon the attraction of the atoms to each other to form chemical bonds. When elements react, each atom tries to achieve a complete outer shell of electrons. For most elements, this outer shell consists of eight electrons, and thus the process is called *the octet rule.*

Ionic bonds are formed by the attraction between oppositely charged ions (cations and anions). In order to obtain a complete outer shell, metal atoms lose electrons to become positively charged *cations* and nonmetal atoms gain electrons to become negatively charged *anions.* An example of an ionic compound is NaCl (sodium chloride). In this compound, sodium has lost an electron to become a cation while chlorine has gained an electron to become an anion. Similarly, in the ionic compound CaF_2 (calcium fluoride), calcium has lost two electrons to become a cation while fluorine has gained an electron to become an anion.

Covalent bonds are formed between two nonmetals sharing electrons to complete their outer shells. These compounds form individual units called *molecules.* Examples of covalent compounds are CO_2 (carbon dioxide) and H_2O (water).

Chemical formulas show the composition of a compound by indicating the number of atoms of each element. Ionic compound formulas show the lowest whole number ratio of atoms, while covalent compound formulas show the actual number of atoms in a single molecule. Examples of ionic compounds are iron (III) oxide or ferric oxide (Fe_2O_3) and tin (II) bromide or stannous bromide ($SnBr_2$). Examples of covalent compounds are glucose ($C_6H_{12}O_6$) and dinitrogen tetroxide (N_2O_4). The ability to determine a chemical substance from both its name and chemical formula is a critical aspect of chemistry. A systematic approach identifies patterns in the naming of chemical substances. This systematic approach is called chemical nomenclature and it is used to assign unambiguous names for chemical substances.

Elements and compounds exhibit physical properties such as color, luster, density, melting point, boiling point and crystal shape. The physical properties of individual elements are different from the physical properties of the compounds that they form when they combine. For example, at room temperature sodium (Na) is a shiny metal and chlorine (Cl_2) is a green gas, but they combine to form sodium chloride (NaCl), a white, crystalline solid.

Lewis dot structures

The outer shell electrons are known as *valence electrons* and they are capable of taking part in the formation of chemical bonds. Also known as an electron-dot structure, a Lewis structure is a convenient method to indicate the valence electrons. When drawing the dots to represent the electrons, only one may be placed on each side of the element symbol until all four sides have one dot. Then the dots may be paired.

Figure 1 ***Examples of Lewis dot structures***

K·   Ca·   ·As:

Metals will lose their valence electrons to form cations and nonmetals will gain electrons to form anions. The gain or loss of electrons to form these ions results in the element having a complete outer shell of electrons.

Table 1 ***Lewis dot structures of atoms and ions***

	Symbol	Electron arrangement	# Protons	# Electrons	Net charge	Lewis dot structure
sodium atom	Na	2–8–1	11	11	0	Na·
sodium *ion*	Na^+	2–8	11	10	+1	$[Na]^+$
fluorine atom	F	2–7	9	9	0	:F:
fluoride *ion*	F^-	2–8	9	10	–1	$[:F:]^-$

Writing ionic formulas

The group number of an element may be used to determine the number of valence electrons. A helpful hint when writing ions for the representative elements is that groups 1A – 3A are metals and will lose their electrons to form ions and groups 5A – 7A are nonmetals and will gain electrons to form ions.

Table 2 ***Group numbers and valence electrons***

Group number	1A	2A	3A	4A	5A	6A	7A	8A
Valence electrons	1 e^-	2 e^-	3 e^-	4 e^-	5 e^-	6 e^-	7 e^-	8 e^- †
Gain/lose electrons	lose 1	lose 2	lose 3	none	gain 3	gain 2	gain 1	no change
Ionic charge	+1	+2	+3	none	–3	–2	–1	none

† Note: helium is in group 8A but it only has two valence electrons. It is an exception.

In an ionic formula, the total positive charges must equal the total negative charges. The overall net charge is zero. Thus, it is necessary to determine the smallest number of cations and anions required to give a neutral compound. For example, the ions of sodium and fluorine (see Table 1) have +1 and –1 charges, respectively, and therefore only one ion of each is required. The resulting formula is NaF. But since calcium is in Group 2A it will form a +2 cation (see Table 2) and two fluoride ions are required for an overall net charge of zero. The resulting formula is CaF_2. Notice that the subscript refers to the number of ions for each element, with "1" being understood. The formulas do NOT show the charges of the ions, only the number of ions each element requires.

A more visual method of determining the correct formula is by stacking boxes. Draw a box for the cation and the anion. The height of the box is the number part of its charge. If the heights are not equal, then additional boxes must be stacked on top. The goal is to make the overall height of the cation side equal to the anion side.

In the example below, the calcium ion box is two units high while the fluoride ion box is one unit high. Because the heights are not equal, additional boxes must be stacked. When stacking additional boxes, the size of the boxes cannot be altered (each calcium ion box and fluoride ion box has a definite size). One additional fluoride ion box makes the overall heights the same.

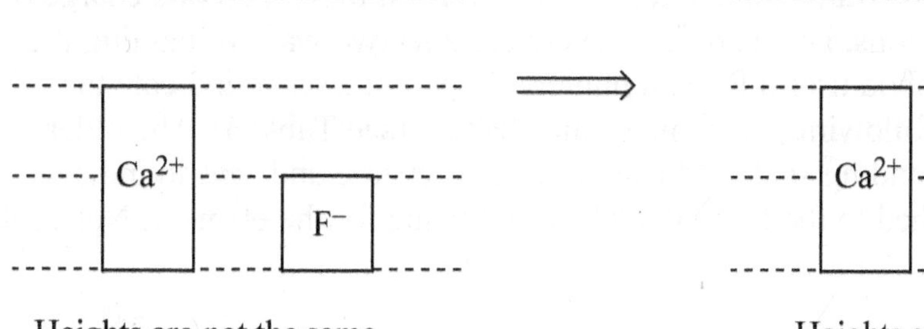

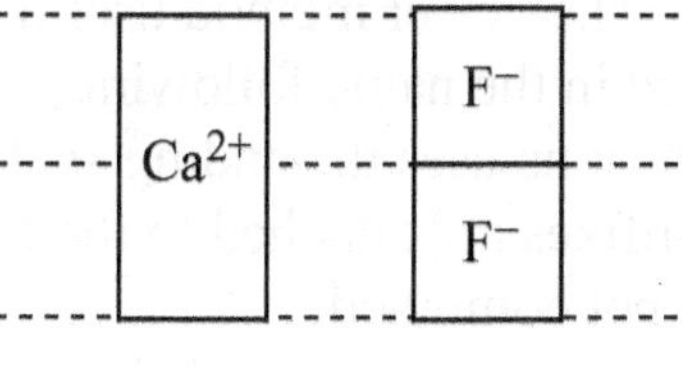

Heights are not the same.
An additional F^- box is needed.

Heights are the same.
Correct formula: CaF_2

Using this method, it is easy to arrive at the correct formula when aluminum and oxygen are brought together to make an ionic compound.

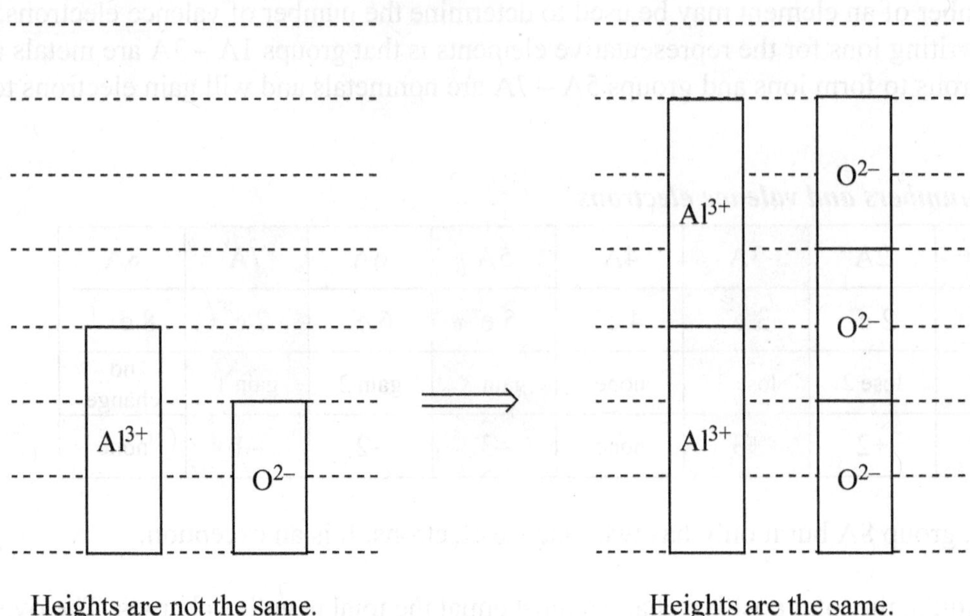

Heights are not the same.
Additional boxes are needed.

Heights are the same.
Correct formula: Al_2O_3

Naming binary ionic compounds

A binary ionic compound is composed of only two elements, a metal and a nonmetal. The metal is always present as the positive ion (cation) and the nonmetal is always present as the negative ion (anion). The name of a binary ionic compound is a two-word name. The first word is the full name of the metal. The second word contains the stem of the nonmetal and the suffix *–ide*. Following this rule, NaCl is sodium chloride, CaF_2 is calcium fluoride, and Li_3N is lithium nitride.

Some metals, especially transition metals, can form more than one type of charged ion. Therefore, a slight modification to the naming rule must be used when the ionic compound contains one of these metals.

Ionic compounds with transition metals

Table 2 shows the ion formation of elements in the representative families but does not address transition metal elements. Many of the transition metals are able to form more than one charge of cation. For example, iron forms two ions, Fe^{2+} and Fe^{3+}. There are also two ways of naming the transition metal ions. The newer method uses a Roman numeral in parentheses to indicate the charge and is included in the name following the name of the element (see Table 3). The older method of naming the ions uses the ending *-ic* to indicate the higher charge and *-ous* to indicate the lower charge. The suffixes are attached to the Latin (usually) root name for the element. Notice that the two methods are not combined.

A few of the transition elements (zinc, cadmium, and silver) form only one charge of cation and do not use either method of naming. The name of the metal ion is simply followed by the name of the nonmetal ion.

zinc ion Zn^{2+} cadmium ion Cd^{2+} silver ion Ag^{+}

Table 3 ***Some transition metal ions***

Ion	Name of ion	Example	Name of compound
Cu^{+}	copper(I) ion or cuprous ion	CuCl	copper(I) chloride or cuprous chloride
Cu^{2+}	copper(II) ion or cupric ion	$CuCl_2$	copper(II) chloride or cupric chloride
Fe^{2+}	iron(II) ion or ferrous ion	$FeCl_2$	iron(II) chloride or ferrous chloride
Fe^{3+}	iron(III) ion or ferric ion	$FeCl_3$	iron(III) chloride or ferric chloride
Ag^{+}	silver ion	AgCl	silver chloride
Zn^{2+}	zinc ion	$ZnCl_2$	zinc chloride

Ionic compounds with polyatomic ions

A *polyatomic ion* is a group of atoms consisting of two or more elements with an overall charge. Table 4 lists some of the more common polyatomic ions. Notice that they are all negatively charged except for the ammonium ion. Most polyatomic ions are formed with nonmetals and oxygen. Replacing the ending of the nonmetal name with *-ate* or *-ite* forms the polyatomic ion name. Similar ions ending in *-ate* have one more oxygen atom than the ion ending in *-ite*. The prefix *bi-* indicates the presence of a hydrogen atom in the ion.

Table 4 ***Some common polyatomic ions***

Formula	Name(s)	Formula	Name(s)
NH_4^+	ammonium ion	CO_3^{2-}	carbonate ion
OH^-	hydroxide ion	HCO_3^-	bicarbonate ion or hydrogen carbonate ion
NO_3^-	nitrate ion	SO_4^{2-}	sulfate ion
NO_2^-	nitrite ion	HSO_4^-	bisulfate ion or hydrogen sulfate ion
PO_4^{3-}	phosphate ion	SO_3^{2-}	sulfite ion

Ionic compounds formed with a metal (or the ammonium ion) and a polyatomic ion must have an overall net charge of zero. When two or more polyatomic ions are needed to obtain equal positive and negative charges, a subscript is used to indicate the number of ions. But the polyatomic ion must be placed inside parentheses and the subscript is written outside the parentheses. Parentheses are NOT used if the subscript is "1." Examples for writing ionic compounds with polyatomic ions are NaOH, sodium hydroxide; $Ca(NO_3)_2$, calcium nitrate; and $Fe_2(SO_4)_3$, iron(III) sulfate or ferric sulfate. Notice that the name of the polyatomic ion is used with no changes and the charges are not shown. The stacking boxes method to determine the correct formula also works with polyatomic ions.

Experimental Procedures

Eye protection and appropriate clothing must be worn at all times.

Wash any contacted skin area immediately with running water and alert the instructor. Discard all wastes properly as directed by the instructor.

A. Physical properties of elements, compounds, and mixtures

1. Observe the display of elements and compounds provided in the laboratory. Record the name and appearance of each on the report sheet.
2. Use a reference book to look up the density and melting point for each of the elements and compounds on display.
3. Obtain a small pea-sized sample of each of the following: sulfur powder, iron filings, iron and sulfur mixture, iron(II) sulfide. Record the appearance of each sample on the report sheet. Record the effect of the magnet on each substance. Indicate whether each substance is an element, compound or mixture.

B. Ionic compounds and formulas

1. Obtain a spot plate. Draw a sketch or label the wells in order to keep track of each reaction performed. Refer to Table B.1 on the Report Sheet. In each block, write the correct name and formula for the compound formed when the cation on the top of the column combines with the anion listed on the left of the row. The cations and anions will be obtained from the bottles labeled:

   Cation #1—Fe^{2+} Anion #1—S^{2-}
   Cation #2—Ag^{+} Anion #2—OH^{-}
   Cation #3—Al^{3+} Anion #3—CO_3^{2-}

2. Place 5 drops of the Cation #1 and 5 drops of the Anion #1 in a well and observe the reaction for the formation of precipitates (solids or cloudiness), or color changes. Look carefully since white precipitates may be difficult to see on a white spot plate. Record your observations in the correct block in Table B.1.
3. Repeat procedure B.2 until you have reacted 5 drops of each cation with 5 drops of each anion in a new well in the spot plate. There are a total of nine combinations (reactions) that need to be done here.
4. Complete Table B.2 by using the periodic table to determine the Lewis dot structure for each element.
5. Complete Tables B.3 – B.9 by following the examples given.

Name ______________________ Date ______________________

Partner(s) ______________________ Section ______________________

______________________ Instructor ______________________

1. Write the formulas for the following polyatomic ions. Be sure to include the correct charges:
   a) hydroxide ____________
   b) sulfate ____________
   c) sulfite ____________
   d) carbonate ____________
   e) phosphate ____________
   f) ammonium ____________

2. Identify the following as elements, compounds, or mixtures:
   a) water, H_2O ____________
   b) sugar (glucose) $C_6H_{12}O_6$ ____________
   c) sugar water ____________
   d) carbon ____________

3. Write the correct formulas for each of the following ionic compounds:
   a) magnesium bromide ____________
   b) aluminum oxide ____________
   c) ammonium hydroxide ____________
   d) copper(II) sulfite ____________

4. List two transition metals that have only one ionic charge. Write their formulas and charges.

5. When obtaining a chemical from a reagent bottle, a student accidently takes too much. What should be done with the excess?

Name ______________________ Date ______________________

Partner(s) ______________________ Section ______________________

______________________ Instructor ______________________

A. Physical properties of elements, compounds, and mixtures

Table A.1 Elements and compounds display

Substance name	Appearance	Density	Melting point
#1			
#2			
#3			
#4			
#5			
#6			
#7			
#8			
#9			

Table A.2 Iron and sulfur comparison

Substance name	Appearance	Effect of magnet	Element, compound, or mixture?

B. Ionic compounds and formulas

Table B.1 Formation of compounds from cations and anions

	S^{2-}	OH^{-}	CO_3^{2-}
Fe^{2+}	FeS iron(II) sulfide or ferrous sulfide black precipitate forms		
Ag^{+}			
Al^{3+}			

Table B.2 Lewis dot structures

Element	Symbol	Atomic number	Electron arrangement	Lewis dot structure	Charge
oxygen	O	8	2–6	$\cdot\ddot{\underset{\cdot}{O}}:$	0
oxide ion	O^{2-}	8	2–8	$[:\ddot{\underset{\cdot\cdot}{O}}:]^{2-}$	2–
sodium					
sodium ion					
nitrogen					
nitride ion					
aluminum					
aluminum ion					
sulfur					
sulfide ion					

Table B.3 Formulas of ionic compounds

Name	Cation	Anion	Formula
potassium chloride	K^+	Cl^-	KCl
magnesium oxide			
sodium nitride			
aluminum sulfide			
calcium fluoride			

Table B.4 Names of ionic compounds

Formula	Name
CaI_2	calcium iodide
Mg_3P_2	
Al_2O_3	
K_3N	
Na_2S	

Table B.5 Formulas of ionic compounds with transition metals

Name	Cation	Anion	Formula
copper(I) oxide	Cu^+	O^{2-}	Cu_2O
iron(III) sulfide			
cupric chloride			
ferrous nitride			
silver iodide			

Table B.6 Names of ionic compounds with transition metals

Formula	Name
ZnO	zinc oxide
FeS	
$CuCl$	
FeI_3	
Cu_3N_2	

Table B.7 Formulas of ionic compounds with polyatomic ions

Name	Cation	Anion	Formula
sodium bicarbonate	Na^+	HCO_3^-	$NaHCO_3$
calcium hydroxide			
iron(II) nitrate			
ammonium sulfate			
magnesium phosphate			

Table B.8 Names of ionic compounds with polyatomic ions

Formula	Name
$Zn(NO_2)_2$	zinc nitrite
$Fe_3(PO_4)_2$	
$(NH_4)_2S$	
$CaCO_3$	
$AgOH$	

Table B.9 Naming compounds and writing formulas of ionic compounds

Chemical formula	Name
$AlCl_3$	**answer:** aluminum chloride
$NiBr_3$	
$Mn(NO_3)_4$	
Na_2CO_3	
CrF_6	
$BaSO_4$	
Li_3N	
$KMnO_4$	
$Fe(HCO_3)_3$	
$LiOH$	
$Zn_3(PO_4)_2$	
answer: $Cu(C_2H_3O_2)_2$	copper(II) acetate
	titanium(IV) oxide
	potassium nitride
	lead(IV) sulfate
	ammonium hydroxide
	barium oxide
	calcium phosphate
	magnesium sulfite
	sodium bromide
	copper(I) nitrite
	silver sulfide

Problems and questions

1. Write the chemical symbol for each of the following ions:

   a) A chlorine atom that has gained one electron ____________

   b) An oxygen atom that has gained two electrons ____________

   c) A cesium atom that has lost one electron ____________

   d) An aluminum atom that has lost three electrons ____________

2. Which noble gas has the same electron arrangement as each of the following ions?

   a) K^+

   b) Mg^{2+}

   c) S^{2-}

3. a) How do the group numbers on the periodic table correspond to the number of valence electrons for representative elements?

   b) How is the charge for transition elements determined?

4. Answer the following questions about the ionic compound formed from the Ca^{2+} and PO_4^{3-} ions.

   a) Is the compound a binary ionic compound or does it contain a polyatomic ion?

   b) What is the name of the compound?

   c) What is the chemical formula of the compound?

Goals

- Identify a compound as ionic or covalent.
- Write Lewis dot structures for compounds.
- Write correct names and formulas for compounds.
- Determine the molecular and electronic shape for compounds.

Materials

- Molecular models

Discussion

An *element* is a pure substance that cannot be broken down by ordinary chemical methods to form simpler substances. *Compounds* are composed of two or more elements chemically bonded in a definite ratio and may be broken down into simpler substances by chemical methods.

Covalent bonds are formed between two nonmetals sharing electrons to complete their outer shells. These compounds form individual units called *molecules.* Examples of covalent compounds are CO_2 (carbon dioxide), H_2O (water), and N_2O_4 (dinitrogen tetroxide).

Naming binary covalent compounds

A binary covalent compound is composed of only two elements, and both are **nonmetals.** Names for binary **covalent** compounds contain numerical prefixes that indicate the number of each type of atom present in the molecule.

Table 1 ***Numerical prefixes for the numbers 1 through 10***

Number	Prefix	Number	Prefix
1	mono-	6	hexa-
2	di-	7	hepta-
3	tri-	8	octa-
4	tetra-	9	nona-
5	penta-	10	deca-

The name of a binary covalent compound has two words. The first word is the full name of the element written on the left in the formula (the one with the lower electronegativity). The second word contains the stem of the element written on the right in the formula (the one with the higher electronegativity) and the suffix *–ide*. Numerical prefixes that indicate the number of each type of atom precede the names of both elements. Following this rule, N_2O_3 is dinitrogen trioxide, S_4N_4 is tetrasulfur tetranitride, and S_2F_{10} is disulfur decafluoride.

The prefix *–mono* is different than the other prefixes. It is not used to modify the name of the first element in a compound's name, but it is used to modify the name of the second element in the compound's name. Thus, the name of the compound CO is carbon monoxide rather than monocarbon monoxide, and the name of the compound CO_2 is carbon dioxide.

Notice that the name of CO is carbon monoxide rather than carbon monooxide. When the numerical prefix ends in "a" or "o" and the element name begins with "a" or "o," the final vowel of the prefix is usually dropped for ease of pronunciation. Thus "monoxide" and "tetroxide" are used.

There is an exception to the use of numerical prefixes when naming binary covalent compounds where hydrogen is the first listed element in the formula. They are named without numerical prefixes. Therefore, H_2S is called hydrogen sulfide instead of dihydrogen monosulfide and HCl is called hydrogen chloride.

Writing covalent formulas

Covalent bonds are formed when two or more nonmetals share their electrons to form a compound. A *single bond* is one pair of shared electrons, a *double bond* is two pairs of shared electrons, and a *triple bond* is three pairs of shared electrons. To write the formula of a covalent compound, it is necessary to determine the number of electrons shared by two atoms. This is most easily done by using Lewis dot structures.

Water is made up of oxygen, which has 6 valence electrons, and hydrogen, which has 1 valence electron. Hydrogen is an element that only needs two electrons to have a complete outer shell so hydrogen needs to share one more electron from oxygen. Oxygen needs to share two more electrons to complete its outer shell of eight electrons. Therefore, there must be two hydrogens to share the electrons that oxygen needs, and oxygen can share one electron with each of those hydrogens. The Lewis dot structure of H_2O is shown in Figure 1. Each pair of "shared" electrons can also be represented with a line, indicating a bond. The lines make the drawing less cluttered.

Figure 1 ***Lewis dot structures for water***

```
  ..          ..
 :O: H       :O—H
  ..          |
  H           H
```

Not all covalent compounds consist solely of single bonds. Since many covalent compounds come in different ratios (i.e. CO, carbon monoxide and CO_2, carbon dioxide) it is possible to determine if double or triple bonds are needed using the following steps:

1) Determine the total number of valence electrons available.
2) Write the symbols for all of the atoms given in the formula.
3) If there are more than two atoms, a "central" atom should be placed in the middle. In simple molecules, the central atom is usually the element with the smallest subscript in the formula. That element is normally found to the left and/or below the other elements on the periodic table. In a compound containing carbon, carbon is generally the central atom.
4) Hydrogen can never be the central element since it can only form one bond.
5) Draw two electrons between each pair of atoms. Two electrons represent a *single bond.*

6) Starting with the outer atoms (not the central atom), draw the remaining electrons around the elements so that each atom has a complete outer shell. Any remaining electrons should be placed around the central atom.
7) If each atom does not have a complete outer shell, move "unshared" electrons so that atoms needing extra electrons share them, thus creating double or triple bonds.
8) The total number of electrons in the Lewis dot structure must equal the number of available electrons.
9) Two pairs of electrons representing a double bond may be indicated with two lines, and three pairs of electrons representing a triple bond may be indicated with three lines.
10) Show any "unshared" electrons around atoms as dots.

Figure 2 ***Lewis structures of carbon monoxide, phosgene, and ammonia***

:C≡O:

:O:
||
:Cl—C—Cl:

H—N—H
|
H

Electron geometry and molecular geometry

Most elements gain, lose, or share electrons in an attempt to complete their outer shell according to the octet rule. These "pairs" of electrons may be shared (bonding electrons) or unshared (nonbonding electrons or lone pairs). In the valence shell electron pair repulsion model (VSEPR), an *electron group* is a set of valence electrons present in a confined region around the central atom. Single, double, and triple bonds are all counted as one electron group because each takes up only one region of space around the central atom. In addition, each lone pair around the central atom is also counted as an electron group. In the VSEPR model, the electron groups move as far apart from each other as possible. Counting the number of electron groups determines the *electron geometry* and bond angles.

In this chemistry course the number of electron groups around the central atom is 2, 3, or 4. Other numbers are possible, but they are studied in more advanced chemistry courses. Each number (2, 3, or 4) gives rise to a different electron geometry as shown in Figure 3.

Figure 3 ***Electron geometries***

O—A—O

Linear
2 groups

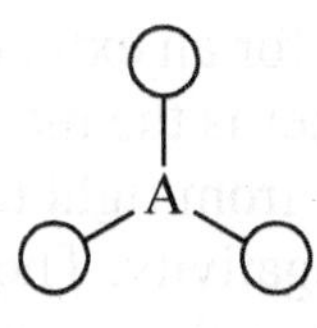

Trigonal planar
3 groups

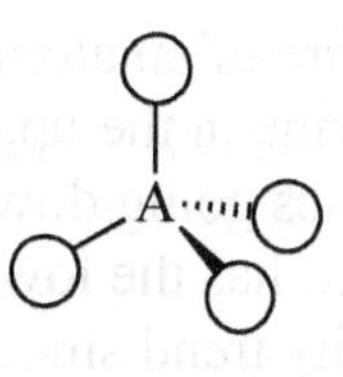

Tetrahedral
4 groups

After determining the electron geometry for a formula, it is then possible to determine the *molecular geometry.* Different molecular geometries result from different number of bonded atoms and lone pairs. The electron and molecular geometries for methane, ammonia, and water are shown in Figure 4. Note that in these examples all of the molecules exhibit tetrahedral *electron* geometry, but each shows a different *molecular* geometry. Common molecular geometries are summarized in Table 2.

Figure 4 ***Electron and molecular geometry examples***

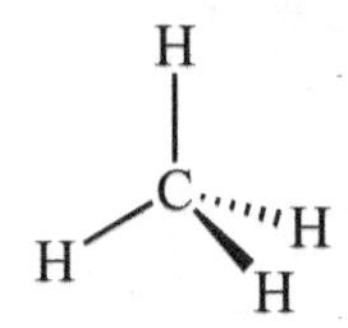

methane

electron geometry: tetrahedral
molecular geometry: tetrahedral

ammonia

electron geometry: tetrahedral
molecular geometry: trigonal pyramidal

water

electron geometry: tetrahedral
molecular geometry: bent

Table 2 ***Electron and molecular geometry for atoms***

Total electron groups	Bonded atoms	Lone pairs	Electron geometry	Molecular geometry	Bond angles	Example
2	2	0	linear	linear	180°	CO_2
3	3	0	trigonal planar	trigonal planar	120°	SO_3
3	2	1	trigonal planar	bent	~120°	SO_2
4	4	0	tetrahedral	tetrahedral	109.5°	CH_4
4	3	1	tetrahedral	trigonal pyramidal	~109.5°	NH_3
4	2	2	tetrahedral	bent	~109.5°	H_2O

Electronegativity

Electronegativity is a measure of an atom's attraction for an extra electron. The general trend on the periodic table is that fluorine in the upper right corner is the most electronegative element and the electronegativity decreases going down a group or from right to left in a period. Therefore, francium, in the lower left corner, has the lowest electronegativity. The noble gas elements are not considered in the electronegativity trend since they are generally unreactive.

The *polarity* of bonds is determined by the difference in electronegativity of the elements. The greater the difference in electronegativity, the more polar the bond will be. A difference in electronegativity between the two atoms of 0.5 – 1.9 makes the bond a *polar* covalent bond. A difference less than 0.5 is classified as a nonpolar bond, while a difference greater than 1.9 is classified as an ionic bond. In a polar covalent bond, the element with the larger electronegativity value will be the negative end of the bond, shown with a δ^- in the Lewis structure. The lowercase Greek letter δ means "partial." The element with the smaller electronegativity value will be the positive end of the bond, shown with a δ^+ in the Lewis structure. Examples of bond polarity are shown in Figure 5.

Figure 5 ***Bond polarity***

electronegativity # of C 2.5
electronegativity # of H 2.1
electronegativity # of O 3.5

H−H	C−H	δ^+ δ^- C−O
difference in electronegativity: (2.1 − 2.1 = 0.0) **nonpolar covalent bond**	difference in electronegativity: (2.5 − 2.1 = 0.4) **nonpolar covalent bond**	difference in electronegativity: (3.5 − 2.5 = 1.0) **polar covalent bond**

Polarity of molecules

Molecules can be classified as polar or nonpolar. If a molecule has polar bonds, it is not necessarily a *polar molecule*. To be a polar molecule, two conditions must be satisfied:

- The molecule must have at least one polar covalent bond.
- The molecule must be asymmetric.

In a perfectly symmetrical molecule, the dipoles cancel. When the dipoles do not cancel, the molecule is asymmetric. Of the five molecular geometries studied in this course, two are always asymmetric: bent and trigonal pyramidal. The other three shapes (linear, trigonal planar, and tetrahedral) may be symmetric or asymmetric. For the purpose of this experiment, each of these three shapes is *symmetric* if all the outer atoms are the same. If they are not all the same, the molecule is *asymmetric*.

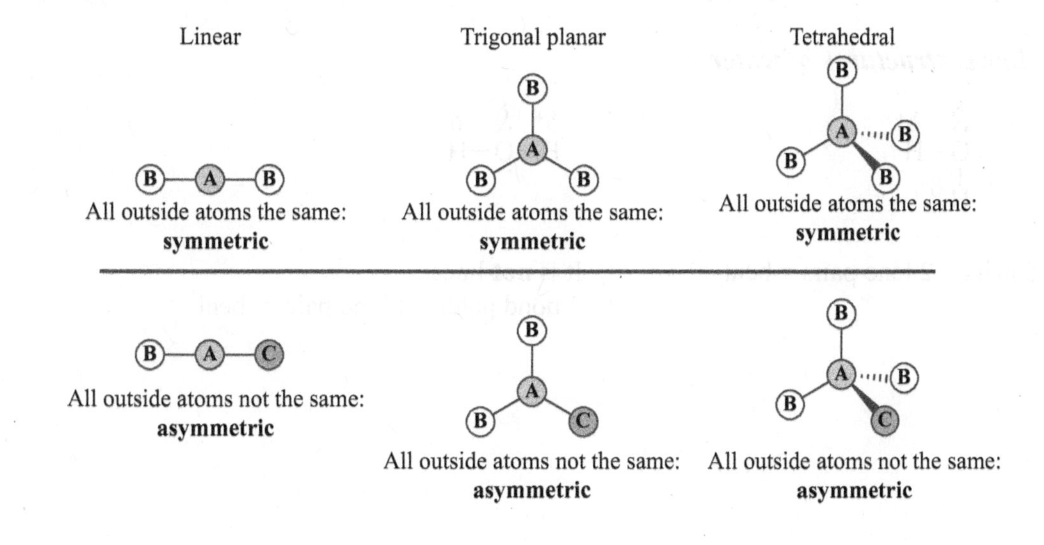

To determine whether a molecule is polar or nonpolar, the polarity of the bonds must be checked. Also, the shape must be determined, which requires the Lewis structure to be drawn. Determination of molecular polarity is shown in Figure 6.

Figure 6 ***Determination of molecular polarity***

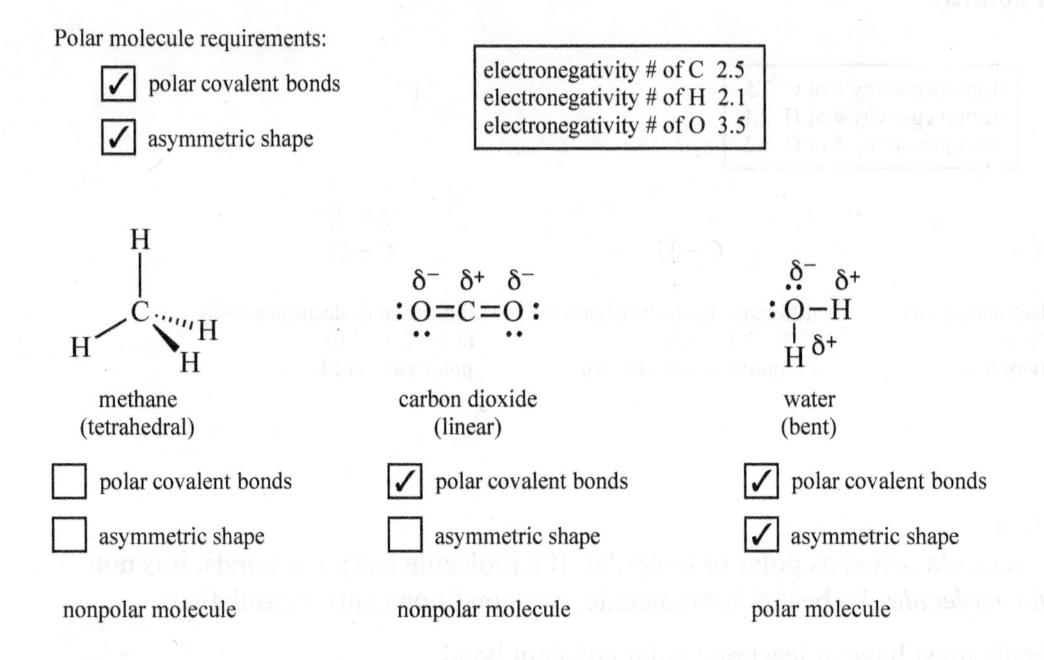

A Lewis structure is only a two-dimensional representation of a molecule. The symmetrical nature of the molecule CANNOT be determined by how the Lewis structure appears on paper. The molecular geometry must be determined based on the number of bonded atoms and lone pairs. For example, water may be drawn two ways (see Figure 7) and each is a perfectly valid Lewis structure. In the structure on the left, the molecule looks bent. In the structure on the right, the molecule looks linear. It is tempting to say that the molecule on the right is symmetric because it is linear and both outer atoms are the same. However, both structures are bent because there are two bonded atoms and two lone pairs. Bent is always asymmetric.

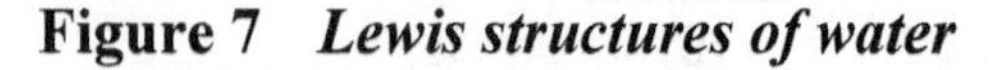

Figure 7 ***Lewis structures of water***

δ^- δ^+ : O – H, H δ^+

δ^+ δ^- δ^+ H – O – H

2 bond pairs + 2 lone pairs = bent

It is **not** linear.
2 bond pairs + 2 lone pairs = bent

Experimental Procedures

A. Covalent compounds and formulas

1. Complete Table A.1 by using the periodic table to determine the Lewis dot structure for each element.
2. Complete Tables A.2 – A.3 by following the examples given.

B. Electron and molecular geometries

1. Using a molecular model kit, build a model of each molecule on Table B and have it checked by the instructor.
2. Complete Table B by following the example given.

C. Bond polarity and molecular polarity

1. Using the table of electronegativities below, put a check mark in the polar bonds box in Table C if the compound listed has polar bonds.
2. Put a check mark in the asymmetric shape box in Table C if the compound listed is asymmetric.
3. Put a check mark in the polar molecule box in Table C if the compound listed is a polar molecule. Complete Table C by following the example given.

Table 3 ***Electronegativity values for main group elements***

1A	2A	3A	4A	5A	6A	7A	8A
H 2.1							
Li 1.0	Be 1.5	B 2.0	C 2.5	N 3.0	O 3.5	F 4.0	
Na 0.9	Mg 1.2	Al 1.5	Si 1.8	P 2.1	S 2.5	Cl 3.0	
K 0.8	Ca 1.0	Ga 1.6	Ge 1.8	As 2.0	Se 2.4	Br 2.8	
Rb 0.8	Sr 1.0	In 1.7	Sn 1.8	Sb 1.9	Te 2.1	I 2.5	

D. Compounds and formulas

1. Complete Table D by identifying each compound as ionic or covalent and writing the correct name or formula. Two examples have been given.

PRE-LAB QUESTIONS: COVALENT COMPOUNDS: NAMES, FORMULAS, AND SHAPES

Name ______________________ Date ______________________

Partner(s) ______________________ Section ______________________

______________________ Instructor ______________________

1. Circle the molecule in each of the following pairs of compounds that would most likely contain covalent bonds.

    a) BaO or CO

    b) ClF or NaF

    c) $FeCl_3$ or NCl_3

    d) CH_4 or $CaCl_2$

2. Which of the compounds N_2O_3, KCl, HF, and CO_2 have names that

    a) End in the suffix *–ide*?

    b) Contain two numerical prefixes?

    c) Contain the prefix *mono–*?

    d) Do not contain numerical prefixes?

3. Circle the correct Lewis structure for the OCS molecule.

    $:\ddot{\underset{\cdot\cdot}{O}}-\ddot{\underset{\cdot\cdot}{C}}-\ddot{\underset{\cdot\cdot}{S}}:$ $\quad :\ddot{O}=C=\ddot{S}: \quad :O\equiv C\equiv S: \quad :\ddot{\underset{\cdot\cdot}{O}}-C=\ddot{S}:$

4. How does the shape of a molecule factor into its molecular polarity?

5. Why are safety goggles not required for this particular lab?

REPORT SHEET: COVALENT COMPOUNDS: THEIR NAMES, FORMULAS, AND SHAPES

Name ______________________ Date ______________________

Partner(s) ______________________ Section ______________________

______________________ Instructor ______________________

A. Covalent compounds and formulas

Table A.1 Lewis dot structures for nonmetallic elements

Hydrogen	Carbon	Nitrogen	Oxygen	Sulfur	Fluorine
H·					

Table A.2 Lewis dot structures for covalent compounds

Compound	Name	Lewis dot structure
CCl_4	carbon tetrachloride	:C̈l: / | / :C̈l—C—C̈l: / | / :C̈l:
NF_3		
H_2O		
CO		
PCl_3		

Table A.3 Formulas and names for covalent compounds

Name	Formula
nitrogen triiodide	**answer:** NI_3
dinitrogen trioxide	
chlorine monofluoride	
sulfur dioxide	
carbon tetrabromide	
sulfur trioxide	
diphosphorus pentoxide	
iodine heptafluoride	
hydrogen bromide	
tetraphosphorus decoxide	
answer: sulfur hexafluoride	SF_6
	ClO_2
	P_4O_6
	SO_2
	BrCl
	NCl_3
	H_2Se
	N_2O_5
	SiF_4
	PI_3

B. Electron and molecular geometries

Formula	Model checked	Lewis dot structure	Electron geometry	Molecular geometry
CBr_4	✓	$\begin{array}{c} :\ddot{Br}: \\ \vert \\ :\ddot{Br}-C-\ddot{Br}: \\ \vert \\ :\underset{\cdot\cdot}{Br}: \end{array}$	tetrahedral	tetrahedral
PH_3				
H_2S				
CO_2				
$CHCl_3$				
CH_2O				

C. Bond polarity and molecular polarity

Lewis dot structure	Polar bonds or nonpolar bonds?	Symmetric or asymmetric?	Polar molecule or nonpolar molecule?
:F: H–C–F: H	polar bonds	asymmetric	polar molecule
:Cl–P–Cl: :Cl:			
:Cl: :Cl–Si–Cl: :Cl:			
F–O–F (O with two lone pairs)			
:O: ‖ H–C–H			
:S=C=S:			

D. Compounds and formulas

Ionic or covalent?	Name	Formula
ionic	nickel(II) oxide	NiO
	calcium nitride	
	sulfur dichloride	
	aluminum nitrate	
	carbon tetrafluoride	
	diphosphorus pentoxide	
	lithium nitride	
	silicon dioxide	
	chlorine pentafluoride	
	lead(IV) chloride	

Ionic or covalent?	Name	Formula
covalent	nitrogen trichloride	NCl_3
		SO_2
		Mg_3N_2
		N_2O_4
		AgCl
		Na_2SO_4
		$CrBr_2$
		Cl_2O
		Cu_3P

Problems and questions

1. Contrast ionic and covalent compounds. Describe their differences regarding bonding, naming, and types of elements involved.

IONIC	COVALENT

2. Identify each of the following terms as being associated with either a "bond" or a "molecule."

   a) polar covalent ______________

   b) nonpolar covalent ______________

   c) polar ______________

   d) nonpolar ______________

3. Draw the Lewis dot structures for the ammonium ion and the hydroxide ion and predict their molecular shapes.

4. The correct name for the compound Na_2CO_3 is not disodium monocarbon trioxide. Explain why this name is incorrect.

Goals

- List the physical property information for elements and compounds using a reference book in the laboratory.
- Observe physical and chemical changes in reactions.
- Identify a reaction type as combination, decomposition, single replacement, or double replacement.
- Predict the products of a reaction.
- Write a balanced equation for a chemical reaction.

Materials

- Bunsen burner
- Deflagration spoon
- Crucible tongs
- Small test tubes
- Wood splints
- Stopper
- 10 mL graduated cylinder
- Red litmus paper
- Metal samples of Mg, Ca, Zn, Cu
- $CuSO_4 \cdot 5H_2O$ crystals
- Solid $CaCO_3$
- Solid $KClO_3$
- 3 M solutions of HCl, H_2SO_4, NaOH
- 1 M solutions of Na_2CO_3, NH_4Cl
- 0.1 M solutions of $AgNO_3$, $CuSO_4$, $ZnSO_4$, $NaNO_3$, $Pb(NO_3)_2$, KI, NaCl, KCl
- *Merck Index* or *CRC Handbook of Chemistry and Physics*

Discussion

Substances undergo physical changes and chemical changes. A *physical change* does not change the formula of the substance. A *chemical change* results in a substance with a new formula and different chemical properties. Some examples of physical and chemical changes are shown in Table 1.

Chemical changes occur during a chemical reaction. A balanced equation can be written for the reaction showing the reactants (on the left side of the equation) and the products (on the right side of the equation). Since the atoms of the reactants are rearranged to form new combinations for the products, it is necessary to ensure that equal numbers of atoms of each element are present on both sides of the equation. To write a balanced equation, a few simple rules are necessary:

1) Write the correct chemical formula for each reactant and product.
2) Remember to write the correct formulas for the diatomic elements.
3) Balance the number of atoms for each element using *coefficients* in front of the formulas.

4) The state of each substance may be indicated with a subscript such as (*g*) for gas, (*aq*) for aqueous, (*s*) for solid, and (*l*) for liquid. *Aqueous* refers to a solution consisting of a solute dissolved in water.

Table 1 ***Examples of physical and chemical changes***

Physical changes	Chemical changes
Change in state: Boiling, Freezing, Melting, Condensation, Sublimation	Formation of a gas (bubbles formed in a reaction)
	Formation of a solid (precipitate formed in a reaction)
	Change in color
Change in size: Tearing, Grinding	Change in temperature in a reaction
	Change in pH

Types of reactions

Most chemical reactions can be classified into one of four general categories: combination, decomposition, single replacement, and double replacement.

1. *Combination reactions* occur when two or more substances combine to form one substance and can be represented by the general equation:

$$A + B \longrightarrow AB$$

Examples of combination reactions include:

$$2\,Na_{(s)} + Cl_2(g) \longrightarrow 2\,NaCl_{(s)}$$

$$S_{(s)} + O_2(g) \longrightarrow SO_2(g)$$

$$SO_3(g) + H_2O(l) \longrightarrow H_2SO_4(aq)$$

$$CaO_{(s)} + CO_2(g) \longrightarrow CaCO_3(s)$$

2. *Decomposition reactions* occur when one substance breaks down into two or more substances and can be represented by the general equation:

$$AB \longrightarrow A + B$$

Examples of decomposition reactions include:

$$CaCO_{3(s)} \longrightarrow CaO_{(s)} + CO_{2(g)}$$

$$2\,HgO_{(s)} \longrightarrow 2\,Hg_{(l)} + O_{2(g)}$$

$$CoCl_2 \cdot 6\,H_2O_{(s)} \longrightarrow CoCl_{2(s)} + 6\,H_2O_{(g)}$$

$$Ca(HCO_3)_{2(aq)} \longrightarrow CaCO_{3(s)} + H_2O_{(l)} + CO_{2(g)}$$

3. *Single replacement reactions* are reactions in which an ion (or atom) in a compound is replaced by an ion (or atom) of another element and can be represented by the general equation:

$$AX + B \longrightarrow A + BX \quad \text{or} \quad AX + Y \longrightarrow AY + X$$

Examples of single replacement reactions include:

$$Fe_{(s)} + 2\,HCl_{(aq)} \longrightarrow FeCl_{2(aq)} + H_{2(g)}$$

$$Zn_{(s)} + CuSO_{4(aq)} \longrightarrow ZnSO_{4(aq)} + Cu_{(s)}$$

$$Br_{2(l)} + 2\,NaI_{(aq)} \longrightarrow 2\,NaBr_{(aq)} + I_{2(s)}$$

$$V_2O_{5(s)} + 5\,Ca_{(l)} \longrightarrow 2\,V_{(l)} + 5\,CaO_{(s)}$$

Single replacement reactions are also examples of redox reactions in which electron transfer occurs (combination and decomposition reactions may or may not be redox reactions).

4. *Double replacement reactions* are reactions in which cations and anions that were partners in the reactants are interchanged in the products and can be represented by the general equation:

$$AX + BY \longrightarrow AY + BX$$

If all substances in the reaction are water-soluble ionic compounds, no overall reaction takes place and no signs of chemical changes are observed. If one of the products (AY or BX) is a precipitate, a gas, or a weak electrolyte, then a chemical change has taken place.

a) To determine if a precipitate will form, the solubility rules must be applied. Examples of formation of a precipitate include:

$$AgNO_3(aq) + NaCl(aq) \longrightarrow AgCl(s) + NaNO_3(aq)$$

$$BaCl_2(aq) + K_2SO_4(aq) \longrightarrow BaSO_4(s) + 2KCl(aq)$$

b) The common gases are CO_2, SO_2, H_2S and NH_3. Examples of formation of a gas include:

$$Na_2CO_3(aq) + H_2SO_4(aq) \longrightarrow Na_2SO_4(aq) + H_2O(l) + CO_2(g)$$

$$CaSO_3(aq) + 2HCl(aq) \longrightarrow CaCl_2(aq) + H_2O(l) + SO_2(g)$$

c) An electrolyte is a substance that ionizes or dissociates in water to form an electrically conductive solution. Electrolytes can be divided into two categories: strong electrolytes and weak electrolytes. A strong electrolyte completely (or almost completely) ionizes or dissociates in water, while a weak electrolyte only partially ionizes/dissociates. Soluble ionic compounds, strong acids, and strong bases are strong electrolytes. Weak acids and weak bases are weak electrolytes. Water is a nonelectrolyte (or maybe a VERY weak electrolyte). Examples of formation of electrolytes include:

$$HNO_3(aq) + KOH(aq) \longrightarrow KNO_3(aq) + H_2O(l)$$

$$CaO(s) + 2HCl(aq) \longrightarrow CaCl_2(aq) + H_2O(l)$$

$$SO_3(g) + 2NaOH(aq) \longrightarrow Na_2SO_4(aq) + H_2O(l)$$

All of these reactions are *exothermic reactions* (give off heat) and will increase the temperature of the solution. Double replacement reactions are also examples of nonredox reactions.

Experimental Procedures

Eye protection and appropriate clothing must be worn at all times.

All of these chemicals are corrosive to the skin and toxic if ingested. Wash any contacted skin area immediately with running water and alert the instructor.

Observe proper safety precautions when handling the Bunsen burner and other hot objects.

For experimental procedures involving heating a test tube, hold the test tube with a test tube clamp and gently move the test tube back and forth through the flame.

Dispose of all solutions, metal pieces, and cooled/dampened wooden splints as directed by the lab instructor, NOT in the sinks.

A. Physical properties of reactants

Observations of chemical reactions often require noticing color changes or changes in other physical properties.

1. Using a *Merck Index*, *CRC Handbook of Chemistry and Physics*, or other chemistry reference books or sources, complete the table by looking up the information as directed by the instructor.
2. Include appropriate units. Close attention may be needed to distinguish between similar compounds.

For sections B and C, use Table 2 to determine if a chemical reaction has taken place after mixing two aqueous solutions or after heating a reactant.

Table 2 ***Evidence of chemical reactions***

Chemical change	Observation
Formation of a precipitate	Solid or cloudiness forms
Formation of a gas (SO_2, H_2S, NH_3)	Moist litmus paper turns red for an acid or blue for a base
Formation of a gas (H_2)	Glowing wood splint will "pop" audibly
Formation of a gas (O_2)	Glowing wood splint will glow brightly or ignite
Formation of a gas (non-flammable N_2, CO_2, Cl_2)	Glowing wood splint will be extinguished
Change in color	Color change from reactants to products — it is helpful to hold a piece of white paper behind the container to observe colors
Change in temperature	Touch the exterior of the container

A splint test is used to determine the identity of a gas product. Collect the gas in the test tube by **carefully** holding a stopper over the open end of the test tube for approximately 5 seconds. Then refer to Figure 1 for instructions on how to complete the splint test.

Figure 1 ***How to perform a splint test***

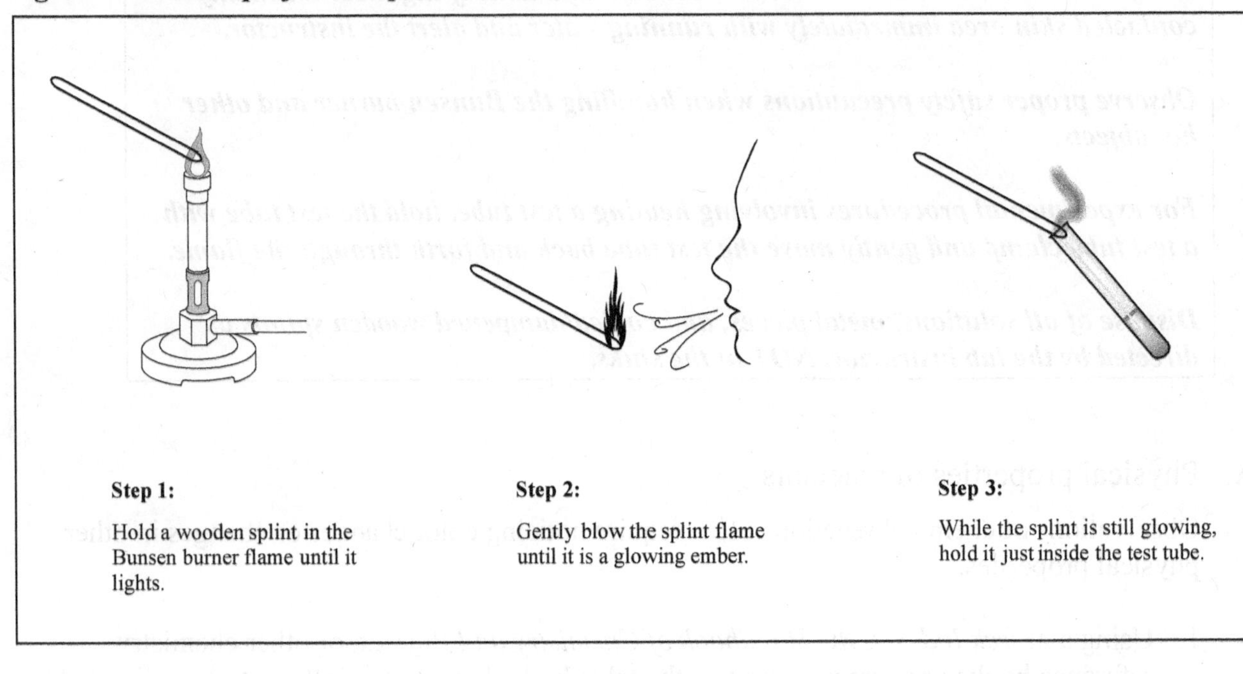

B. Experiments involving mixing solutions

> ***Note:*** ***1 mL is approximately 20 drops.***
>
> ***Dispose of all solutions and metal pieces as directed by the lab instructor, NOT in the sinks.***

For each of the following reactions, record your observations, classify the type of reaction (combination, decomposition, single replacement, or double replacement), and write a balanced equation for the reaction. If no reaction occurs, write "NR" for "no reaction."

1. Mix 1 mL of 3 M HCl with 1 mL of 1 M Na_2CO_3 in a test tube. Perform a splint test, and refer to Table 2 to determine the identity of the gas. For this first experiment, the balanced equation has been given as an example.
2. To a test tube containing 2 mL of 1 M NH_4Cl, add 2 mL of 3 M NaOH. Note the odor of the gas evolved. Perform a moist red litmus paper test by wetting a piece of litmus paper with distilled water and holding it in the mouth of the test tube, being careful not to touch the paper to the glass.
3. Mix 1 mL of 3 M H_2SO_4 with 2 mL of 3 M NaOH in a test tube. Note any change in temperature of the reaction by touching the exterior of the test tube with your hand.

4. To a test tube containing 3 mL of 0.1 M $CuSO_4$, add a pea-sized piece of mossy zinc metal or a 2 cm length of zinc ribbon.
5. To a test tube containing 3 mL of 0.1 M $ZnSO_4$, add a pea-sized piece of copper metal or a 2.5 cm long copper wire.
6. To a test tube containing a pea-sized amount of zinc metal, add 3 mL of 3 M HCl. What evidence is there that a gas is produced? Perform a splint test, and refer to Table 2 to determine the identity of the gas.
7. To a test tube containing a pea-sized amount of copper metal, add 3 mL of 3 M HCl.
8. Mix 1 mL of 0.1 M $Pb(NO_3)_2$ with 2 mL of 0.1 M KI in a test tube.
9. Mix 1 mL of 0.1 M $AgNO_3$ with 1 mL of 0.1 M NaCl in a test tube.
10. Mix 1 mL of 0.1 M KCl with 1 mL of 0.1 M $NaNO_3$ in a test tube.

C. Experiments involving heating and burning

Note: ***Remove the test tube from the flame and collect the gas in the test tube by carefully holding a stopper over the open end of the test tube for approximately 5 seconds in preparation for performing a splint test.***

Dispose cooled/dampened wooden splints as directed by the lab instructor, NOT in the sinks.

For each of the following reactions, record your observations, classify the type of reaction (combination, decomposition, single replacement, or double replacement), and write a balanced equation for the reaction. If no reaction occurs, write "NR" for "no reaction."

In a chemistry context, burning means to add oxygen, therefore oxygen gas (O_2) is written as a reactant. Conversely, heating means to make the substance warmer and does not involve a chemical reaction with oxygen.

1. Note the color of calcium metal. Burn a pea-sized amount of calcium on a deflagration spoon and hold it in the flame of a Bunsen burner for 2 – 3 minutes. Note the color after burning. Record your observations of the reaction. Note: the calcium will not "burn" by "catching on fire," but there will be a visual change.
2. Obtain a 3 cm strip of magnesium ribbon. Using crucible tongs, burn the ribbon in the hottest part of the flame of a Bunsen burner. **DO NOT** look directly at the bright light.
3. Heat a test tube containing 0.1 g $CaCO_3$ with a Bunsen burner for 1 – 2 minutes. Perform a splint test, and refer to Table 2 to determine the identity of the gas.
4. Heat a test tube containing 0.1 g $KClO_3$ with a Bunsen burner for 1 minute. Perform a splint test, and refer to Table 2 to determine the identity of the gas.
5. Gently heat a test tube containing 1 g $CuSO_4 \cdot 5\,H_2O$ with a low flame from a Bunsen burner. Carefully observe the crystals and the cooler upper part of the test tube while continuing to heat for several minutes.

Name ______________________ Date ______________________

Partner(s) ______________________ Section ______________________

______________________ Instructor ______________________

1. Label each of the following as an example of a physical or chemical change:
   a) grinding pepper ______________
   b) burning wood ______________
   c) freezing water ______________
   d) rusting of an iron nail ______________

2. How should a wooden splint be prepared in order to perform a splint test?

3. Write the names and correct formulas for the diatomic elements.

4. What are the guidelines in the "Special Directions" boxes for the disposal of metal pieces and wooden splints?

Name ______________________ Date ______________________

Partner(s) ______________________ Section ______________________

______________________ Instructor ______________________

A. Physical properties of reactants

Substance	Melting point	Boiling point	Density	Color
Cu				
Zn				
$CaCO_3$				
NaOH				
$CuSO_4$				
$CuSO_4 \cdot 5\,H_2O$				

B. Experiments involving mixing solutions

Balanced equation for reaction
hydrochloric acid (HCl) and sodium carbonate $2HCl(aq) + Na_2CO_3(s) \longrightarrow 2NaCl(aq) + H_2O(l) + CO_2(g)$
ammonium chloride and sodium hydroxide
sulfuric acid (H_2SO_4) and sodium hydroxide
copper(II) sulfate and zinc
zinc sulfate and copper
hydrochloric acid and zinc
hydrochloric acid and copper
lead(II) nitrate and potassium iodide
silver nitrate and sodium chloride
potassium chloride and sodium nitrate

B. Experiments involving mixing solutions (continued)

Observations and test results	Type of reaction

C. Experiments involving heating and burning

Balanced equation for reaction	Observations and test results	Type of reaction
Burning calcium		
Burning magnesium		
Heating calcium carbonate		
Heating potassium chlorate		
Heating copper(II) sulfate pentahydrate		

Problems and questions

1. Balance the following equations:

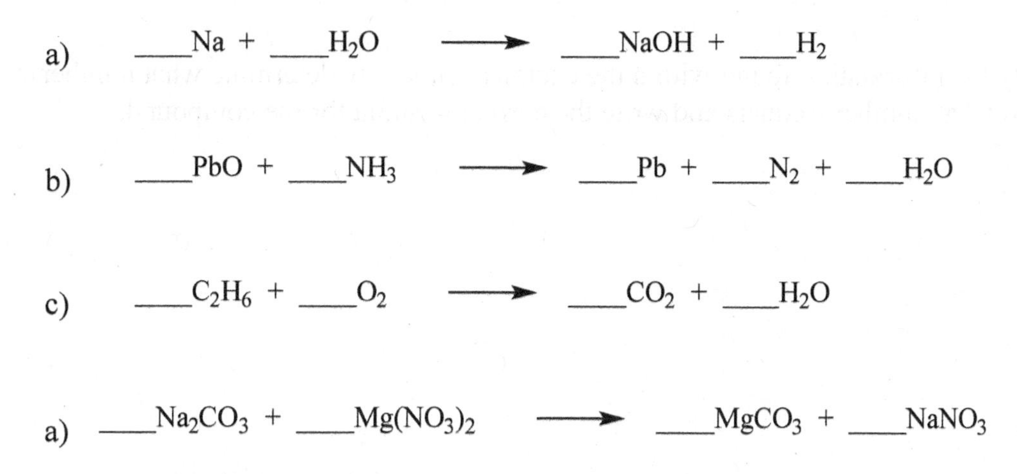

a) ___Na + ___H_2O ⟶ ___NaOH + ___H_2

b) ___PbO + ___NH_3 ⟶ ___Pb + ___N_2 + ___H_2O

c) ___C_2H_6 + ___O_2 ⟶ ___CO_2 + ___H_2O

a) ___Na_2CO_3 + ___$Mg(NO_3)_2$ ⟶ ___$MgCO_3$ + ___$NaNO_3$

2. Write the correct balanced equation for each of the following reactions. Include the proper subscripts to identify the state of matter for all reactants and products. Write the type of reaction on the blank line.

   a) solid aluminum reacts with oxygen gas to form solid aluminum oxide
   type of reaction: ____________________

   b) aqueous solutions of calcium chloride and sodium phosphate react to form solid calcium phosphate and aqueous sodium chloride
   type of reaction: ____________________

   c) solid magnesium carbonate decomposes into solid magnesium oxide and carbon dioxide gas
   type of reaction: ____________________

   d) solid iron and aqueous copper(II) bromide react to form solid copper and aqueous iron(II) bromide
   type of reaction: ____________________

3. A compound composed of carbon and hydrogen with the formula C_xH_y was burned. The balanced equation is shown below.

$$2C_xH_y + 15\,O_{2(g)} \longrightarrow 12\,CO_{2(g)} + 6\,H_2O_{(g)}$$

Using only the information found within the chemical equation, determine what number x equals and what number y equals and write the correct formula for the compound.

4. Predict the product(s) that would form from the reaction of the following reactants. Use product coefficients as needed to balance the equations.

a) single replacement $Zn_{(s)} + Cu(NO_3)_{2(aq)} \longrightarrow$ ________ + ________

b) combination $2\,Na_{(s)} + Cl_{2(g)} \longrightarrow$ ________

c) double replacement $Ba(OH)_{2(aq)} + 2\,HCl_{(aq)} \longrightarrow$ ________ + ________

Goals

- Convert from grams to moles and moles to grams using the molar mass from the periodic table.
- Use experimental data to determine the empirical formula of a compound.
- Use experimental data to determine the percent water in a hydrate.
- Determine the formula of a hydrate.

Materials

- Crucible and cover
- Crucible tongs
- Clay triangle
- Ring stand
- Iron ring
- Bunsen burner
- Watch glass
- Glass stirring rod
- Magnesium ribbon
- Steel wool
- Dropper
- Wire gauze
- Electronic balance
- $MgSO_4$ (hydrate) solid

Discussion

The *Law of Definite Composition* states that any pure chemical compound is made up of two or more elements in the same proportion by mass. For example, water is composed of 89% oxygen and 11% hydrogen by mass, and has the formula H_2O. Water always shows this same ratio, no matter where it is found — the ocean, the kitchen faucet, or a teardrop. The *empirical formula* is the simplest whole number ratio of atoms in a compound and can be determined experimentally. The *molecular formula* is the actual number of atoms in each molecule of a compound. Water has the same empirical formula and molecular formula. However, glucose, a sugar found in the blood, has an empirical formula of CH_2O but a molecular formula of $C_6H_{12}O_6$.

Hydrates are ionic compounds that include a fixed number of water molecules in the structure of the compound. The formula for a hydrate is written with a dot indicating that a specific number of water molecules are associated with the ionic formula. For example, $CuSO_4 \cdot 5\,H_2O$ represents one formula unit of copper(II) sulfate associated with five water molecules and is a beautiful blue hydrate. The name of this compound is copper(II) sulfate pentahydrate. Some compounds may form more than one hydrate. For example, cobalt(II) chloride forms a red hydrate with the formula $CoCl_2 \cdot 6\,H_2O$ and also forms a violet hydrate with the formula $CoCl_2 \cdot 2\,H_2O$. These compounds are named cobalt(II) chloride hexahydrate and cobalt(II) chloride dihydrate, respectively.

Empirical formulas

This experiment will verify the empirical formula for the product formed when magnesium metal is burned in the presence of oxygen in the air. The balanced equation for the reaction is:

$$2\,Mg(s) + O_2(g) \longrightarrow 2\,MgO(s)$$

Magnesium may also combine with nitrogen in the air. The magnesium nitride formed in this reaction can be converted to magnesium oxide and ammonia gas by adding water and reheating. As a result, all of the original mass of magnesium is converted to magnesium oxide.

$$3\,Mg(s) + N_2(g) \longrightarrow Mg_3N_2(s)$$

$$Mg_3N_2(s) + 3\,H_2O(l) \xrightarrow{\Delta} 3\,MgO(s) + 2\,NH_3(g)$$

The original mass of the magnesium will be obtained by weighing the magnesium ribbon. To determine the mass of the oxygen used to form the magnesium oxide, the mass of the magnesium can be subtracted from the mass of the final magnesium oxide product.

$$\text{mass}_{(\text{oxide product})} - \text{mass}_{(\text{magnesium})} = \text{mass}_{(\text{oxygen})}$$

The following steps are used to determine the empirical formula of a compound:

1. Calculate the moles of each element using the molar mass from the periodic table.
2. Divide each of those mole amounts by the smallest number of moles.
3. Use the whole number ratio to determine the formula subscripts.

Why must masses be converted to moles? Suppose one feather can be glued to one lead bar. Does it then follow that 1 kg of feathers are needed for 1 kg of lead? No, because 1 kg of lead could be only one bar while 1 kg of feathers would be a very large amount. One must use item amounts rather than item masses because the masses of the items involved are different. Similarly, because elements have different masses, one must use the amount of each in calculations. In chemistry, the metric unit for amount is the *mole*.

For example, when determining the empirical formula for magnesium oxide if the mass of magnesium is 3.147 g and the mass of oxygen is 2.072 g, the steps outlined would be as follows:

1. To calculate moles from grams, the molar mass of each element from the periodic table is used:

   1 mole Mg = 24.31g Mg

   1 mole O = 16.00 g O

   a) $\text{mass of magnesium ribbon (g)} \times \dfrac{1 \text{ mole Mg}}{24.31 \text{ g Mg}} = \text{moles of magnesium}$

   $$3.147 \text{ g} \times \frac{1 \text{ mol}}{24.31 \text{ g Mg}} = 0.1295 \text{ mol Mg}$$

   b) $\text{mass of oxygen used (g)} \times \dfrac{1 \text{ mole O}}{16.00 \text{ g O}} = \text{moles of oxygen}$

   $$2.072 \text{ g} \times \frac{1 \text{ mol}}{16.00 \text{ g O}} = 0.1295 \text{ mol O}$$

2. Divide the mole amount of each element by the smallest mole amount. In this example, the mole amounts for magnesium and oxygen are the same:

   $$\frac{0.1295 \text{ mol Mg}}{0.1295} = 1.000 \qquad \frac{0.1295 \text{ mol O}}{0.1295} = 1.000$$

3. In this example, the mole ratio is 1:1 and the empirical formula is MgO.

Not all compounds have a 1:1 ratio. For example if sulfur and oxygen react to form a compound with 0.140 mol S and 0.289 mol O, dividing each of those numerical values by the smaller value gives:

$$\text{sulfur} = \frac{0.140 \text{ moles}}{0.140} = 1.00$$

$$\text{oxygen} = \frac{0.289 \text{ moles}}{0.140} = 2.06$$

Usually the numbers are close to whole numbers. Use the whole number ratio (1 to 2) to determine the subscripts for the empirical formula of the compound: SO_2.

If the ratio does not result in a reasonable "whole number" subscript after rounding off, both numbers may have to be multiplied by a factor of two or three. For example, if the ratio is 1 to 1.5, then multiplying both numbers by two will give whole number coefficients of 2 and 3.

$$\left.\begin{array}{l} \text{aluminum} = \dfrac{0.450 \text{ moles}}{0.450} = 1.00 \\ \\ \text{oxygen} = \dfrac{0.675 \text{ moles}}{0.450} = 1.50 \end{array}\right\} \times 2 \Longrightarrow \left.\begin{array}{l} 2.00 \\ \\ 3.00 \end{array}\right\} Al_2O_3$$

Hydrates

The percent of water in a hydrate can be determined experimentally by heating the *hydrate* to remove the water. This reaction is called *dehydration* and the product is called an *anhydrate*. The mass of the water removed is calculated by subtracting the mass of the final product from the mass of the original hydrate.

$$\text{mass}_{(\text{hydrate})} - \text{mass}_{(\text{anhydrate})} = \text{mass}_{(\text{water})}$$

The formula of the hydrate may then be determined by calculating the number of moles of anhydrate and the number of moles of water using the molar mass conversions. The "whole number" mole ratio is then used to write the formula of the hydrate.

$$\text{moles of anhydrate product} = \text{grams of anhydrate product} \times \frac{1 \text{ mole of anhydrate}}{\text{molar mass (g) of anhydrate}}$$

$$\text{moles of water} = \text{grams of water} \times \frac{1 \text{ mole of } H_2O}{18.02 \text{ g of } H_2O}$$

Dividing each of the numerical values obtained for the moles of anhydrate and moles of water by the smaller numerical value will give the whole number ratio. The whole number obtained for water will be the coefficient written for H_2O in the formula of the hydrate.

The percent of water in a hydrate may be calculated with the formula:

$$\%\text{ water} = \frac{\text{mass of water}}{\text{mass of hydrate}} \times 100$$

Note: mass of "hydrate" and **not** "anhydrate."

Experimental Procedures

Eye protection and appropriate clothing must be worn at all times.

Record all measurements with the correct number of significant figures and units.

Observe proper safety precautions when using the Bunsen burner and handling hot objects.

Use crucible tongs whenever moving crucible and lid. Never place hot objects on the balance.

Wash any contacted skin area immediately with running water and alert the instructor. Discard all wastes properly as directed by the instructor.

A. Empirical formula

1. Obtain a clean, dry crucible and cover. Dark stains may not be removable. Check the crucible for cracks by tapping the side of the crucible with a hard plastic pen (see Figure 1). If the crucible "rings," it is acceptable to use. If a "thunk" is heard, it has a crack. Do not use a cracked crucible.

Figure 1 ***How to check the crucible for cracks***

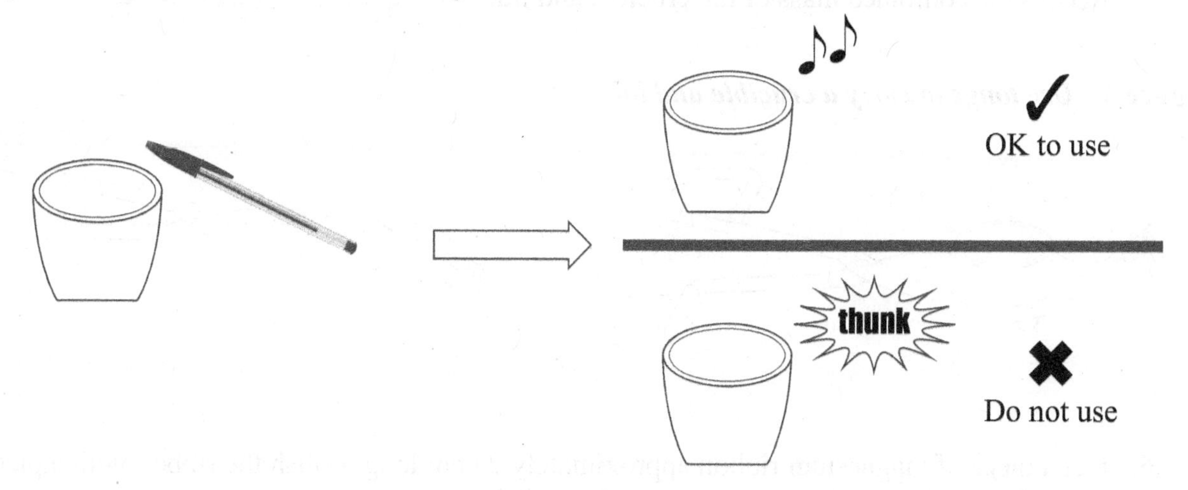

2. Place the crucible and cover in a clay triangle that is sitting on an iron ring attached to a ring stand (see Figure 2). Notice that the lid is slightly askew.

Figure 2 ***Setup for heating a crucible and cover***

3. Heat the crucible and cover for 1 minute.
4. Cool for 5 – 10 minutes until the crucible and lid are at room temperature. NEVER PLACE HOT OBJECTS ON A BALANCE.
5. Use crucible tongs whenever moving the crucible and lid to avoid transferring skin oils (see Figure 3). Follow this practice for the entire experiment. Carry the crucible and lid to the balance. For extra safety, hold the crucible and lid over a watch glass or paper towel. Record the combined mass of the crucible and lid.

Figure 3 ***Use tongs to carry a crucible and lid***

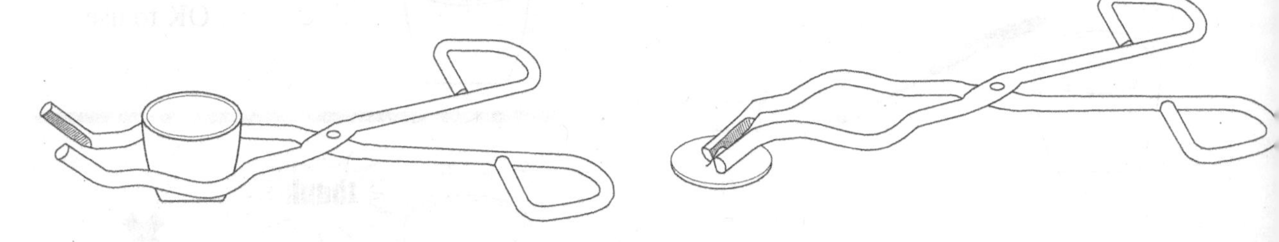

6. Cut a piece of magnesium ribbon approximately 23 cm long. Polish the ribbon with a piece of steel wool. Record the appearance of the polished ribbon.
7. Wind the ribbon into a small tight coil that will fit in the bottom of the crucible. Place the coil in the crucible and weigh the crucible, lid, and magnesium ribbon. Record the combined mass.
8. Place the ring stand and Bunsen burner in the fume hood as directed by the instructor. **Caution: Avoid breathing fumes during the experiment.**
9. Place the crucible containing the magnesium ribbon on the clay triangle. DO NOT COVER UNTIL STEP 10. Heat the crucible *strongly* with the Bunsen burner. Be sure to use the

hottest part of the flame (the tip of the inner cone) to heat the crucible (see Figure 4). The bottom of the crucible should glow red-hot.

Figure 4 ***Heating strongly versus heating gently***

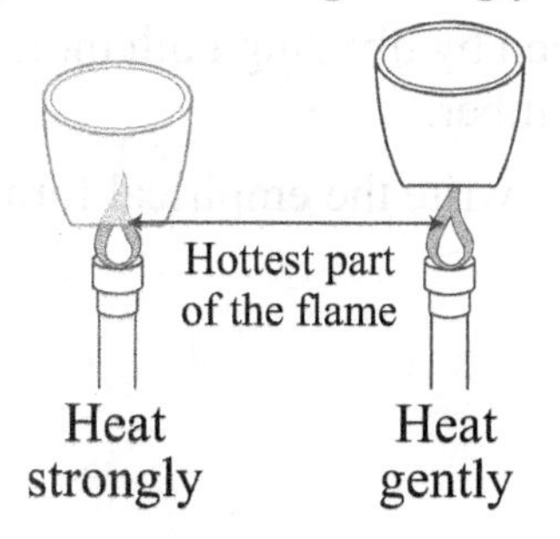

10. The reaction begins when smoke or fumes are visible. When the magnesium sparks, use the crucible tongs to place the lid on the crucible. Remember to leave the lid slightly askew. **Caution: Avoid looking at the bright flame of the burning magnesium.**
11. Continue heating the covered crucible until the reaction no longer produces smoke or flames and a white ash is observed.
12. Remove the lid and place it on a wire gauze. Continue to heat the crucible *strongly* for another 5 minutes.
13. Turn off the Bunsen burner and let the crucible and its contents cool to room temperature. DO NOT weigh anything yet! The magnesium nitride product must first be removed.
14. **Caution: Be sure the crucible and its contents are cooled to room temperature before proceeding.** To remove any magnesium nitride product formed in the reaction, carefully add 15 – 20 drops of water to the crucible.
15. Replace the lid (slightly askew) on the crucible and heat the crucible *gently* (see Figure 4) for 5 minutes. **Caution: Some water may splatter.**
16. Then heat *strongly* for an additional 5 minutes. **Caution: Avoid breathing the fumes from the reaction.**
17. Remove the lid and place it on a wire gauze to cool. Allow the crucible and its contents to cool completely.
18. Describe the appearance of the product.
19. Reweigh the crucible, lid, and its contents. Record the combined mass.

Calculations for Procedure A

20. Calculate the mass of the original magnesium ribbon. This also equals the mass of the magnesium in the magnesium oxide product.
21. Calculate the mass of the magnesium oxide product.

22. Calculate the mass of oxygen that combined with the magnesium to form the magnesium oxide product.
23. Determine the number of moles of magnesium in the product.
24. Determine the number of moles of oxygen in the product.
25. Determine the whole number mole ratio of magnesium to oxygen by dividing both mole values by the smaller one and rounding to the nearest whole number.
26. Use the whole numbers obtained in the mole ratio calculation to write the empirical formula for the magnesium oxide product.

B. Formula and percent water of a hydrate

Note: ***The crucible lid is not used in part B.***

1. Obtain a clean dry crucible and check it for cracks (see Figure 1). Heat it for 2 – 3 minutes and then let cool to room temperature. Use crucible tongs whenever moving the crucible (see Figure 3).
2. Weigh and record the mass of the crucible.
3. Fill the crucible about $^1/_3$ full with the hydrate of $MgSO_4$.
4. Weigh and record the combined mass of the crucible and hydrate.
5. Place the crucible and hydrate on a clay triangle (see Figure 2). Heat *gently* for 5 minutes (see Figure 4).
6. Heat *strongly* for an additional 15 minutes. The bottom of the crucible should be red hot.
7. Turn off the burner and allow the crucible and its contents to cool to room temperature.
8. Weigh and record the mass of the crucible and its contents.

Calculations for Procedure B

9. Calculate the mass of the hydrate used.
10. Calculate the mass of the anhydrate obtained after heating.
11. Calculate the mass of water driven off from the hydrate.
12. Calculate the percent H_2O in the hydrate.
13. Calculate the moles of H_2O driven off from the hydrate.
14. Calculate the moles of anhydrate in the product.
15. Calculate the mole ratio of anhydrate to water by dividing both mole values by the smaller one and rounding to the nearest whole number.
16. Determine the empirical formula of the hydrate.

Name ______________________ Date ______________________

Partner(s) ______________________ Section ______________________

______________________ Instructor ______________________

1. Why is the crucible tapped with a plastic pen before using it?

2. What color should the bottom of the crucible be when heating magnesium ribbon?

3. Compare the composition of a hydrate to an anhydrate.

4. Describe the placement of the crucible lid on the crucible when heating the magnesium. Why is it important that this be done correctly?

5. a) How many molecules of water, called the *water of hydration*, surround each molecule of sodium carbonate decahydrate, $Na_2CO_3 \cdot 10\,H_2O$?

   b) Calculate the molar mass of sodium carbonate decahydrate.

Name ______________________ Date ______________________

Partner(s) ______________________ Section ______________________

______________________ Instructor ______________________

Show the correct number of significant figures and units for all measurements and calculations.

A. Empirical formula

Mass of crucible and lid ______________

Mass of crucible, lid, and magnesium ______________

Appearance of magnesium ribbon:

Appearance of product:

Mass of crucible, lid, and product ______________

Calculations (Show your work)

Mass of magnesium ribbon
(show calculations) ______________

Mass of magnesium oxide product
(show calculations) ______________

Mass of oxygen
(show calculations) ______________

Number of moles of magnesium
(show calculations) ______________

Number of moles of oxygen
(show calculations) ______________

Mole ratio of magnesium to oxygen
(show calculations) ______________

Empirical formula for the magnesium oxide product produced ______________

B. Formula and percent water of a hydrate

Mass of crucible ______________

Mass of crucible and hydrate ______________

Mass of crucible and anhydrate after heating ______________

Calculations (Show your work)

Mass of hydrate used
(show calculations) ______________

Mass of anhydrate after heating
(show calculations) ______________

Mass of water driven off from hydrate
(show calculations) ______________

Percent water in the hydrate
(show calculations) ______________

Number of moles of water driven off from the hydrate
(show calculations) ______________

Number of moles of anhydrate in the product
(show calculations) ______________

Mole ratio of anhydrate to water
(show calculations) ______________

Empirical formula of the hydrate ______________

Problems and questions (show all calculations, correct significant figures and units)

1. Why is it important to test the crucible for cracks before heating it?

2. Write the correct balanced equation for the dehydration (removal of water) of the magnesium sulfate hydrate obtained in Procedure B of this experiment.

3. How many moles are in 70.0 grams of potassium hydroxide?

4. A compound is formed when 0.300 moles of barium combine with 0.600 moles of chlorine. What is the empirical formula for this compound?

5. A compound contains 3.24 grams of sodium, 2.26 grams of sulfur, and 4.51 grams of oxygen. What is the empirical formula for this compound?

6. Calculate the percent water in iron(II) sulfate heptahydrate.

Goals

- Calculate the heat gain for a given amount of water.
- Use a calorimeter to determine the specific heat of a metal.
- Identify an unknown metal from its measured specific heat.
- Determine the energy content per gram of a candle.
- Use conversion factors to calculate the number of calories, kilocalories and Joules in a nutritional Caloric measurement.
- Use the nutritional label on food products to calculate the energy in one serving.

Materials

- Thermometer
- Clamp with thermometer holder
- Hot plate
- Matches
- Small votive candle in metal container (a.k.a. tea light)
- Ring stand
- Iron ring
- Wire screen
- Balance
- Calorimeter (Styrofoam cup and cover)
- Aluminum can with adaptations as shown in Figure 2
- Metal object
- Three food products with nutritional labels

Discussion

Energy is defined as the capacity to do work and is involved in all physical and chemical changes. Changes that absorb energy are referred to as endothermic, and changes that release energy as exothermic. The Law of Conservation of Energy states that energy cannot be created nor destroyed, but may be transferred from one object to another.

<u>Measuring physical energy change</u>

Specific heat is the amount of heat that raises the temperature of 1 gram of a substance by 1°C. Specific heat is a physical property and varies depending on the substance. The equation for specific heat is:

$$\text{specific heat} = \frac{\text{amount of heat}}{\text{mass} \times \Delta \text{T}} = \frac{\text{calories } (\textbf{or } \text{Joules})}{\text{grams} \times {}^\circ\text{C}}$$

In many chemistry textbooks, specific heat is given the symbol *c* while the amount of heat is given the symbol *Q* or *q*. Therefore, the above equation may be simplified as:

$$c = \frac{Q}{m \times \Delta \text{T}}$$

The specific heat of water is used to define the value of a *calorie* since it takes 1.00 cal to raise the temperature of 1 gram of water 1°C. The measurement units for the amount of heat may be converted using the conversion factor: 1 calorie = 4.184 Joules.

The specific heats of several substances are shown in Table 1. Notice that liquid water has a very large specific heat compared to most other substances. Because each substance has a unique specific heat value, specific heat can be used to identify a substance.

Table 1 ***Specific heat values***

Substance	Specific heat (cal/g•°C)	Substance	Specific heat (cal/g•°C)
Water (liquid)	1.00	Aluminum	0.214
Water (solid)	0.486	Iron	0.107
Water (gas)	0.481	Copper	0.0920
Ethyl alcohol	0.583	Lead	0.0311
Air	0.239	Zinc	0.0927
Carbon (graphite)	0.169	Brass	0.0908

Calorimeter

In theory, to measure the specific heat of a solid such as a metal, the temperature of the metal can be measured before and after heating. Together with the mass of the metal, the specific heat can be calculated. However, this method requires that the amount of heat (calories) transferred to the metal is known. It is often difficult to know how many calories the metal absorbed.

Instead, if a solid piece of metal is heated and then put into a container of cool water, the metal cools down while the water heats up. Eventually, both reach the same temperature. The metal transferred its energy to the water and the amount of energy the metal lost is **exactly** the same amount of energy the water gained.

Rearranging the specific heat formula allows for the amount of energy to be calculated when the specific heat is known:

equation: amount of heat = mass × specific heat × ΔT

$$\text{units:} \qquad \text{calories} = (grams) \times \left(\frac{cal}{g \cdot {}^\circ C}\right) \times ({}^\circ C)$$

Again, the simplification for this equation is:

$$Q = m \times c \times \Delta T$$

The specific heat of water is known. If the mass of the water is measured as well as the beginning and ending temperatures, the number of calories that the water gained can be calculated. As stated earlier, this is the same number of calories the metal lost. The mass of the metal and its temperature change can be easily measured. Together with the calories lost, the specific heat of the metal may be calculated.

As the metal cools down, it is important that it transfers its heat to the water and not the surroundings. A *calorimeter* is an insulated container that prevents rapid heat exchange between the contents inside the container and the exterior surroundings. The transfer of heat from the metal to the water will be contained in the calorimeter. In this experiment, a Styrofoam cup will be used as the calorimeter.

Measuring chemical energy change

Different types of bonds, such as carbon–hydrogen or oxygen–oxygen, have different energies associated with them. When molecules react, bonds are broken in the reactants and new bonds are formed in the products. Breaking a bond always requires an input of energy, while forming a bond always releases energy. The difference in energy between the products and the reactants determines whether there is a net loss or a net gain of energy. When energy is absorbed, the reaction is said to be *endothermic*. When energy is released, the reaction is said to be *exothermic*. Examples of exothermic chemical reactions include the burning of fuels such as oil, gas, and even candles.

Food products can also be considered as "fuels" for our bodies. When any of these fuels react, bonds are broken and energy is released. The energy released can then be expressed as cal/g. In food energy measurements, nutritionists use calorimeters to establish the caloric values for the food types as shown in Table 2. The energy values for foods are expressed in Calories (with a capital C) per gram. The equalities for the different units are:

$$1 \text{ Calorie} = 1000 \text{ calories} = 1 \text{ kilocalorie} = 4184 \text{ Joules}$$

Table 2 ***Caloric values for food types***

Food type	Calories/g
Fat	9
Carbohydrate	4
Protein	4
Fiber	0

Nutritional values are required by law to be printed on food labels (see Figure 1). Food labels list the grams of each food type rather than the Calories from each type. A simple calculation is needed to convert the grams to Calories. To do some of the Calories problems, round your answer to the nearest whole Calorie. However, the Calories on food labels are typically rounded to the nearest tens place.

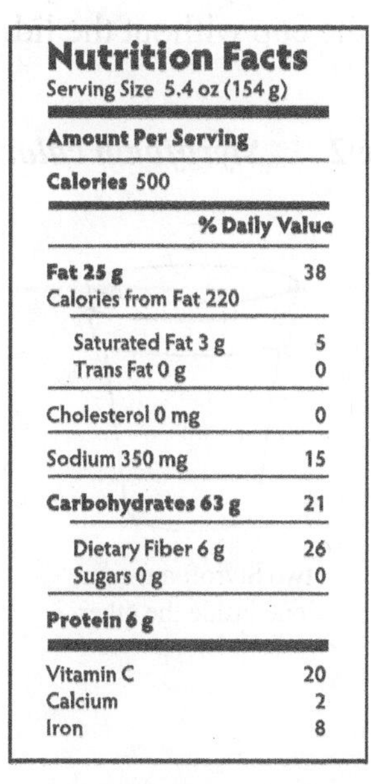
Nutrition Facts

Serving Size 5.4 oz (154 g)

Amount Per Serving	
Calories 500	
	% Daily Value
Fat 25 g	38
Calories from Fat 220	
Saturated Fat 3 g	5
Trans Fat 0 g	0
Cholesterol 0 mg	0
Sodium 350 mg	15
Carbohydrates 63 g	21
Dietary Fiber 6 g	26
Sugars 0 g	0
Protein 6 g	
Vitamin C	20
Calcium	2
Iron	8

Figure 1 ***Nutrition label for French fries***

Sample problem:

What is the Caloric value of a serving of meat containing 3 grams of protein and 1 gram of fat?

$$\underbrace{(3\text{ g})(4\,{}^{\text{Cal}}\!/\!_{\text{g}})}_{\text{protein}} + \underbrace{(1\text{ g})(9\,{}^{\text{Cal}}\!/\!_{\text{g}})}_{\text{fat}} = 21\text{ Cal}$$

On a food label, this would be listed as 20 Cal.

Experimental Procedures

Eye protection and appropriate clothing must be worn at all times.
Record all measurements with the correct number of significant figures and units.
NEVER leave a boiling water bath unattended.
Discard all wastes properly as directed by the instructor.

A. Specific heat of a metal

1. Prepare a boiling water bath by filling a 400 mL beaker with approximately 250 mL of water. Set the beaker and water on a hotplate with the temperature control turned to high and heat the water to boiling.
2. Obtain a metal object. Use a balance to determine the mass of the metal object. Record the mass and identification number of the metal object on the report sheet.
3. Tie a length of string around the metal object and carefully lower it into the boiling water bath. Boil the water bath with the metal object submerged for 10 minutes.
4. Obtain a Styrofoam cup calorimeter apparatus as shown in Figure 2 and determine the mass of the cup without the lid. Record.

Figure 2 ***Styrofoam calorimeter apparatus***

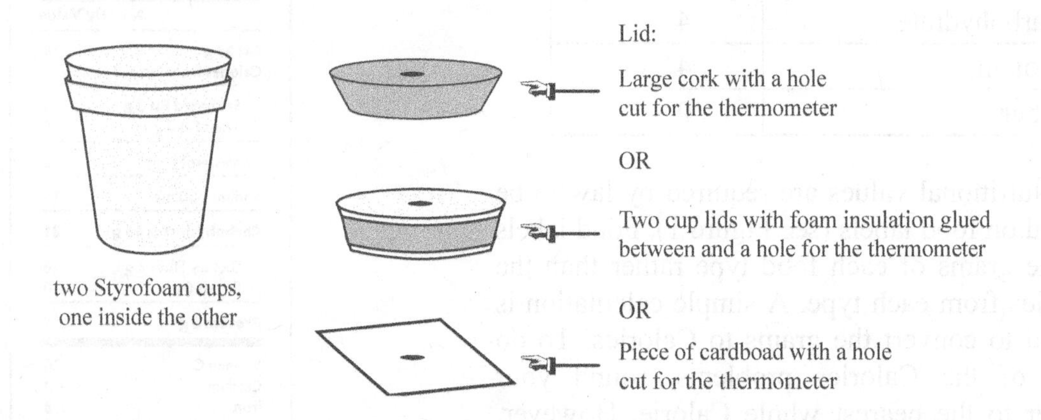

5. Pour approximately 50–100 mL of water into the Styrofoam cup. There should be just enough water to cover the metal object when it is transferred to the cup. Determine the combined mass of the cup and the water. Record the combined mass.
6. After the metal object has been in the boiling water bath for 10 minutes, use a thermometer to measure the temperature of the boiling water. NEVER SHAKE THE THERMOMETER. ***This is also the initial temperature of the metal object***. Record.

7. Use the thermometer to measure the initial temperature of the water in the calorimeter. Record.
8. Carefully and quickly use the string to remove the metal object from the boiling water bath and transfer it to the calorimeter. Quickly place the lid on the calorimeter and use the thermometer to gently stir the water, being careful not to hit the metal object. Stirring with a thermometer is not the best practice because the thermometer may be broken, but it is allowed here for convenience. Record the highest temperature of the water in the calorimeter. ***This is also recorded as the final temperature of the metal object.***

Calculations for Procedure A

9. Calculate the temperature change for the water in the calorimeter.

$$\Delta T_{water} = \text{final highest temperature} - \text{initial temperature}$$

10. Calculate the heat energy (in calories) gained by the water in the calorimeter.

$$\underset{\substack{\uparrow \\ \text{water}}}{\text{calories}} = \underset{\substack{\uparrow \\ \text{water}}}{\text{mass}} \times \underset{\substack{\uparrow \\ \text{water}}}{\text{specific heat}} \times \underset{\substack{\uparrow \\ \text{water}}}{\Delta T}$$

11. Record the heat energy (in calories) lost by the metal object (same as the amount gained by the water).
12. Calculate the ΔT of the metal.

$$\Delta T_{metal} = \text{initial temperature of metal} - \text{final temperature of metal}$$

13. Calculate the specific heat of the metal

$$\text{specific heat} = \frac{\text{calories lost by metal}}{\text{mass of metal} \times \Delta T \text{ of metal}}$$

14. Convert the specific heat of the metal to Joules/g•°C.
15. Use Table 1 to identify the unknown metal object. Additional specific heat values may be provided by the instructor.

B. Energy content of fuel

1. Use a balance to determine the mass of an aluminum can. Record.
2. Pour approximately 100 mL of water into the can. Weigh and record the combined mass of the can and the water.

3. The instructor will demonstrate how to set up the can on a ring stand as shown in Figure 2 and Figure 3.

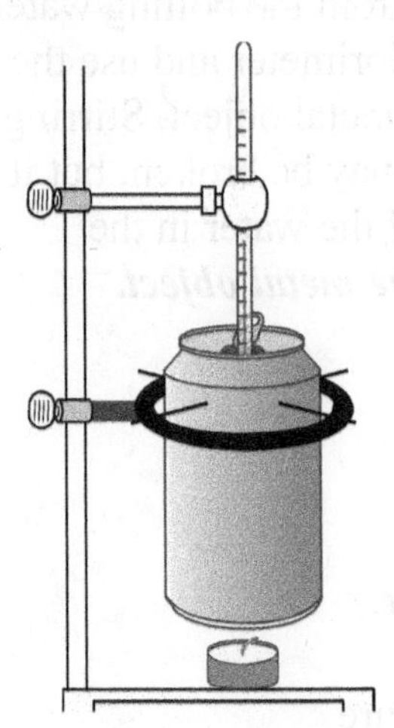

Figure 3 ***Apparatus setup for Procedure B showing heavy wire or straightened paper clip inserted through an aluminum can to secure it to the iron ring***

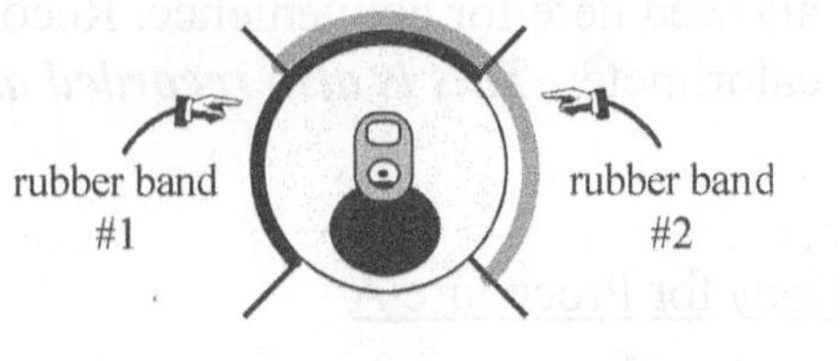

Figure 4 ***Top view showing two rubber bands holding the wires in place***

4. Obtain a votive candle and use a balance to determine its initial mass. Record.
5. Place the votive candle on the ring stand. Adjust the height of the can by moving the ring clamp so that the bottom of the can just touches the wick of the candle.
6. Attach a clamp with a thermometer in a stopper to the ring stand.
7. Determine the initial temperature of the water. Leave the thermometer in the water. Do not let the thermometer touch the bottom or sides of the can.
8. Light the candle and let it burn until the water temperature increases by at least 5°C. The larger the change in temperature, the better the results will be.
9. Extinguish the candle. Be sure not to blow any of the molten wax off the candle. Record the final temperature of the water.
10. When the candle is cool, use a balance to determine its final mass.

Calculations for Procedure B

11. Calculate the heat energy (in calories) gained by the water using the formula from Procedure A.

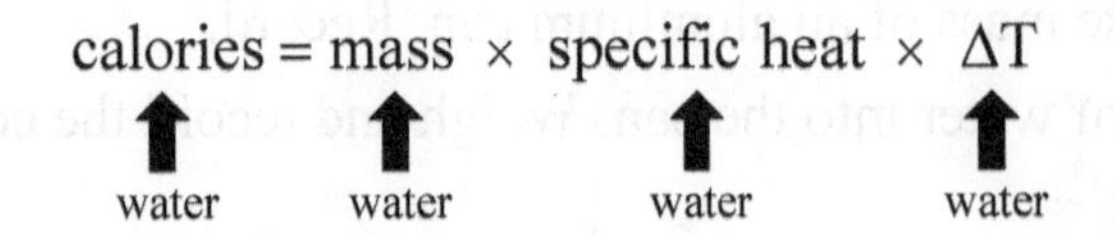

12. Record the heat energy gained by the water and the heat energy lost by the candle.
13. Calculate the heat energy per gram released from the candle. The energy released comes from the burning wax molecules breaking their bonds to form the new products:

$$C_{46}H_{92}O_{(s)} + 69\,O_{2(g)} \longrightarrow 46\,CO_{2(g)} + 46\,H_2O_{(g)} + \text{heat}$$

The equation for the heat energy per gram released from the candle is:

$$\text{heat energy per gram} = \frac{\text{heat energy lost by candle}}{\text{mass of candle burned}}$$

C. Food Calories

1. Choose three food items from the display provided.
2. Complete the chart on the report sheet.
3. Use the mass of each food type and the appropriate Cal/g values to calculate the Calories in one serving.
4. Compare your calculated values with those provided on the product label.

Name ______________________ Date ______________________

Partner(s) ______________________ Section ______________________

______________________ Instructor ______________________

1. Water has a large specific heat compared to other substances. How is this property beneficial to the human body?

2. Why must the metal object be transferred and covered with the lid quickly when using a calorimeter?

3. Why is a Styrofoam cup a good choice for a calorimeter?

4. How many calories are required to increase the temperature of a 27.3 g sample of copper from 36.2°C to 74.4°C?

5. List two reasons why a boiling water bath should never be left unattended.

REPORT SHEET: **CALORIMETRY**

Name ____________________ Date ____________________

Partner(s) ____________________ Section ____________________

____________________ Instructor ____________________

Read the procedures before completing the following tables. Show the correct number of significant figures and units for all measurements and calculations.

A. Specific heat of a metal

Mass of the metal ______________ ID# __________

Mass of the Styrofoam cup ______________

Mass of Styrofoam cup + water ______________

Mass of water ______________

Initial temperature of metal ______________

Initial temperature of water in Styrofoam cup ______________

Final highest temperature of water in Styrofoam cup ______________

Final temperature of metal ______________

<u>Calculations (Show your work)</u>

Temperature change of water in calorimeter ______________

Heat energy gained by water in calorimeter ______________

Heat energy lost by metal object ______________

Temperature change of the metal ______________

Specific heat of the metal ______________

Specific heat of the metal converted to Joules/g·°C ______________

Identity of metal object (ID#__________) ______________

B. Energy content of fuel

Mass of aluminum can ______________

Mass of can + water ______________

Mass of water ______________

Initial mass of candle ______________

Final mass of candle ______________

Mass of candle burned ______________

Initial temperature of water ______________

Final temperature of water ______________

<u>Calculations (Show your work)</u>

Heat energy gained by water ______________

Heat energy lost by candle ______________

Heat energy per gram burned from candle ______________

C. Food Calories

	Brand__________	Brand__________	Brand__________
Serving size			
Fat grams			
Carbohydrate grams			
Protein grams			
Fat Calories			
Carbohydrate Calories			
Protein Calories			
Calculated total Calories per serving (sum of above three numbers)			
Total Calories per serving listed on label			
Is the calculated value close to the listed value?			
Calculated total Calories converted to kilocalories, calories, and Joules	kcal	kcal	kcal
	cal	cal	cal
	J	J	J

Problems and questions (show all calculations, correct significant figures and units)

1. Consider a pot used for cooking. Would you expect the handle to have a small or a large specific heat? Why?

2. What are the units for specific heat?

3. A potato chip with a mass of 0.20 grams is burned. The heat released is used to warm 75 grams of water from 27°C to 37°C. Calculate the caloric value of the potato chip in kcal/g.

4. A 12.5 g sample of lead absorbs 28.0 calories of heat. If the initial temperature is 23.0°C, what is the final temperature?

5. An unknown metal has a mass of 4.67 g. It is heated to 95.1°C and then placed in a calorimeter that contains 24.3 g of water at 21.7°C. The metal and water both reach a final temperature of 24.6°C. What is the specific heat of this metal? What is the unknown metal?

Goals

- Graph a heating curve.
- Graph a cooling curve.
- Determine the boiling/condensation point of a substance.
- Determine the melting/freezing point of a substance.
- Determine the heat of fusion of a substance.
- Purify a substance by sublimation.

Materials

- 400 mL beaker
- 250 mL beaker
- 100 mL beaker
- Hot plate
- Ring stand
- Iron ring
- Wire gauze
- Thermometer with rubber stopper and clamp
- Timer
- 100 mL graduated cylinder
- Freezing point apparatus with Salol (phenyl salicylate), stopper, thermometer, and wire stirrer
  DO NOT ATTEMPT TO REMOVE THERMOMETER
- Calorimeter
- Ice
- Spoon or scoopula
- Electronic balance
- Evaporating dish to fit on 100 mL beaker
- Naphthalene
- Bunsen burner

Discussion

Physical properties can be used to describe substances and include observable characteristics such as color, shape, odor, and taste. Other physical properties are measurable, such as density and hardness. Numerical values for these measurable properties can be looked up in reference materials and are often used to help identify unknown samples. In this experiment, some physical properties that are associated with states of matter and the changes between states will be measured.

Three physical states exist for matter: solid, liquid, and gas. Changes in physical state are commonly observed in the chemistry lab. In a physical change, the substance changes its physical appearance but not its chemical composition. Melting, freezing, and vaporization are examples of physical changes (see Figure 1).

Figure 1 ***Changes of state***

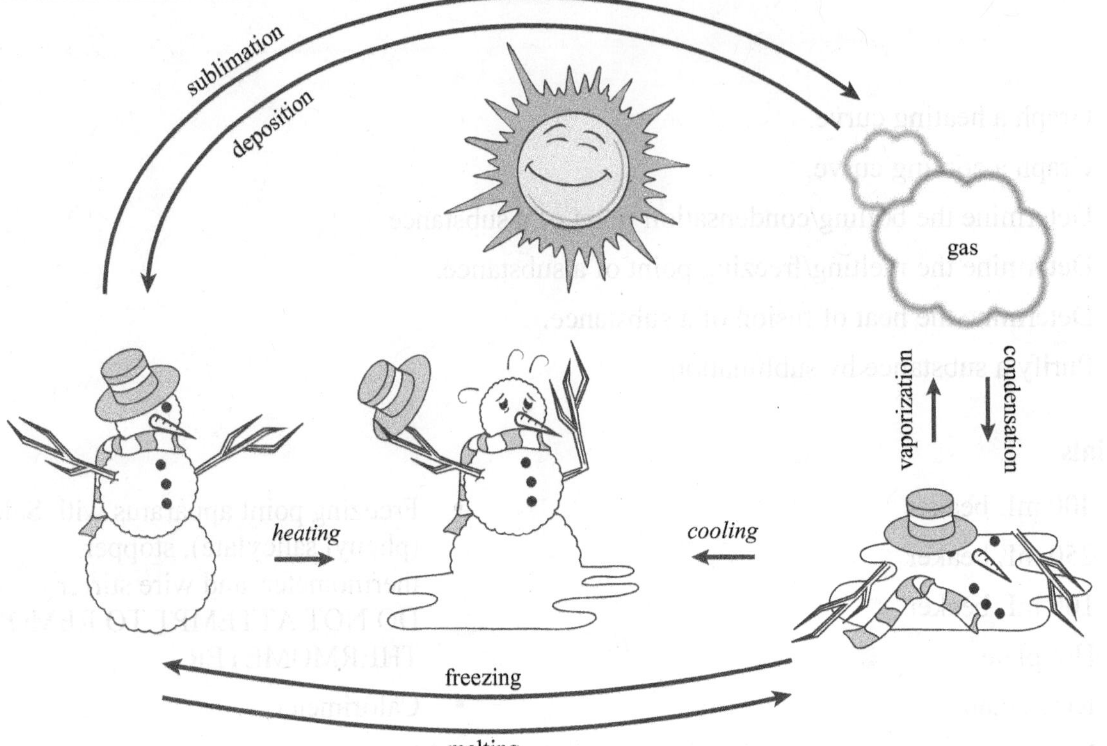

Heating curves and *cooling curves* are graphs that show the change in temperature of a substance relative to the amount of energy added to or removed from the substance. In this experiment, the amount of energy will be plotted as the amount of time since it is easier to measure passing time than actual amount of energy gained or lost. Even though the temperatures and states of matter are changing, the formula for the substance does not change.

Specific heat is the amount of heat that raises the temperature of 1 gram of a substance by 1°C. The specific heats for a substance vary, depending upon its state (solid, liquid, or gas) and each is unique. The sloping lines on a graph for the heating curve represent specific heats for solids, liquids and gases. The equation for specific heat is:

$$\text{specific heat} = \frac{\text{amount of heat}}{\text{mass} \times \Delta T} = \frac{\text{calories } (\textbf{or} \text{ Joules})}{\text{grams} \times {}^\circ\text{C}}$$

In many chemistry textbooks, specific heat is given the symbol *c* while the amount of heat is given the symbol *Q* or *q*. Therefore, the above equation may be simplified as:

$$c = \frac{Q}{m \times \Delta T}$$

Although specific heat is not measured in this experiment, an important equation can be obtained by rearranging the specific heat equation to solve for amount of heat:

$$\text{equation:} \quad \text{amount of heat} = \text{mass} \times \text{specific heat} \times \Delta T$$

$$\Downarrow \qquad \Downarrow \qquad \Downarrow$$

$$\text{units:} \quad \text{calories} = (grams) \times \left(\frac{cal}{g \cdot {}^\circ C}\right) \times ({}^\circ C)$$

Again, the simplification for this equation is:

$$Q = m \times c \times \Delta T$$

Notice that in these equations there is a change in temperature, ΔT.

Heat of fusion is the amount of energy required to melt 1 gram of a substance from a solid state to a liquid state. Although energy is added to the substance, the temperature of the substance does not change, only the state of matter changes. The amount of energy absorbed to melt the substance is given by the following equation:

$$\text{equation:} \quad \text{amount of heat} = \text{mass} \times \text{heat of fusion}$$

$$\Downarrow \qquad \Downarrow$$

$$\text{units:} \quad \text{calories} = (grams) \times \left(\frac{cal}{g}\right)$$

The symbol for heat of fusion is H_f, and this equation may be simplified to:

$$Q = m \times H_f$$

The numerical value for the heat of fusion is also the amount of energy released when a substance freezes, changing from a liquid to a solid. *Melting point* or *freezing point* refers to the temperature at which the state change between solid and liquid occurs. This constant temperature appears as a horizontal area (plateau) on a graph for the heating curve. *Supercooling* occurs when heat is removed from a liquid so rapidly that the molecules have no time to assume the ordered structure of a solid. The temperature of the liquid drops temporarily below the freezing point. But as the liquid continues to freeze, the temperature will return to the freezing point. This dip in the graphed data points is known as the area of supercooling.

Heat of vaporization is the amount of energy required to vaporize 1 gram of a substance from a liquid state to a gaseous state. Although energy is added to the substance, the temperature of the substance does not change, only the state of matter changes. The amount of energy absorbed to vaporize the substance is given by the following equation:

$$\text{equation:} \quad \text{amount of heat} = \text{mass} \times \text{heat of vaporization}$$

$$\Downarrow \qquad \Downarrow$$

$$\text{units:} \quad \text{calories} = (grams) \times \left(\frac{cal}{g}\right)$$

The symbol for heat of vaporization is H_v, and this equation may be simplified to:

$$Q = m \times H_v$$

The numerical value for heat of vaporization is also the amount of energy released when a substance condenses, changing from a gas into a liquid. *Boiling point* or *condensation point* refers to the temperature at which the state change between liquid and gas occurs. This constant temperature appears as a horizontal area (plateau) on a graph for the heating curve.

A typical heating curve for any substance is shown in Figure 2. The numerical values for the specific heats and heats of fusion and vaporization would differ depending upon the substance. For water, those values are found in Table 1.

Figure 2 ***Heating curve***

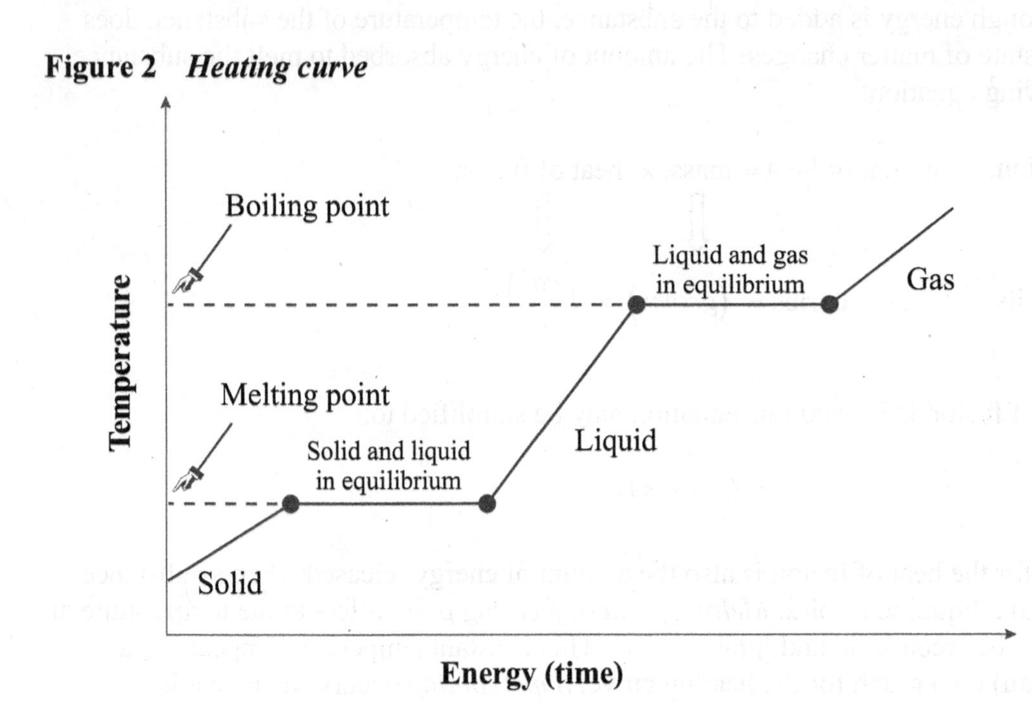

Table 1 ***Energy values for water***

Substance state	Numerical value
Specific heat: $H_2O_{(solid)}$	0.51 cal/g•°C
Specific heat: $H_2O_{(liquid)}$	1.00 cal/g•°C
Specific heat: $H_2O_{(gas)}$	0.48 cal/g•°C
Heat of fusion: H_2O	80. cal/g
Heat of vaporization: H_2O	540. cal/g

Sublimation is a process that involves direct conversion of a solid to a gas without passing through a liquid state. Two examples of substances that can undergo sublimation are solid carbon dioxide (dry ice) and naphthalene (moth balls). Sublimation is a method that can be used to purify a substance. Impurities in the substance often do not sublimate. Sublimating the substance and then redepositing the gas as a solid can remove the impurities.

Summary

There are three important equations in this experiment:

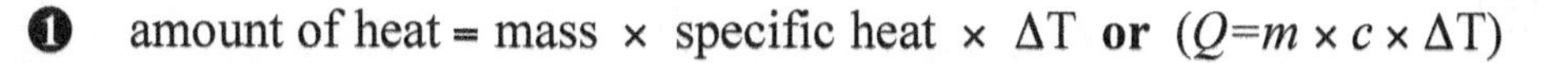

❶ amount of heat = mass × specific heat × ΔT **or** ($Q = m \times c \times \Delta T$)

❷ amount of heat = mass × heat of fusion **or** ($Q = m \times H_f$)

❸ amount of heat = mass × heat of vaporization **or** ($Q = m \times H_v$)

To solve a problem, use the following flowchart to plot a path from the start of the problem to the end:

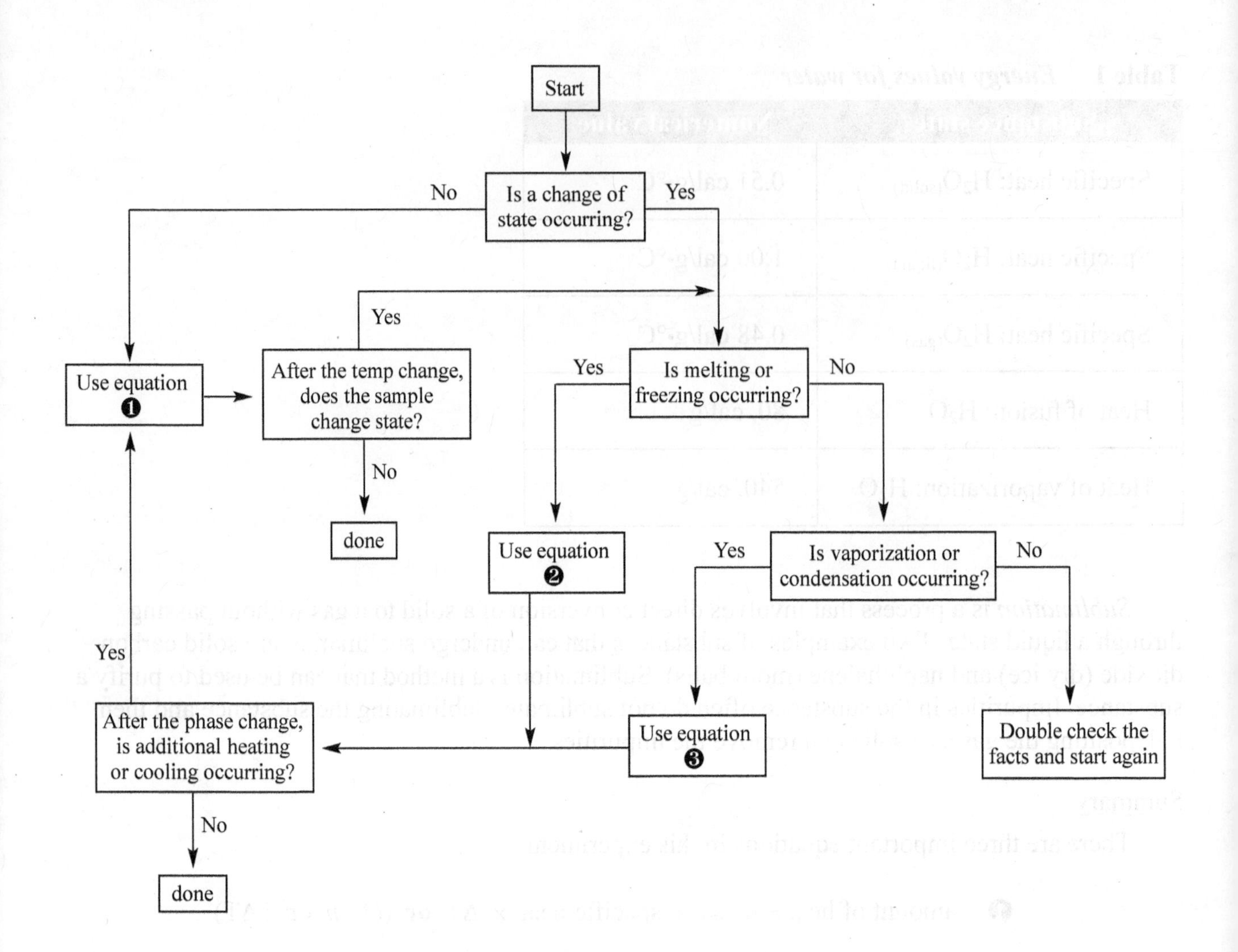

Example 1:

If a hot water bottle contains 500. g of water at 65°C, how much heat, in calories, will it have supplied to a person's "aching back" by the time it has cooled to 37°C (assuming all the heat energy goes into the person's back)?

Solution:

No change of state occurs, therefore, only equation #1 is needed.

❶ amount of heat = mass × specific heat × ΔT **or** ($Q = m \times c \times \Delta T$)

facts:
- mass of water = 500. g
- specific heat of $H_2O_{(liquid)} = 1.00$ cal/g•°C
- $\Delta T = 65°C - 37°C = 28°C$

$$\text{amount of heat} = (500.\ g)\left(1.00\ {}^{cal}/_{g\cdot°C}\right)(28°C) = 1.4 \times 10^4\ cal$$

Example 2:

Calculate the heat required to melt 14.7 g of ice at 0°C.

Solution:

A change of state (melting) occurs, therefore, equation #2 is needed.

❷ amount of heat = mass × heat of fusion **or** ($Q = m \times H_f$)

facts: { mass of ice = 14.7 g
heat of fusion of H_2O = 80. cal/g

$$\text{amount of heat} = (14.7\ g)\left(80.\ {}^{cal}\!/_{g}\right) = 1.2 \times 10^3\ cal$$

Example 3:

Calculate the heat released when 2.45 g of steam at 100°C condenses to liquid water at 100°C.

Solution:

A change of state (condensation) occurs, therefore, equation #3 is needed.

❸ amount of heat = mass × heat of vaporization **or** ($Q = m \times H_v$)

facts: { mass of steam = 2.45 g
heat of vaporization of H_2O = 540. cal/g

$$\text{amount of heat} = (2.45\ g)\left(540.\ {}^{cal}\!/_{g}\right) = 1.32 \times 10^3\ cal$$

Example 4:

Calculate the heat required to take an 11.8 g ice cube at –13.2°C and melt it (change it to liquid water at 0.0°C).

Solution:

This is a two-step problem.

In the first step, no phase change occurs (the ice is just being warmed from –13.2°C to 0.0°C). Equation #1 is needed.

❶ amount of heat = mass × specific heat × ΔT **or** ($Q = m \times c \times \Delta T$)

facts:
- mass of ice = 11.8 g
- specific heat of $H_2O_{(solid)}$ = 0.51 cal/g•°C
- $\Delta T = 0.0°C - (-13.2°C) = 13.2°C$

$$\text{amount of heat} = (11.8\ g)\left(0.51\ \frac{cal}{g \cdot °C}\right)(13.2°C) = 79\ cal$$

In the second step, a phase change (melting) occurs and equation #2 is needed.

❷ amount of heat = mass × heat of fusion **or** ($Q = m \times H_f$)

facts:
- mass of ice = 11.8 g
- heat of fusion of H_2O = 80. cal/g

$$\text{amount of heat} = (11.8\ g)\left(80.\ \frac{cal}{g}\right) = 9.4 \times 10^2\ cal$$

The total energy required is the sum of the calories from the two steps:

$$\text{total calories} = 9.4 \times 10^2\ cal + 79\ cal = 1019\ cal$$

Experimental Procedures

> ***Eye protection and appropriate clothing must be worn at all times.***
>
> ***Observe proper safety precautions when working with flames and other hot objects.***
>
> ***Record all measurements with the correct number of significant figures and units.***
>
> ***Wash any contacted skin area immediately with running water and alert the instructor. Discard all wastes properly as directed by the instructor***

Graphs should include the following:

- **Label and units on each axis**
- **Numerical values on each axis**
- **Data plot should cover as much of the graph as possible**
- **Axis scales should be equally spaced**
- **A smooth line or curve that best fits the data points**

A. Heating curve for water

1. Use a graduated cylinder to measure approximately 100 mL of cool water. Record the actual volume as accurately as possible. Pour the water into a 250 mL beaker.
2. Place the beaker with water on a hot plate (see Figure 3).
3. Insert a thermometer into a rubber stopper and secure with a clamp onto a ring stand. Position the thermometer so that the bulb is in the center of the water. Do not let the thermometer touch the bottom or sides of the beaker.

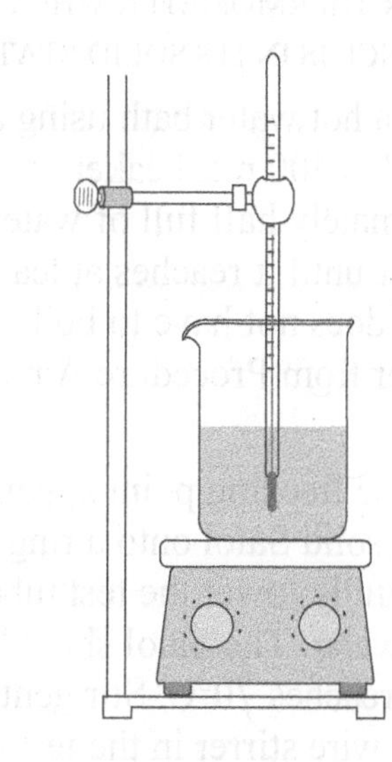

Figure 3 ***Hot water bath***

4. Measure the initial temperature (T_i) of the water. Record.
5. Turn the thermostat on the hotplate to high. Record the temperature at one-minute intervals, using a second hand on a clock or a timer.
6. Once the water has come to a *full boil* (bubbles reaching the surface of the water), continue to boil for 5 more minutes and continue recording the temperature at one-minute intervals. Save this set-up for Procedure B.

7. Graph the results of the heating curve. Plot the temperature on the *y*-axis and the energy added (time) on the *x*-axis. The plateau indicates the boiling point (T_f) of the water. Record.
8. Calculate the $\Delta T = T_f - T_i$ and record.
9. Calculate the mass of the water using the original actual recorded volume and the density of water (1.00 g/mL).
10. Calculate the energy used to heat the water from the original temperature to the boiling point.

$$\text{equation: amount of heat} = \text{mass} \times \text{specific heat} \times \Delta T$$

$$\text{units: calories} = (grams) \times \left(\frac{cal}{g \cdot {}^{\circ}C}\right) \times ({}^{\circ}C)$$

B. Cooling curve for Salol (phenylsalicylate)

1. Observe the demonstration set up for the use of the freezing point apparatus (see Figure 4). DO NOT ATTEMPT TO REMOVE THE STOPPER, WIRE, OR THERMOMETER WHILE THE SUBSTANCE IS IN ITS SOLID STATE.
2. Prepare a hot water bath using a hot plate. Fill a 400 mL beaker approximately half full of water. Heat the water until it reaches at least 80°C. It does not have to boil. The hot water from Procedure A may be used.
3. Clamp the freezing point apparatus with the solid Salol onto a ring stand and carefully lower the test tube into the hot water. The Salol should melt when it reaches 70°C. Stir gently with the wire stirrer in the test tube until all of the Salol is liquid.

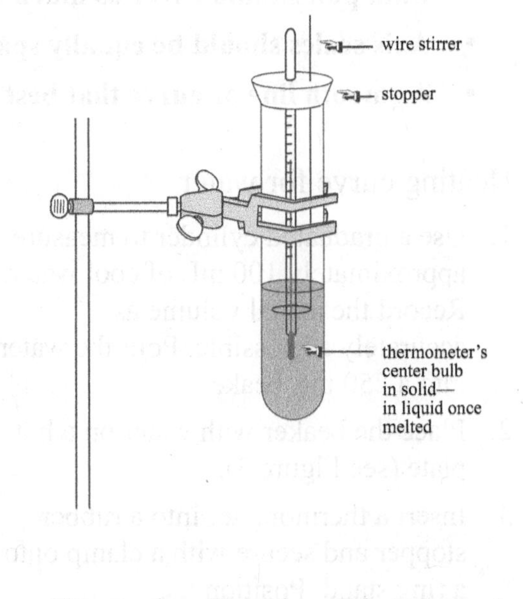

Figure 4 ***Freezing point apparatus***

4. Turn off the hot water bath and remove the freezing point apparatus with the liquid Salol from the hot water.
5. Remove the hot plate and reclamp the freezing point apparatus to the ring stand. Continue to gently stir with the wire stirrer. Observe and record the temperature at one-minute intervals as the liquid Salol cools.
6. Stop stirring when the stirring wire becomes frozen in the Salol. Continue to observe and record the temperature at one-minute intervals for five additional minutes.

7. Return the freezing point apparatus to the instructor. Do not attempt to remove the stopper, wire or thermometer.
8. Graph the results of the cooling curve. Plot the temperature on the *y*-axis and the energy removed (time) on the *x*-axis.

C. Heat of fusion

1. Weigh an empty calorimeter cup. Record the mass.
2. Add approximately 100 mL of water to the cup. Reweigh and record the combined mass of the cup and water.
3. Calculate the mass of the water in the cup (used in the equation in C.11).
4. Record the initial temperature of the water in the calorimeter cup.
5. Fill a 100 mL beaker with crushed ice and add the ice to the water in the calorimeter cup.
6. Stir vigorously and check the temperature of the water. It should be approximately 2–3°C.
7. If the temperature is above 3°C, continue to add ice and stir. Once the temperature reaches 2–3°C, remove all of the unmelted ice using a spoon or scoopula.
8. Record the temperature of the water.
9. Weigh the cup, water, and melted ice. Record the combined mass.
10. Calculate the ΔT of the water.
11. Calculate the energy lost by the water using the equation:

$$\text{calories lost by water} = \text{mass of water} \times \text{specific heat of water} \times \Delta T$$

(This is also the number of calories gained by the ice when it melted.)

12. Calculate the mass of the melted ice (used in the denominator of the equation in C.13).

$$\text{mass of melted ice} = \text{mass of cup, water, and melted ice} - \text{mass of cup and water}$$

13. Calculate the experimental value for the heat of fusion for ice by using the equation:

$$\text{Heat of fusion}\left(\frac{cal}{g}\right) = \frac{\text{energy gained to melt ice}\ (cal)}{\text{mass of melted ice}\ (g)}$$

D. Purification of naphthalene by sublimation

Note: ***Be sure to work under the ventilation hoods.***

1. Weigh out approximately 0.5 g of impure naphthalene. Record the appearance of the substance.
2. Place the impure naphthalene in the bottom of a 100 mL beaker.
3. Set an evaporating dish on top of the beaker and fill the dish with ice.

4. Place the beaker and evaporating dish on a wire gauze, iron ring and ring stand (see Figure 5).
5. Using a low flame on a Bunsen burner, gently heat the bottom of the 100 mL beaker by passing the flame back and forth beneath the wire gauze.
6. Observe the solid flakes of sublimated naphthalene reform on the bottom of the evaporating dish.
7. Turn off the burner and carefully remove the evaporating dish, pouring out the water and ice.
8. Scrape off the solid flakes and record their appearance. Note the appearance of any substance remaining in the beaker and record.
9. Dispose of the naphthalene in the appropriate waste container.

Figure 5 ***Sublimation apparatus***

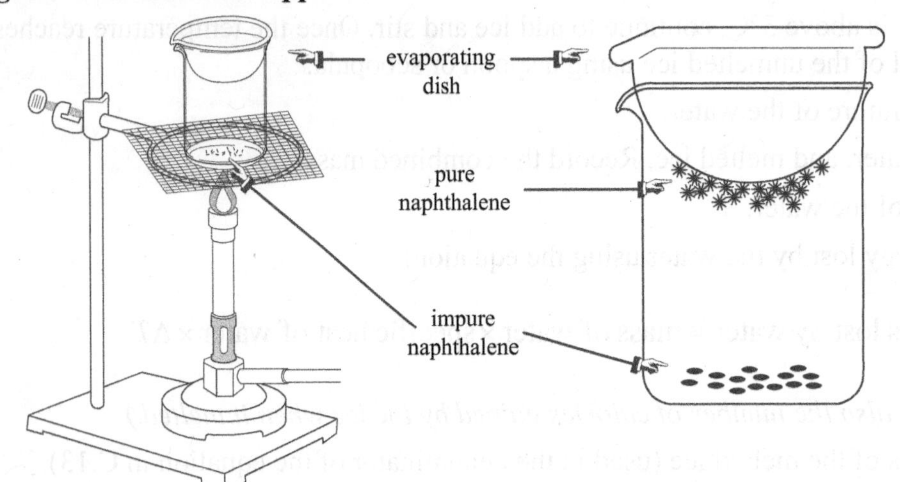

Name ______________________ Date ______________________

Partner(s) ______________________ Section ______________________

______________________ Instructor ______________________

1. Draw the heating curve for water. Label the areas for the states of matter and state changes. Insert the numerical values for all specific heats, heat of fusion, and heat of vaporization for the correct areas of the curve.

2. Which chemical substance used in this lab must be disposed of in a designated waste container and **NOT** placed in the trash or the sink?

3. The procedure for using the freezing point apparatus carries a warning, "Do not attempt to remove the stopper, wire, or thermometer while the substance is in its solid state." What would happen if you disregard this warning?

Name ______________________ Date ______________________

Partner(s) ______________________ Section ______________________

______________________ Instructor ______________________

A. Heating curve for water

Table A.1 Water temperatures

Time (minutes)	Temperature (°C)	Time (minutes)	Temperature (°C)

Graph A.1 Heating curve

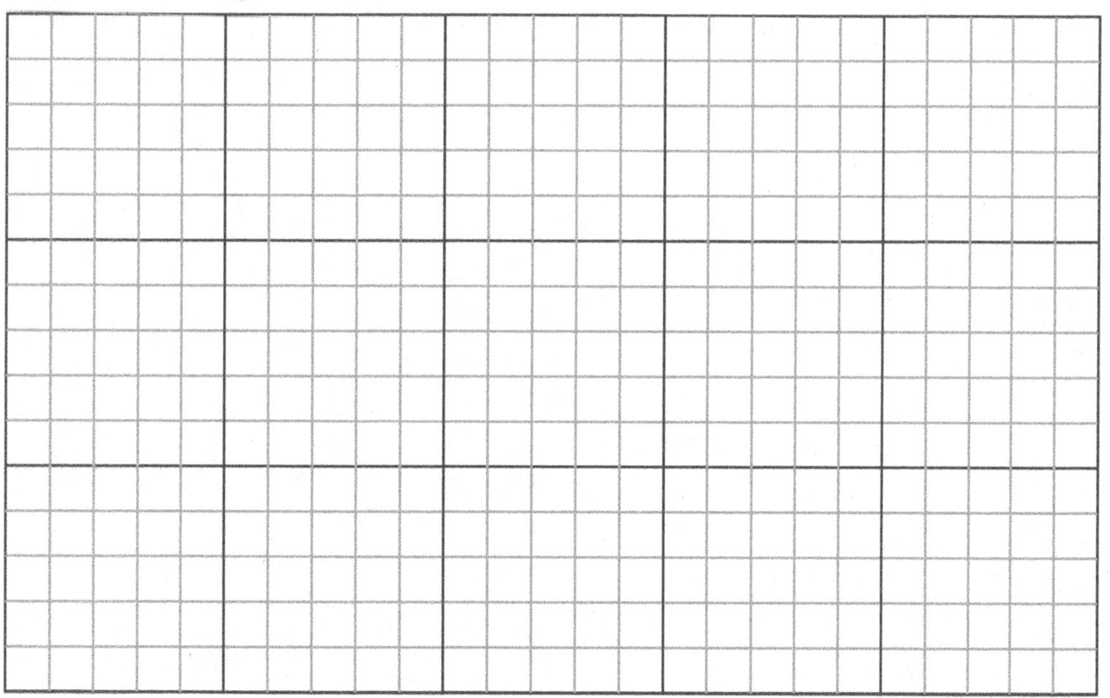

Volume of cool water ____________

Change in water temperature (ΔT)
(show calculations) ____________

Mass of water
(show calculations) ____________

Energy (calories) used to heat water
(show calculations) ____________

B. Cooling curve for Salol

Table B.1 Salol temperatures

Time (minutes)	Temperature (°C)	Time (minutes)	Temperature (°C)

Graph B.1 Salol cooling curve

C. Heat of fusion

Mass of empty calorimeter cup ________________

Mass of cup and water ________________

Mass of water ________________

Initial water temperature ________________

Final water temperature ________________

Mass of cup, water, and melted ice ________________

Mass of melted ice ________________

Change in water temperature (ΔT) ________________

Energy lost by water (calories)
(show calculations) ________________

Experimental value for heat of fusion
(show calculations) ________________

D. Purification of naphthalene by sublimation

Appearance of impure naphthalene

Appearance of solid flakes collected after sublimation

Appearance of substance remaining in beaker

<u>Problems and questions</u> (show all calculations, correct significant figures, and units)

1. Calculate the number of calories required to melt 35 g of ice at 0.0°C.

2. Calculate the number of calories required to condense 45.0 g of steam at 100.°C.

3. Calculate the amount of heat released by 35 g of water at 100.°C as it touches the skin and cools to a body temperature of 37°C.

4. Calculate the amount of heat released when 35 g of steam at 100.°C as it touches the skin and cools to a body temperature of 37°C. (Hint: this calculation requires two steps.)

5. Compare the two values and explain why steam burns are more severe than boiling water burns.

6. Calculate the number of calories released when 20.0 g of steam condenses at 100.°C, cools to 0.0°C, and then freezes. (Hint: this calculation requires three steps.)

7. A Styrofoam container filled with solid carbon dioxide (dry ice) was left open. When examined later, there was no solid or liquid in the container. What happened to the carbon dioxide?

Goals

- Examine the effect of pressure on the volume of a gas at constant temperature and formulate the relationship between the two.
- Examine the effect of temperature on the volume of a gas at constant pressure and formulate the relationship between the two.
- Examine the effect of temperature on the pressure of a gas at constant volume and formulate the relationship between the two.

Materials

The website address for the experiment is
http://highered.mheducation.com/sites/0073511250/student_view0/gas_laws_online_lab.html.

Discussion

Solids have a definite shape and a definite volume. Liquids have an indefinite shape (they take the shape of the container they are in) but a definite volume. Gases, on the other hand, differ from the other two states of matter because they have an indefinite shape (they take the shape of the container they are in) and an indefinite volume (a gas's volume is the volume of the container it is in).

The behavior of an ideal gas can be described reasonably well by simple relationships called *gas laws*. A gas law describes, in mathematical terms, the relationships among the amount, pressure, temperature, and volume of a gas.

Boyle's law

The first gas law to be discovered relates the gas pressure to the gas volume. It was formulated more than 300 years ago, in 1662, by the British chemist and physicist Robert Boyle. Boyle's law states that the volume of a fixed amount of a gas is inversely proportional to the pressure applied to the gas if the temperature is kept constant. This means that if the pressure on the gas increases, the volume decreases. Conversely, if the pressure decreases, the volume increases. The mathematical equation for Boyle's law is

$$P_1 \times V_1 = P_2 \times V_2$$

where P_1 and V_1 are the pressure and volume of a gas at an initial set of conditions, and P_2 and V_2 are the pressure and volume of the gas under a new set of conditions.

Charles's law

The French scientist Jacques Alexandre César Charles discovered the relationship between the temperature and the volume of a gas at constant pressure in 1787. Charles's law states that the volume of a fixed amount of a gas is directly proportional to its Kelvin temperature if the pressure is kept constant. Directly proportional means that one quantity increases when the other one increases. If one quantity decreases, the other one also decreases. In the laboratory, temperatures are often measured in degrees Celsius, but this gas law requires the temperature to be measured in (or converted to) Kelvin. The mathematical equation for Charles's law is

$$\frac{V_1}{T_1} = \frac{V_2}{T_2}$$

where V_1 and T_1 are the volume and Kelvin temperature of a gas at an initial set of conditions, and V_2 and T_2 are the volume and Kelvin temperature of a gas under a new set of conditions.

Gay-Lussac's law

Toward the end of the 1600s, the French physicist Guillaume Amontons built a thermometer based on the fact that the pressure of a gas is directly proportional to its temperature (Amontons' law). However, the French chemist and physicist Joseph Louis Gay-Lussac is credited for rediscovering this law in 1802 and it is known today as Gay-Lussac's law. Gay-Lussac's law states that the pressure of a fixed amount of a gas is directly proportional to its Kelvin temperature if the volume is kept constant.

The mathematical equation for Gay-Lussac's law is

$$\frac{P_1}{T_1} = \frac{P_2}{T_2}$$

where P_1 and T_1 are the pressure and Kelvin temperature of a gas at an initial set of conditions, and P_2 and T_2 are the pressure and Kelvin temperature of the gas under a new set of conditions.

Measurements

Barometers, manometers, and gauges are the instruments most commonly used to measure gas pressures. Atmospheric pressure is expressed in terms of the height of the barometer's mercury column, usually in millimeters of mercury (mm Hg). Another name for millimeters of mercury is *torr*, named after Evangelista Torricelli, the Italian physicist who invented the barometer.

Atmospheric pressure varies with the weather and the altitude, but it averages about 760 mm Hg at sea level. By definition, this is also 760 torr. The gauges used in this experiment measure gas pressure in torr.

The gases in this experiment are contained in cylindrical tubes. The volume of a cylinder is

$$V = \pi r^2 h$$

The volume is directly proportional to the height of the column. As the height increases, so does the volume. In this experiment, the height of the column will be measured and it will be converted to volume by a Microsoft Excel worksheet.

As was mentioned previously, the temperature needs to be in Kelvin. In this experiment, temperature will be measured in degrees Celsius and it will be converted to Kelvin by a Microsoft Excel worksheet.

Experimental Procedures

1. Go to the website provided by the instructor.
2. Each of the three experiments has a tab for "Lab Manual" and a tab for "Worksheet."
3. Open the "Lab Manual" tab and follow the directions.
4. Open the "Worksheet" tab and enter all data into the Excel worksheet.
    a. In Experiment 1, Excel will calculate the volume as you enter each height value.
    b. In Experiment 2, Excel will calculate the volume as you enter each height value and will also calculate the Kelvin temperature as you enter each Celsius value.
    c. In Experiment 3, Excel will calculate the Kelvin temperature as you enter each Celsius value.
5. Print out the computer-generated worksheets OR transfer all data and graphs to the printed report sheets.
6. Turn in the data sheets and the Problems and Questions pages.

PRE-LAB QUESTIONS: GAS LAWS

Name ______________________ Date ______________________

Partner(s) ______________________ Section ______________________

______________________ Instructor ______________________

1. What temperature units are required for all gas law calculations?

2. Describe in words what happens to the volume of a container of gas at constant temperature when the pressure is changed.

3. If one variable increases as the other decreases, what term is used to describe the relationship?

4. Write the mathematical formula for Charles's law.

5. Rearrange Gay-Lussac's law to solve for final temperature.

6. Using all three variables (pressure, volume, temperature), write the mathematical formula for the combined gas law.

Name ______________________ Date ______________________

Partner(s) ______________________ Section ______________________

______________________ Instructor ______________________

Experiment 1: Boyle's law

Number of weights	Temp (°C)	Temp (K)	Height (cm)	Pressure (torr)	Volume (L)
0	24.1	297			
1	24.1	297			
2	24.1	297			
3	24.1	297			
4	24.1	297			
5	24.1	297			

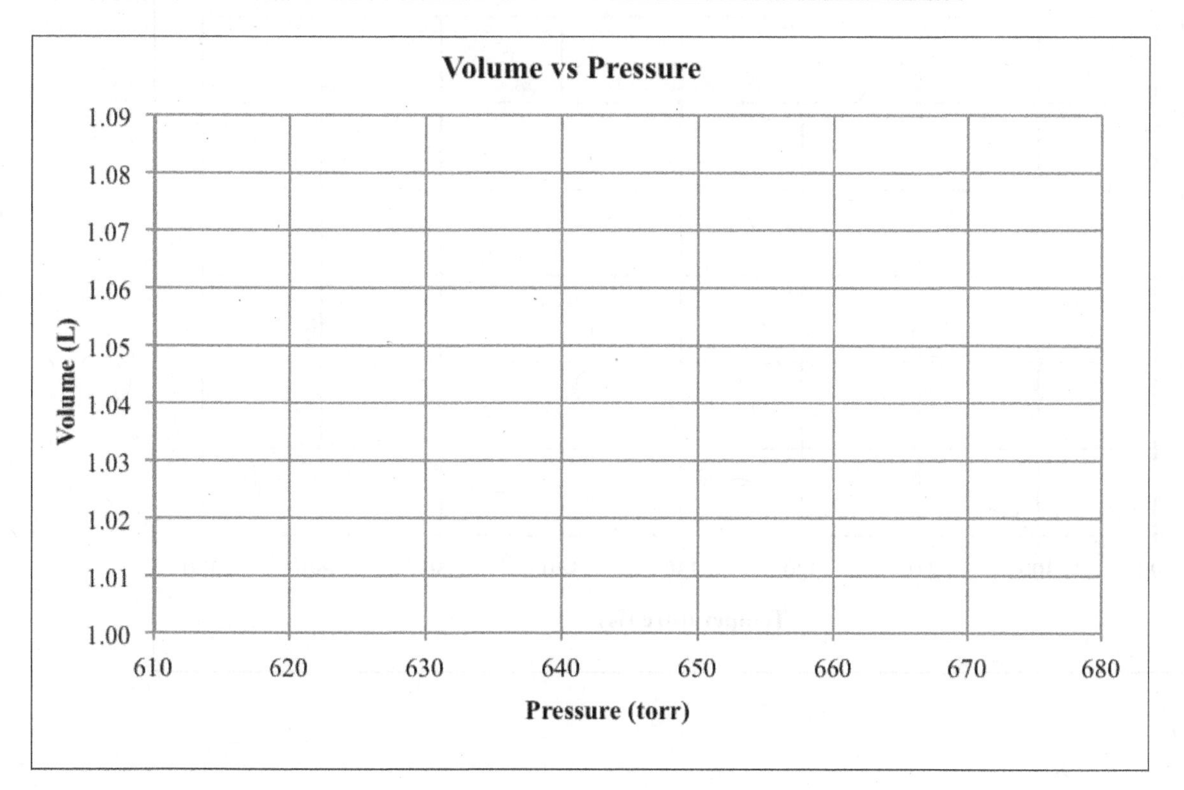

Experiment 2: Charles's law

Dial setting	Pressure (torr)	Height (cm)	Temp (°C)	Temp (K)	Volume (mL)
0					
1					
2					
3					
4					
5					
6					
7					

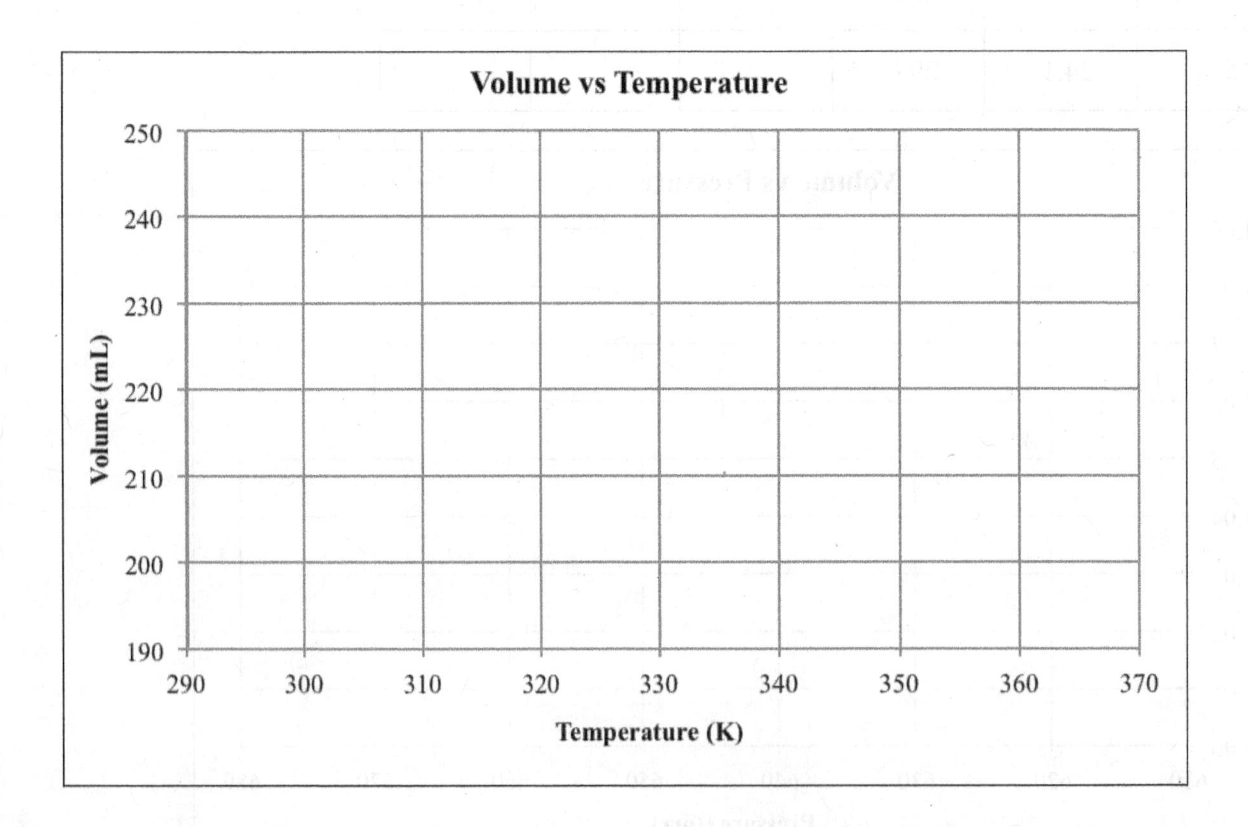

Experiment 3: Gay-Lussac's law

Dial setting	Volume (mL)	Temp (°C)	Temp (K)	Pressure (torr)
0	391			
1	391			
2	391			
3	391			
4	391			
5	391			
6	391			
7	391			

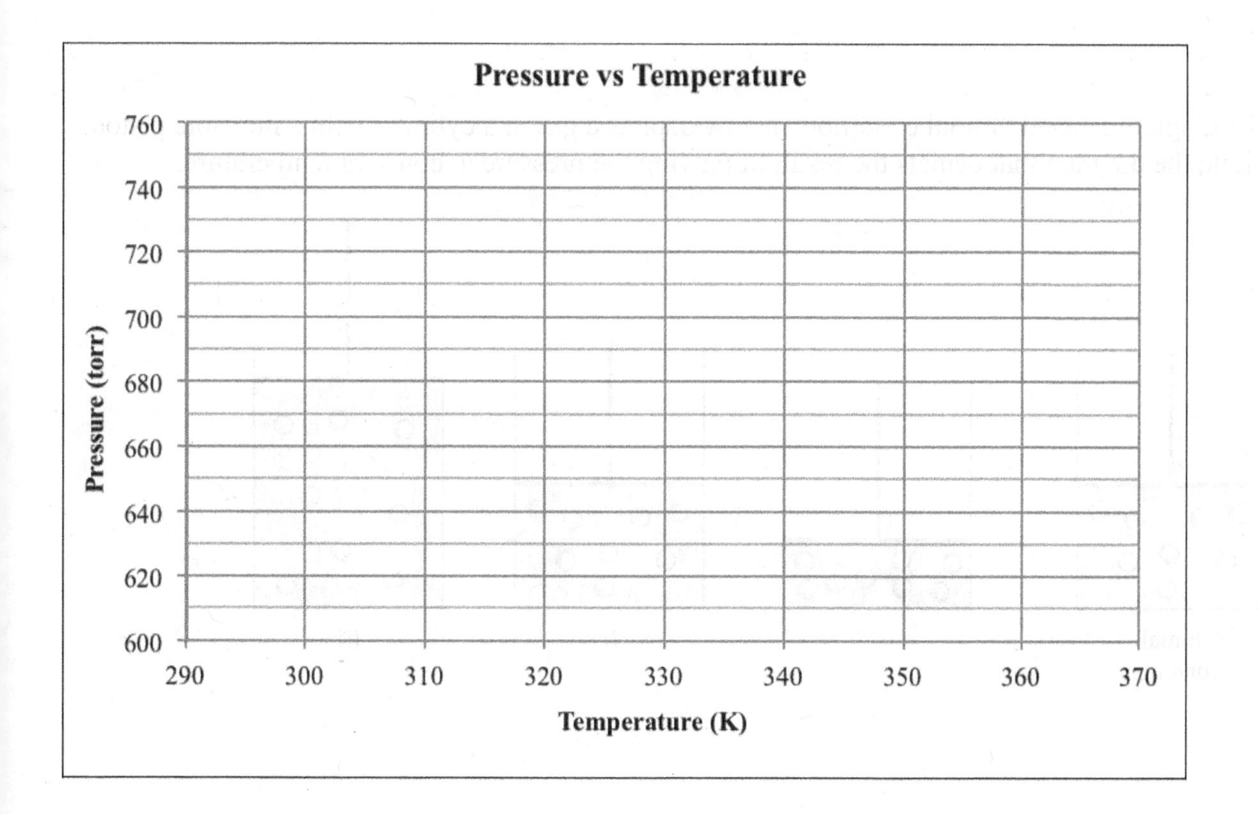

Problems and questions

Experiment 1 analysis:

1. How does the volume of a gas change with pressure?

2. Write the mathematical formula for the relationship between V_1 and V_2 (the volumes of a gas) and P_1 and P_2 (the pressures of a gas).

3. If a sample of gas has a volume of 536 mL at 637 torr, what will its volume be if the pressure is increased to 712 torr at constant temperature?

4. The diagram labeled "Initial condition" below depicts a gas in a cylinder with a movable piston. Circle the diagram that depicts the result of halving the pressure at constant temperature.

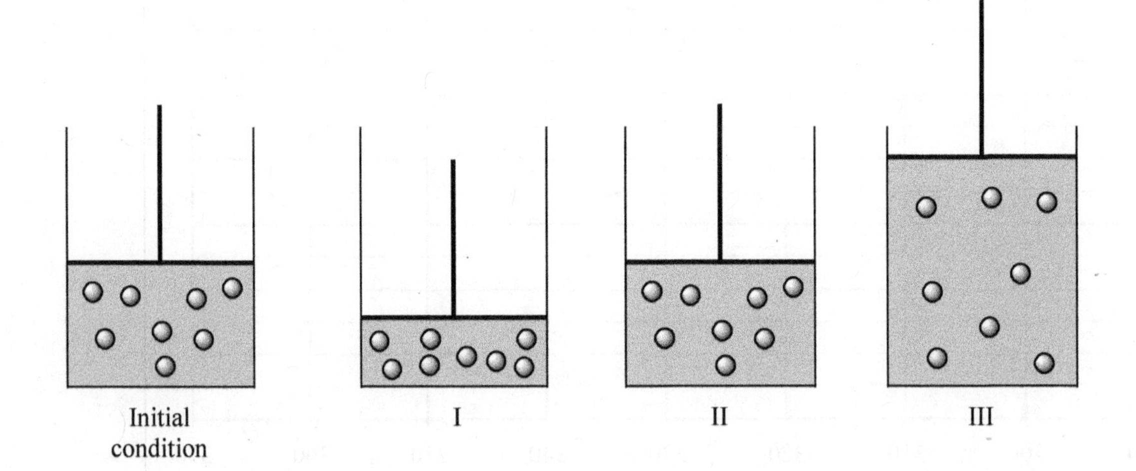

Experiment 2 analysis:

1. How does the volume of a gas change with temperature?

2. Write the mathematical formula for the relationship between V_1 and V_2 (the volumes of a gas) and T_1 and T_2 (the temperatures of a gas).

3. If a sample of gas at constant pressure has a volume of 417 mL at 32.4°C, what will its volume be if the temperature is increased to 64.8°C.

4. The diagram labeled "Initial condition" below depicts a gas in a cylinder with a movable piston. Circle the diagram that depicts the result of decreasing the Kelvin temperature by a factor of two at constant pressure.

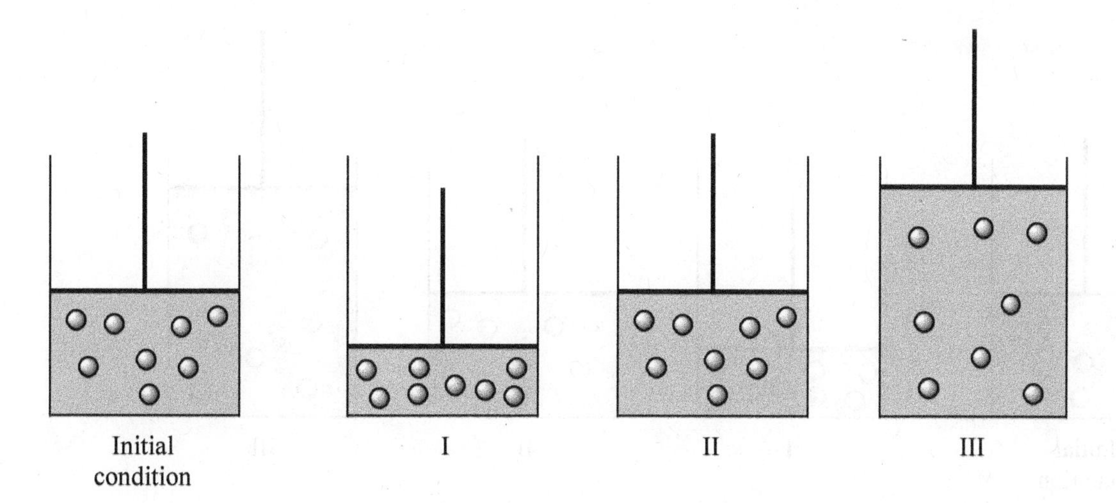

Experiment 3 analysis:

1. How does the pressure of a gas change with temperature?

2. Write the mathematical formula for the relationship between P_1 and P_2 (the pressures of a gas) and T_1 and T_2 (the temperatures of a gas).

3. If a sample of gas in a container of fixed volume is initially at 26.3°C and 652 torr, what will its pressure be if the temperature is increased to 52.6°C?

4. The diagram labeled "Initial condition" below depicts a gas in a cylinder with a movable piston. Circle the diagram that depicts the result of doubling the Kelvin temperature and doubling the pressure.

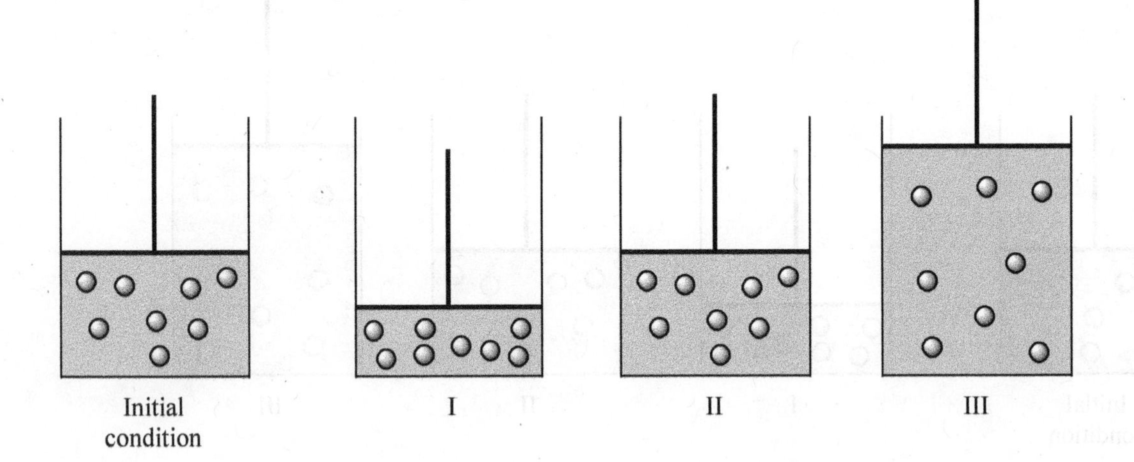

Goals

- Observe the conductivity of strong, weak, and nonelectrolytes.
- Observe the solutes and their concentrations in intravenous (IV) solutions.
- Perform chemical tests for chloride ions, glucose, and starch.
- Use a semipermiable membrane to separate colloids from a solution.

Materials

Instructor demonstration:

- Conductivity apparatus
- 150 mL beakers half-filled with: 0.5 M NaCl, 0.05 M NaCl, 0.005 M NaCl, 0.1 M HCl, 0.1 M NaOH, 0.1 M $HC_2H_3O_2$, 0.1 M NH_4OH, 0.1 M sucrose, 0.1 M glucose, ethanol, deionized water, tap water
- Empty 250 mL beaker
- Rinse bottle with deionized water

Student experiment:

- 20 cm cellophane tubing (pre-cut and soaking in a beaker of distilled water)
- Dental floss
- Scissors
- 100 mL beaker
- 250 mL beaker
- 400 mL beaker
- Funnel
- Test tubes
- Test tube rack
- Test tube holder
- Stirring rod
- 50 mL graduated cylinder
- 10% NaCl
- 10% glucose
- 1% starch
- 0.1 M $AgNO_3$
- Benedict's solution
- Iodine reagent
- Hot plate
- Crucible tongs
- Wire gauze
- Droppers
- Bags of IV solutions: saline, Ringer's, dextrose, etc.
- Disposable nitrile gloves

Discussion

The solubility and the concentration of a solute in a solution are two important factors that will be studied in this experiment. A *solution* is a homogeneous mixture of two or more substances. The *solvent* is the substance that does the dissolving and is usually present in the larger amount. The *solute* is the substance that is dissolved and is usually present in the smaller amount. Solutes and solvents may be solids, liquids, or gases. The state of matter of the solvent determines the state of matter of the solution.

Polarity of solutes and solvents

The general rule for determining the attractive forces between solutes and solvents is "Like dissolves like." Ionic and polar covalent solutes are attracted to ionic and polar covalent solvents. Nonpolar solutes are attracted to nonpolar solvents. When an ionic solute such as NaCl dissolves in water, the individual ions separate as positive ions (cations) and negative ions (anions). The Na^+ ion is attracted to the partially negative oxygen atom of water and the Cl^- ion is attracted to the partially positive hydrogen atom of water. The water molecules surrounding each ion hold the dissolved particles in solution in a process called *hydration*. The polar water molecules are also attracted to polar solutes such as sucrose and are attracted to the polar sites on the molecules.

Nonpolar solutes dissolve in nonpolar solvents. Examples of nonpolar solvents include paint thinner, acetone, and gasoline. These compounds will dissolve nonpolar solutes such as oil-based paints, nail polish, and grease.

Electrolytes

An electrolyte is a substance that conducts electricity in its molten state or in an aqueous solution. Substances may be classified as strong, weak, or nonelectrolytes by testing them with a conductivity apparatus. A strong electrolyte completely (or almost completely) ionizes or dissociates in water to form ions in water. A weak electrolyte only partially ionizes/dissociates to form a few ions, but mostly forms molecules in water. Nonelectrolytes dissolve as molecules in water (see Figure 1). The light bulb of a conductivity apparatus will glow when enough ions are present in the aqueous solution to complete the electrical circuit between the two submerged electrodes. The more ions in solution, the brighter the light bulb glows.

Figure 1 ***Strong electrolyte, weak electrolyte, and nonelectrolyte***

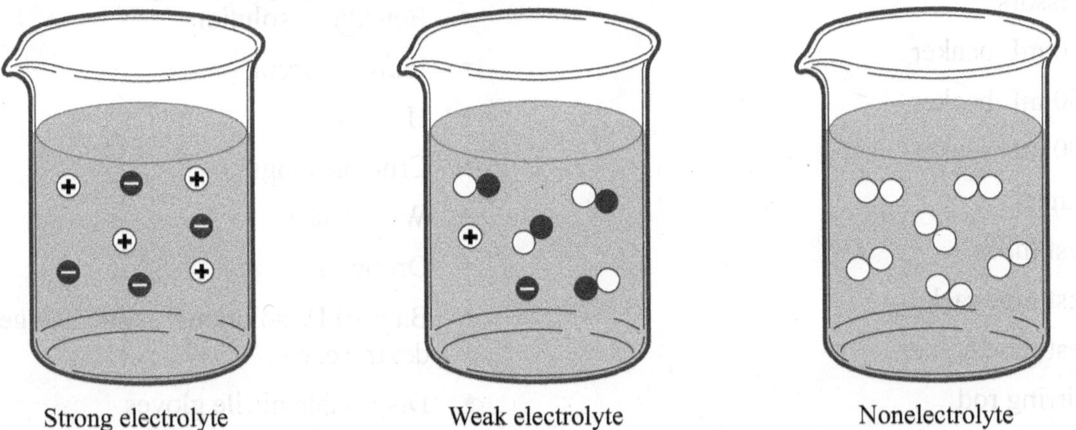

Electrolytes in the medical field

Electrolytes are present in body fluids and play an important role in cellular activity. Replacement fluids are usually given in the form of intravenous (IV) solutions. The concentration of the ions is expressed in milliequivalents per liter (mEq/L). An equivalent (Eq) is the amount of an ion equal to one mole of positive or negative charge. For example, one mole of Na^+ ions equals one equivalent. But one mole of Ca^{2+} ions equals two equivalents. One equivalent (Eq) is equal to 1000 milliequivalents (mEq).

All solutions must have equal numbers of positive and negative charges. Specific IV solutions are prescribed depending upon the nutritional, electrolyte, and fluid needs of the patient. The names and concentrations of the solutes are available on the IV label. Some of the effects of electrolyte imbalance are shown in Table 1.

Table 1 ***Effects of electrolyte imbalance on the human body***

Electrolyte	Low levels	High levels
Na^+, Cl^-	Dry tongue, fever	Edema
K^+	Cardiac failure	Cardiac failure
Ca^{2+}	Numbness in extremities	Poor cardiac function
HCO_3^-	Metabolic acidosis (low pH)	Metabolic alkalosis (high pH)
Mg^{2+}	Muscle weakness, insomnia	Cardiac arrest

Osmosis and dialysis

Osmosis occurs when water moves through a semipermeable membrane in an attempt to equalize the concentration of solutes. The semipermeable membrane of a red blood cell allows osmosis to occur. *Isotonic* solutions exert the same osmotic pressure. For example, a 0.90% (m/v) saline (NaCl) solution and a 5.0% (m/v) glucose solution exert the same osmotic pressure as the fluids in the red blood cell. A red blood cell placed into an isotonic solution will remain unchanged because the flow of water into and out of the red blood cell is equal. A *hypotonic* solution has a lower osmotic pressure than the red blood cell and a net flow of water into the cell will result in *hemolysis*. This expansion of the cell may cause the cell to rupture. A *hypertonic* solution has a higher osmotic pressure than the red blood cell and a net flow of water out of the cell will result in *crenation*. The red blood cell volume will decrease and form small clumps.

If the pores in the membrane are large enough to allow small molecules and ions to also pass through, the process is called *dialysis*. A schematic of dialysis is shown in Figure 2. The lining of the intestinal tract is a dialyzing membrane. This allows the dissolved particles from digestion to pass into the blood and lymph systems, while the larger, undigested particles remain inside the digestive tract.

Figure 2 ***Schematic of dialysis***

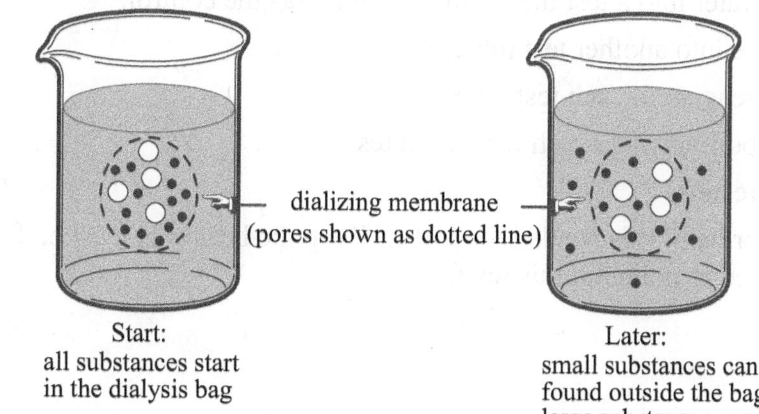

Experimental Procedures

Eye protection and appropriate clothing must be worn at all times.

Wash any contacted skin area immediately with running water and alert the instructor. Discard all wastes properly as directed by the instructor

Time management will be critical in this experiment. The lab instructor will perform the electrolyte demonstration.

A. Electrolytes—conductivity apparatus

1. The instructor will explain and demonstrate the conductivity apparatus. Draw and label a sketch of the apparatus on the report sheet.
2. Record the intensity of the light bulb for each substance tested.
3. Identify each substance as a strong, weak, or nonelectrolyte. Record.

B. Identification tests

Caution: Gloves should be worn since $AgNO_3$ and iodine reagent will stain skin, clothing, and lab reports.

Chloride ion test

1. Pour 3 mL of deionized water into a test tube. This will serve as the control.
2. Pour 3 mL of 10% NaCl into another test tube.
3. Add 2 drops of 0.1 M $AgNO_3$ to each test tube.
4. Compare and record the results.

Dispose of all solutions containing $AgNO_3$ as directed by the instructor.

Glucose test

1. Prepare a boiling water bath using a 250 or 400 mL beaker and tap water on a hot plate.
2. Pour 3 mL of deionized water into a test tube. This will serve as the control.
3. Pour 3 mL of 10% glucose into another test tube.
4. Add 3 mL of Benedict's reagent to each test tube.
5. Heat both test tubes in a boiling water bath for 5 minutes.
6. Compare and record the results.
7. Maintain the boiling water bath. It is needed for part C. Keep the beaker at least half full of water. Add water as needed to maintain this level.

Starch test

1. Pour 3 mL of deionized water into a test tube. This will serve as the control.
2. Pour 3 mL of 1% starch solution into another test tube.
3. Add 2–3 drops of iodine reagent to each test tube.
4. Compare and record the results.

C. Osmosis and dialysis

1. Obtain 12 clean, dry test tubes and label them #1–#12. These will be needed to perform the identification tests.
2. In a small beaker, combine 10 mL of 10% NaCl, 10 mL of 10% glucose, and 10 mL of starch solution. Stir.
3. Prepare the dialysis bag by obtaining a piece of cellophane tubing and two pieces of dental floss. The cellophane tubing must be soaked in deionized water until softened. Use a piece of dental floss to tie one end of the bag closed.
4. Use a funnel to pour about 20 mL of the prepared saline/glucose/starch solution into the bag. Use another piece of dental floss to tie the other end of the bag closed.
5. Rinse the outside of the bag thoroughly with deionized water.
6. Fill a 250 mL beaker $^{2}/_{3}$ full with deionized water. Carefully submerge the dialysis bag in the beaker water.
7. Using test tubes #1, #2, and #3, immediately, pour 3 mL of water from the beaker into each test tube. Perform the following identification tests:
   - In test tube #1, add 2 drops of 0.1 M $AgNO_3$. Record the observations in Table C. Compare the results with those in Table B to determine whether Cl^- is present and record in Table C.
   - In test tube #2, add 3 mL of Benedict's reagent. Heat the test tube in a boiling water bath for 5 minutes. Record the observations in Table C. Compare the results with those in Table B to determine whether glucose is present and record in Table C.
   - In test tube #3, add 2–3 drops of iodine reagent. Record the observation in Table C. Compare the results with those in Table B to determine whether starch is present and record in Table C.
8. After 15 minutes, pour 3 mL of water from the beaker into test tubes #4, #5, and #6. Perform the following identification tests:
   - In test tube #4, add 2 drops of 0.1 M $AgNO_3$. Record the observations in Table C. Compare the results with those in Table B to determine whether Cl^- is present and record in Table C.
   - In test tube #5, add 3 mL of Benedict's reagent. Heat the test tube in a boiling water bath for 5 minutes. Record the observations in Table C. Compare the results with those in Table B to determine whether glucose is present and record in Table C.

- In test tube #6, add 2–3 drops of iodine reagent. Record the observation in Table C. Compare the results with those in Table B to determine whether starch is present and record in Table C.

9. After a total of 30 minutes, pour 3 mL of water from the beaker into test tubes #7, #8, and #9. Perform the following identification tests:
   - In test tube #7, add 2 drops of 0.1 M $AgNO_3$. Record the observations in Table C. Compare the results with those in Table B to determine whether Cl^- is present and record in Table C.
   - In test tube #8, add 3 mL of Benedict's reagent. Heat the test tube in a boiling water bath for 5 minutes. Record the observations in Table C. Compare the results with those in Table B to determine whether glucose is present and record in Table C.
   - In test tube #9, add 2–3 drops of iodine reagent. Record the observation in Table C. Compare the results with those in Table B to determine whether starch is present and record in Table C.
10. After the 30 minutes, remove the dialysis bag from the beaker. Open the bag by cutting one end with scissors. Pour the contents of the bag into a small, clean beaker.
11. Pour 3 mL of the contents of the bag into test tubes # 10, #11, and #12. Perform the following identification tests:
    - In test tube #10, add 2 drops of 0.1 M $AgNO_3$. Record the observations in Table C. Compare the results with those in Table B to determine whether Cl^- is present and record in Table C.
    - In test tube #11, add 3 mL of Benedict's reagent. Heat the test tube in a boiling water bath for 5 minutes. Record the observations in Table C. Compare the results with those in Table B to determine whether glucose is present and record in Table C.
    - In test tube #12, add 2–3 drops of iodine reagent. Record the observation in Table C. Compare the results with those in Table B to determine whether starch is present and record in Table C.

D. Electrolytes in intravenous solutions

1. Examine the display of IV bags.
2. Complete the table on the report sheet. List each electrolyte with the ionic symbol and the concentration. For example, record "sodium 47" on the label as "Na^+" and "47 mEq/L."
3. Calculate and record the total number of cation mEq.
4. Calculate and record the total number of anion mEq.
5. Calculate and record the overall sum of positive and negative charges.

Name ______________________ Date ______________________

Partner(s) ______________________ Section ______________________

______________________ Instructor ______________________

1. Explain why paintbrushes used with oil-based paints cannot be cleaned effectively with water.

2. What is the difference between osmosis and dialysis?

3. What is the difference between strong electrolytes and weak electrolytes?

4. What three identification tests are used repeatedly throughout this experiment?

5. Why are gloves required specifically for this experiment?

Name ______________________ Date ______________________

Partner(s) ______________________ Section ______________________

______________________ Instructor ______________________

A. Electrolytes

Draw and label a sketch of the conductivity apparatus.

Complete the table for the demonstration of the conductivity apparatus.

Substance	Light intensity: "bright, dim, or none"	Strong, weak, or nonelectrolyte?
0.5 M NaCl		
0.05 M NaCl		
0.005 M NaCl		
0.1 M HCl		
0.1 M NaOH		
0.1 M Acetic acid ($HC_2H_3O_2$)		
0.1 M Ammonium hydroxide (NH_4OH)		
0.1 M Sucrose ($C_{12}H_{22}O_{11}$)		
0.1 M Glucose ($C_6H_{12}O_6$)		
Ethanol (C_2H_5OH)		
Deionized water		
Tap water		

B. Identification tests

Complete the table for the identification tests.

Test	Appearance of water control	Appearance of positive test
Chloride ion		
Glucose		
Starch		

C. Osmosis and dialysis

Complete the table for the tests.

Time	Cl^- present? (observation)	Glucose present? (observation)	Starch present? (observation)
0 Beaker water 0 minutes	Test tube #1	Test tube #2	Test tube #3
15 Beaker water 15 minutes	Test tube #4	Test tube #5	Test tube #6
30 Beaker water 30 minutes	Test tube #7	Test tube #8	Test tube #9
30 Contents from **inside** the bag 30 minutes	Test tube #10	Test tube #11	Test tube #12

Which substance(s) were found <u>outside</u> the dialysis bag after 30 minutes?
Which substances remained <u>inside</u> the dialysis bag after 30 minutes? Explain both results.

D. Electrolytes in intravenous solutions

Complete the table for the display of the IV bags.

Name of IV solution	Cations listed (include concentrations)	Anions listed (include concentrations)	Total cation mEq	Total anion mEq	Overall charge sum

Problems and questions

1. Circle the solvent in each mixture:

   a) 30 mL of water and 5 mL of ethanol

   b) 15 mL of water and 30 mL of ethanol

   c) toluene dissolved in benzene

   d) an alloy of brass containing 20% Zn and 80% Cu

2. Predict whether each of the following solutes will be strong, weak, or nonelectrolytes:

   a) potassium iodide ____________________

   b) isopropyl alcohol (C_3H_7OH) ____________________

   c) dextrose ($C_6H_{12}O_6$) ____________________

   d) carbonic acid (H_2CO_3) ____________________

3. Describe what will happen to a red blood cell (crenation, hemolysis, or nothing) if it is placed in each of the following solutions:

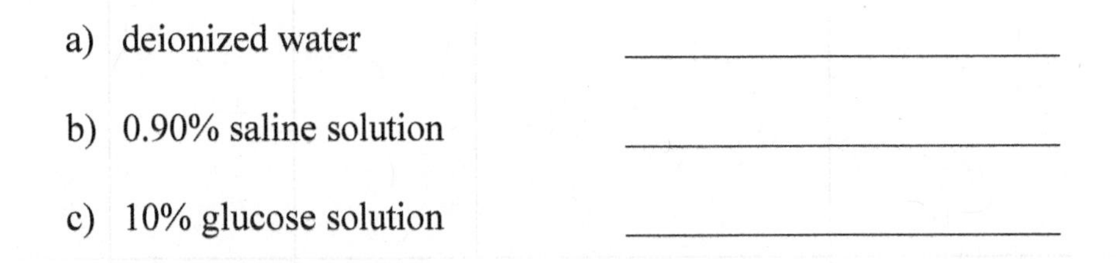

   a) deionized water ____________________

   b) 0.90% saline solution ____________________

   c) 10% glucose solution ____________________

4. Compare the results of the conductivity tests performed on each of the three NaCl solutions. Explain any differences.

Goals

- Identify a chemical reaction as endothermic or exothermic.
- Identify factors that affect the rate of a chemical reaction.
- Determine if a reaction is reversible or irreversible.
- Identify factors that shift the equilibrium of a chemical reaction.
- Use Le Châtelier's Principle to predict the direction of equilibrium shift.

Materials

- Test tube rack
- Test tubes
- NH_4NO_3 crystals
- Anhydrous $CaCl_2$ crystals
- Thermometer
- Magnesium ribbon
- 1.0 M HCl
- 2.0 M HCl
- 3.0 M HCl
- Stirring rod
- 50 mL beakers
- 100 mL beakers
- 250 mL beakers
- 400 mL beaker
- 10 mL graduated cylinder
- Ice
- Vinegar ($HC_2H_3O_2$)
- Baking soda ($NaHCO_3$)
- 0.024 M KIO_3
- 0.016 M $NaHSO_3$
- Freshly made 2% starch solution
- 0.1 M $CuSO_4$
- Droppers
- Rubber stoppers to fit test tubes
- 0.01 M $Fe(NO_3)_3$
- 1.0 M $Fe(NO_3)_3$
- 0.01 M KSCN
- 1.0 M KSCN
- Hot plate
- Beaker tongs
- Grease pencil

Discussion

Every chemical reaction has two factors that should be considered:

1. How much and what type of product is formed?
2. How fast is it formed?

The amount of product formed is determined by the dynamic equilibrium of the system, and how fast it is formed is determined by the rate of reaction.

Exothermic and endothermic reactions

In a chemical reaction, bonds are broken in the reactants and new bonds are formed in the products. Bond-breaking requires energy while bond-forming releases energy. The energy can be in a number of different forms (e.g., heat, electricity, light, kinetic). Considering these two factors, a reaction can have an overall gain or loss of energy. The most common form of energy exchanged is heat. In an *exothermic* chemical reaction, heat is released as the reaction occurs. For example, when a fuel such as methane (CH_4) is burned with oxygen, the release of heat causes the temperature of the surroundings to increase. Often, chemists use the generic word "heat" in the chemical equation instead of specifying exactly how much heat is released (213 kcal/mol in this reaction):

$$CH_{4(g)} + 2O_{2(g)} \longrightarrow CO_{2(g)} + 2H_2O_{(l)} + \text{heat}$$

Cells in the human body produce heat as a byproduct of metabolism when they "burn" carbohydrates and fats. This gives rise to body heat:

$$\underset{\substack{\text{glucose} \\ \text{(a simple carbohydrate)}}}{C_6H_{12}O_{6(s)}} + 6O_{2(g)} \longrightarrow 6CO_{2(g)} + 6H_2O_{(l)} + \text{heat}$$

Conversely, in an *endothermic* chemical reaction, heat is absorbed (from the surroundings) as the reaction occurs. When no other sources of heat are available, this causes a drop in the temperature of the surroundings. An example of an endothermic reaction is melting ice:

$$H_2O_{(s)} + \text{heat} \longrightarrow H_2O_{(l)}$$

When ice is put into a beverage, the ice absorbs heat from the liquid around it and melts. The surrounding liquid becomes colder because it loses heat.

Another example of an endothermic reaction is protein synthesis. Proteins are polymers of amino acids. The stepwise synthesis of a protein molecule involves the joining of individual amino acids. This process is an endothermic one. Where does the body get the energy it needs to assemble proteins? In the cell, the glucose that is metabolized is not reacted in a single step; rather, it is broken down with the aid of enzymes in a series of steps. Much of the released energy is used to synthesize adenosine triphosphate (ATP) from adenosine diphosphate (ADP). When ATP is converted back to ADP, the energy released can be used in protein synthesis.

It is important to note that in an endothermic reaction, heat is a reactant and it is written on the left side of the chemical equation. In an exothermic reaction, heat is a product and is written on the right side of the chemical equation.

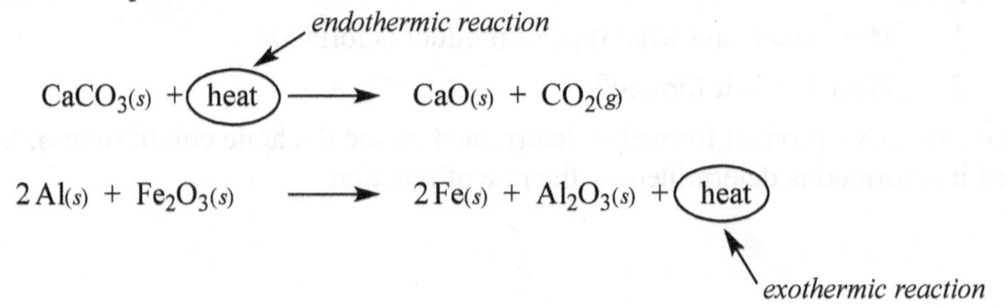

Reaction rates

The *rate of reaction* can be affected by the concentration of the reactants, the temperature, and the presence of a catalyst. If the concentration of a reactant is increased, the product will form more quickly because it is easier for the reactant molecules to "find one another and react." The combustion of fuels is visibly faster when the concentration of oxygen is increased and explains why patients on oxygen therapy should not be near open flames or cigarettes.

Conversely, if the concentration of a reactant is lowered, the product forms more slowly. At high altitudes, the concentration of oxygen is lower and the ability of hemoglobin to form the oxygenated hemoglobin molecule is slowed. This results in a lower level of oxygen transported to the brain and may result in hallucinations and confusion for mountain climbers as they ascend to higher elevations. The use of bottled oxygen by climbers provides an increased concentration of oxygen. This allows for a faster rate of formation of the oxygenated hemoglobin molecules to transport oxygen to the brain.

Next, if the temperature of the reaction is increased, the additional energy typically increases the rate of reaction by providing increased kinetic energy to the reactant molecules, thus allowing for more numerous and more efficient collisions between the molecules. Generally, a 10°C increase in temperature results in a doubling of the reaction rate.

Consider the chemical reactions that occur within the human body. When the body temperature increases due to a high fever, for example, the rates of chemical reactions increase too. Because the reactions are happening faster, more oxygen is needed for the reactions. Rapid breathing occurs and the heart must pump faster to supply the tissues with the extra oxygen, via the bloodstream. The increase in temperature speeds up reactions designed to kill invading organisms such as bacteria. Above 41°C (about 106°F), the condition called *hyperthermia* (heat stroke) occurs. It results from the body's inability to dissipate the extra internal heat generated by the reactions occurring at an increased rate and it is a life-threatening situation.

At temperatures below 35°C, *hypothermia* occurs. Hypothermia is most often caused by exposure to cold weather or immersion in a cold body of water. It results from the body's inability to generate enough internal heat to maintain the normal body temperature because the reaction rates have decreased. Although hypothermia can be life-threatening, it is an integral part of present-day cardiopulmonary bypass surgery. By cooling the body, reaction rates decrease and the demand for oxygen decreases to such an extent that blood flow can be completely interrupted. In some cases, the body is cooled to 15–20°C.

The last factor affecting reaction rates is *activation energy*. Activation energy is the minimum combined kinetic energy that colliding reactant molecules must possess in order for their collision to result in a chemical reaction. The larger the activation energy, the slower the reaction rate because few molecules have the necessary energy. Any substance that lowers the activation energy speeds up the reaction. *Catalysts* are substances that increase the rate of a reaction but are not consumed in the reaction. Think of them as "helper molecules" that help the reactants do what they need to do to react. Catalysts provide an alternative pathway for the reaction to occur, and they may form an intermediate structure with a reactant. This alternative pathway results in a lower activation energy for the reaction, thus increasing the rate of reaction. Upon completion of the reaction, the reactant will have formed new products and the catalyst will return to its original structure. Catalysts may be reused in a reaction and are usually only needed in small quantities. In living organisms, enzymes are used as catalysts in metabolic reactions.

Reversible and irreversible reactions

Some reactions are *irreversible reactions* because the forward reaction goes to completion and no reverse reaction occurs. The combustion of methane (shown earlier) forms two gaseous products that do not recombine under normal conditions. This is the reaction that occurs when natural gas is burned in the presence of oxygen in the air in a gas stove, furnace, or Bunsen burner.

Other reactions are *reversible* and do not go to completion. The reaction of the yellow Fe^{3+} ion with the thiocyanate ion (SCN^-) to form the deep red ion $FeSCN^{2+}$ is often used to demonstrate a reversible reaction according to the net ionic equation:

$$\underset{\textit{yellow}}{Fe^{3+}(aq)} + \underset{\textit{colorless}}{SCN^-(aq)} \rightleftharpoons \underset{\textit{deep red}}{FeSCN^{2+}(aq)}$$

In a reversible reaction, the forward reaction (conversion of reactants to products) and the reverse reaction (conversion of products to reactants) occur simultaneously. This is shown in the chemical equation using double arrows or double half-arrows.

Equilibrium

In a reversible reaction, the rate of the forward reaction slows down as time goes on and the rate of the reverse reaction speeds up. Eventually, the rate of the forward reaction becomes equal to the rate of the reverse reaction. At this point, the reaction has reached *equilibrium*. Thus, one definition for equilibrium is when the rate of the forward reaction is equal to the rate of the reverse reaction.

Once the reaction has reached equilibrium, there is no effective change in the concentrations of reactants and products. Imagine, just as fast as the reactants are being consumed in the forward reaction, they are being produced in the reverse reaction. This serves as a second definition of equilibrium in a reversible reaction.

Equilibrium constant

Because the concentrations of the reactants and products do not change once equilibrium has been reached, an equilibrium constant may be written for the reaction. An *equilibrium constant* is a numerical value that relates the equilibrium concentrations of reactants and products at equilibrium.

Consider the following hypothetical chemical equation:

$$a\mathrm{A} + b\mathrm{B} \rightleftharpoons c\mathrm{C} + d\mathrm{D}$$

The equilibrium constant expression for this reaction is

$$K_{\mathrm{eq}} = \frac{[\text{products}]}{[\text{reactants}]} = \frac{[\mathrm{C}]^c[\mathrm{D}]^d}{[\mathrm{A}]^a[\mathrm{B}]^b}$$

Note the following points about this general equilibrium constant expression:

1) The square brackets refer to molar (moles/liter) concentrations.
2) Product concentrations are always placed in the numerator and they are multiplied together.

3) Reactant concentrations are always placed in the denominator and they are multiplied together.
4) The coefficients in the balanced equation determine the powers to which the concentrations are raised.
5) The abbreviation K_{eq} is used to denote an equilibrium constant. In some textbooks the abbreviation K or K_c is used.

Equilibria shifts

According to *Le Châtelier's Principle*, if a chemical system at equilibrium is disturbed or stressed, the system will react in the direction that counteracts the disturbance or relieves the stress. Factors that can cause the equilibrium to shift include a change in concentration by the addition or removal of a reactant or product, a change in pressure if a gas is involved, or a change in temperature. The body has many complex reactions that are dependent upon each other and thus the shift in the equilibrium of one reaction may affect one or more other reactions. For example, even a small change in the pH of blood caused by the change in concentration of the hydronium ion $[H_3O^+]$ may result in death.

Le Châtelier's Principle can be summarized by:

1) When a component of the equilibrium is added to the mixture, the shift is to the opposite side.
2) When a component of the equilibrium is removed from the mixture, the shift is to the same side.

Consider the following hypothetical chemical reaction at equilibrium:

$$A + B \rightleftharpoons C + D$$

If more A is *added* to the mixture, the equilibrium shift is to the right. If more C is *added* to the mixture, the equilibrium shift is to the left.

$$\text{(A)} + B \rightleftharpoons C + D$$

A is on the ***left***... ➡ ... *shift is opposite (to the* ***right****)*

$$A + B \rightleftharpoons \text{(C)} + D$$

... *shift is opposite (to the* ***left****)* ⬅ *C is on the* ***right***...

If B is *removed* from the mixture, the equilibrium shift is to the left. If D is *removed* from the mixture, the equilibrium shift is to the right.

$$A + (B) \rightleftharpoons C + D$$

*B is on the **left**...*
*... shift is to the same side (to the **left**)*

$$A + B \rightleftharpoons C + (D)$$

*D is on the **right**...*
*... shift is to the same side (to the **right**)*

Experimental equilibria shifts

This experiment will use the iron/thiocyanate reaction to observe the equilibria shifts when different changes are applied to the reaction. The chemical equation and equilibrium constant for the reaction are:

$$Fe^{3+}_{(aq)} + SCN^-_{(aq)} \rightleftharpoons FeSCN^{2+}_{(aq)} \qquad K_{eq} = \frac{[FeSCN^{2+}]}{[Fe^{3+}][SCN^-]}$$

yellow *colorless* *deep red*

Shift to the right:

When the reaction system contains mostly reactants, the solution color is yellow. As the system shifts and more products are formed, the solution color changes to a deep red. Adding additional reactants (Fe^{3+} or SCN^- ions) **increases** the concentration of the reactants and the equilibrium will shift to the opposite side (to the right).

One way to increase the concentration of Fe^{3+} ions is to add a soluble salt such as $Fe(NO_3)_3$. When it dissolves, it dissociates into its ions:

$$Fe(NO_3)_{3(s)} \xrightarrow{H_2O} Fe^{3+}_{(aq)} + 3\,NO_3^-{}_{(aq)}$$

The Fe^{3+} ions released will then disturb the equilibrium.

Similarly, one way to increase the concentration of SCN^- ions is to add a soluble salt such as KSCN. When it dissolves, it dissociates into its ions:

$$KSCN_{(s)} \xrightarrow{H_2O} K^+_{(aq)} + SCN^-_{(aq)}$$

The SCN^- ions released will then disturb the equilibrium.

Shift to the left:

Removing some of the reactants **decreases** the concentration of the reactants and the equilibrium will shift to the same side (to the left). One way to remove some of the reactants is to add the chloride ion (Cl^-) in the form of $HCl_{(aq)}$. The chloride ion decreases the concentration of the Fe^{3+} ion by reacting with it and removing it from the solution to form a colorless ion according to the following equation:

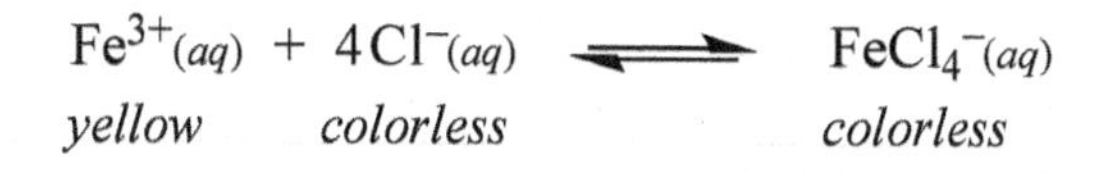

$$Fe^{3+}(aq) + 4Cl^-(aq) \rightleftharpoons FeCl_4^-(aq)$$

yellow *colorless* *colorless*

Add or remove heat:

As mentioned earlier in the discussion, heat is a form of energy. It is possible to determine whether the iron/thiocyanate reaction is endothermic or exothermic by observing the color change when the temperature of the system is changed. If the reaction is endothermic, "heat" will be written as a reactant in the net ionic equation and adding "heat" would shift the equilibrium to the right, making the solution more red. Removing heat by cooling the system would shift the equilibrium to the left, causing the solution to become more yellow.

If the reaction is endothermic: $Fe^{3+}(aq) + SCN^-(aq) + \text{heat} \rightleftharpoons FeSCN^{2+}(aq)$

yellow *colorless* *deep red*

However, if the reaction is exothermic, "heat" will be written as a product in the net ionic equation and adding "heat" would shift the equilibrium to the left, making the solution more yellow. Removing heat by cooling the system would shift the equilibrium to the right, causing the solution to become more red.

If the reaction is exothermic: $Fe^{3+}(aq) + SCN^-(aq) \rightleftharpoons FeSCN^{2+}(aq) + \text{heat}$

yellow *colorless* *deep red*

It is important to remember that increasing the temperature will increase the rate of a reaction (how fast it occurs), whether it is endothermic or exothermic. However, this is a different concept than predicting the direction of equilibrium shift in a reversible reaction.

Increasing the temperature (adding heat) of an exothermic equilibrium reaction will shift the equilibrium to the left. Thus, the reverse reaction will be more favored. Increasing the temperature (adding heat) of an endothermic equilibrium reaction will shift the equilibrium to the right, making the forward reaction more favored.

Experimental Procedures

Eye protection and appropriate clothing must be worn at all times.

Stoppers, thermometers, and glassware should be clean and dry for each experiment.

Wash any contacted skin area immediately with running water and alert the instructor. Discard all wastes properly as directed by the instructor.

A. Endothermic and exothermic reactions

1. Obtain two test tubes and label them #1 and #2.
2. Pour approximately 5 mL of water into each test tube.
3. Record the temperature of the water.
4. In test tube #1 add one scoop (approximately 0.5 g) of $NH_4NO_{3\,(s)}$ crystals. Stir and record the temperature of the solution.

$$NH_4NO_{3(s)} \xrightarrow{H_2O} NH_4^+{}_{(aq)} + NO_3^-{}_{(aq)}$$

5. In test tube #2 add one scoop of *anhydrous* $CaCl_{2\,(s)}$ crystals. Stir and record the temperature of the solution. Anhydrous refers to a crystalline structure that does not contain water molecules.

$$CaCl_{2(s)} \xrightarrow{H_2O} Ca^{2+}{}_{(aq)} + 2Cl^-{}_{(aq)}$$

6. Describe each reaction as endothermic or exothermic. Write the balanced equation for each reaction, showing "heat" as a reactant (if endothermic) or as a product (if exothermic).

B. Rates of reactions

<u>Concentration of reactants</u>

1. Obtain three test tubes and label them #1, #2, and #3.
2. Place a 2.5 cm piece of magnesium ribbon into each test tube. Complete the next three steps simultaneously and record the start time using a watch with a second hand or a stopwatch.
3. Add 10 mL of 1.0 M HCl to test tube #1.
4. Add 10 mL of 2.0 M HCl to test tube #2.
5. Add 10 mL of 3.0 M HCl to test tube #3.
6. Record the time required for the magnesium to completely react in each test tube.
7. After the reactions are completed, determine whether the reaction is endothermic or exothermic by carefully touching the bottom of the test tubes (or the sides if the magnesium strips were floating on the surface).

Temperature

The following reaction will be studied:

$$NaHCO_3(s) + HC_2H_3O_2(aq) \longrightarrow Na^+(aq) + C_2H_3O_2^-(aq) + CO_2(g) + H_2O(l)$$

8. Prepare an ice bath by filling a 250 mL beaker half full with crushed ice. Add tap water to make an ice/water slush.
9. Prepare a warm water bath by filling a 400 mL beaker half full with tap water. Use a hot plate to heat to approximately 55°C. Use beaker tongs to remove the beaker from the hot plate.
10. Obtain two test tubes and label them #1 and #2.
11. Pour approximately 10 mL of vinegar ($HC_2H_3O_2$) into each test tube.
12. Place test tube #1 in the ice bath and insert a thermometer into the test tube. Remove the test tube from the ice bath when the thermometer reads 10°C. Record the temperature of the vinegar.
13. Place test tube #2 in the warm water bath and insert a thermometer into the test tube. Remove the test tube from the warm water bath when the thermometer reads 50°C. Record the temperature of the vinegar.
14. Save both water baths for the iron/thiocyante procedure in Part C.
15. Using a watch with a second hand or a stopwatch, record the start time. Simultaneously add one scoop (approximately 0.5 g) of baking soda ($NaHCO_3$) to each test tube.
16. Record the time when each reaction is complete.

Catalyst

Without a catalyst

17. Pour 10 mL of 0.024 M KIO_3 (potassium iodate) into a 50 mL beaker.
18. Pour 10 mL of 0.016 M $NaHSO_3$ (sodium bisulfite) into a second 50 mL beaker. Add 3 drops of starch solution to the second beaker.
19. Record the time on a watch with a second hand or use a stopwatch. Pour the contents of the first beaker into the second beaker, stirring the mixed solutions continuously. Record the time when the reaction mixture turns deep blue.

With a catalyst

20. Pour 10 mL of 0.024 M KIO_3 into a 50 mL beaker. Add 8 drops of 0.1 M $CuSO_4$ to the beaker. The copper(II) sulfate will act as the catalyst.
21. Pour 10 mL of 0.016 M $NaHSO_3$ (sodium bisulfite) into a second 50 mL beaker. Add 3 drops of starch solution to the second beaker.
22. Record the time on a watch with a second hand or use a stopwatch. Pour the contents of the first beaker into the second beaker, stirring the mixed solutions continuously. Record the time when the reaction mixture turns deep blue.

C. Iron(III)/thiocyanate equilibrium

1. Prepare a stock equilibrium solution by mixing 10 mL of 0.01 M $Fe(NO_3)_3$ and 10 mL of 0.01 M KSCN in a small beaker. **Pay close attention to the concentrations of the reactants!** Do not use the wrong solutions.
2. Obtain six test tubes and label them #1, #2, #3, #4, #5, and #6. Place them in order in a test tube rack.
3. Add 3 mL of the stock equilibrium solution to each test tube.
4. To test tube #1 add 10 drops of water. This test tube will be the "control" for the rest of the experiment. Record the color of the "control" solution. Compare the color of the "control" test tube to the colors that form in the other test tubes as each experiment is performed. It is helpful to hold a piece of white paper behind the test tubes when making color observations.
5. To test tube #2, slowly add 10 drops of 1.0 M $Fe(NO_3)_3$. **Pay close attention to the concentration!** Do not use the wrong solution. Place a clean and dry stopper in the test tube and invert it several times to mix the solutions. Compare and record the color of the solution in test tube #2.
6. To test tube #3, slowly add 10 drops of 1.0 M KSCN. **Pay close attention to the concentration!** Do not use the wrong solution. Place a clean and dry stopper in the test tube and invert it several times to mix the solutions. Compare and record the color of the solution in test tube #3.
7. To test tube #4, slowly add 10 drops of 3.0 M HCl. Place a clean and dry stopper in the test tube and invert it several times to mix the solutions. Compare and record the color of the solution in test tube #4.
8. To test tube #5, slowly add 10 drops of water and place the test tube in the warm water bath from part B. Maintain the temperature at approximately 55°C. Do not boil. After 10 minutes, compare and record the color of the solution in test tube #5.
9. To test tube #6, slowly add 10 drops of water and place the test tube in the ice/water bath from part B. After 10 minutes, compare and record the color of the solution in test tube #6.

Name ______________________ Date ______________________

Partner(s) ______________________ Section ______________________

______________________ Instructor ______________________

1. In a balanced chemical equation, what symbol is used to indicate that a reaction is reversible?

2. In an exothermic reaction, which side of the equation is "heat" written on?

3. What is the purpose of a catalyst?

4. Why is it important to clean and dry the thermometer after each procedure?

Name ____________________ Date ____________________

Partner(s) ____________________ Section ____________________

____________________ Instructor ____________________

A. Endothermic and exothermic reactions

	Test tube #1	Test tube #2
Initial water temperature		
Final water temperature		
Endothermic or exothermic?		
Balanced equation for the reaction, including the "heat" term		

B. Rates of reactions

Concentration of reactants

	Test tube #1	Test tube #2	Test tube #3
Start time			
Completion time			
Total time for reaction			
Endothermic or exothermic?			

Which concentration of HCl resulted in the fastest rate of reaction? Which was the slowest? On a molecular level, why is there a difference in reaction rates due to the concentration?

Temperature

	Test tube #1	Test tube #2
Temperature of vinegar		
Start time		
Completion time		
Total time required for reaction		

Which temperature resulted in the faster reaction rate? On a molecular level, why is there a difference in reaction rates due to the temperature?

Catalyst

	Beaker #1 (without catalyst)	Beaker #2 (with catalyst)
Start time		
Completion time		
Total time required for reaction		

What effect does the catalyst have on the reaction rate? Explain.

C. Iron(III)/thiocyanate equilibrium

$$Fe^{3+}(aq) + SCN^{-}(aq) \rightleftharpoons FeSCN^{2+}(aq)$$

yellow *colorless* *deep red*

	Test tube #1	Test tube #2	Test tube #3	Test tube #4	Test tube #5	Test tube #6
Color observed						
What component was changed? *	*control*					
Was that [component] increased or decreased? †	*control*					
Which direction did the equilibrium shift? ‡	*control*					

* There are only four components in the equilibrium equation: Fe^{3+}, SCN^{-}, $FeSCN^{2+}$, and *heat*. You should list only one component in each box in the above table, do not write more than one. Based on what you added to the test tube, which component changed? Reread the section *Experimental equilibria shifts* if needed.

† For example, if Fe^{3+} was the component that changed, how did the concentration of Fe^{3+} change? State your answer as *increased* or *decreased.*

‡ Did equilibrium shift to the left, to the right, or no change? Use words or symbols for your answer:

left right no change

Is the reaction exothermic or endothermic? Explain how you arrived at your conclusion.

Problems and questions

1. Give two definitions of when equilibrium is attained in a reversible reaction.

   a)

   b)

2. Nitrogen gas and hydrogen gas react to form ammonia gas and 22 kcal of heat.

   a) Write the correct balanced chemical equation for this reversible reaction. Include a "heat" term and the subscripts for the states of matter of each reactant and product.

   b) Predict the shift in equilibrium when each of the following changes occurs:

   Nitrogen is removed ____________________

   Heat is added ____________________

   Ammonia is added ____________________

   Hydrogen gas is added ____________________

3. Chemical hot packs and cold packs are often used to treat injuries. What type of reaction (exothermic or endothermic) is associated with each of these?

Goals

- Measure the pH of several solutions using a natural indicator, an indicator paper, and a pH meter.
- Calculate the pH of a solution from the $[H^+]$ or $[OH^-]$.
- Observe changes in the pH when acids or bases are added to a buffer solution.

Materials

- Red cabbage leaves
- Knife
- 400 mL beaker
- 10 mL graduated cylinder
- Hot plate
- Beaker clamp
- Wire gauze
- Stirring rod
- Test tubes
- Test tube rack
- Set of buffer solutions with pH range from 1 to 13
- pH meter with small beaker for rinsing electrode
- Calibration buffers for pH meter
- Wash bottle with deionized water
- Samples to test for pH: colorless or lightly colored juices and beverages, shampoo, conditioner, mouthwash, detergents, liquid soap, vinegar, household cleaners, etc.
- Small beakers or vials for test samples above
- Universal pH paper (e.g., pHydrion®) and color chart
- High pH buffer (9 – 11)
- Low pH buffer (3 – 4)
- 0.1 M NaCl with dropper
- 0.1 M HCl with dropper
- 0.1 M NaOH with dropper
- Optional: colored pencils

Discussion

Acids and bases are encountered frequently in everyday life. Foods, cleaning products, and biological fluids are examples of substances that may be categorized as acids or bases. Several different methods are available to determine whether a substance is acidic or basic. In this experiment, a natural indicator extracted from red cabbage, a commercially available universal indicator paper, and an electronic pH meter will be used to test for acids and bases.

There are several definitions of acids and bases, some are narrow while others are broader. One of the narrowest definitions was suggested by Svante Arrhenius. *Acids* dissociate when dissolved in water and produce hydrogen ions (H^+). *Bases* dissociate when dissolved in water and produce hydroxide ions (OH^-). For example, hydrogen chloride (HCl) is an Arrhenius acid, while sodium hydroxide (NaOH) is an Arrhenius base, as shown below:

$$\underset{\textit{Acid}}{HCl(g)} \xrightarrow{H_2O} H^+(aq) + Cl^-(aq) \qquad \underset{\textit{Base}}{NaOH(s)} \xrightarrow{H_2O} Na^+(aq) + OH^-(aq)$$

The Brønsted-Lowry theory is a more general definition of acids and bases. *Acids* are hydrogen-containing compounds that can donate protons (H^+) to another substance. *Bases* are substances that accept a proton (H^+). In an *aqueous* (water) solution, nitric acid (HNO_3) donates a hydrogen ion that reacts with a water molecule to form a *hydronium ion* (H_3O^+). Ammonia (NH_3), a base, receives a hydrogen ion from a water molecule to form a hydroxide ion (OH^-). The balanced equations are:

$$\underset{\textit{Acid}}{HNO_3(aq)} + H_2O(l) \longrightarrow H_3O^+(aq) + NO_3^-(aq) \qquad \underset{\textit{Base}}{NH_3(aq)} + H_2O(l) \longrightarrow NH_4^+(aq) + OH^-(aq)$$

Acids may also be defined as substances that increase the hydronium ion (H_3O^+) concentration, while bases are substances that increase the hydroxide ion (OH^-) concentration. Water also produces hydronium ions and hydroxide ions in a process called *self-ionization*. The balanced equation showing the reaction of the very few water molecules undergoing this reaction is:

$$H_2O(l) + H_2O(l) \rightleftharpoons H_3O^+(aq) + OH^-(aq)$$

It is not surprising that water is *amphiprotic* (can act as an acid or a base). In the reaction with nitric acid above, water acts as the base and receives a proton from the acid. In the reaction with ammonia, water acts as the acid and donates a proton to the base.

In pure water, the concentration of the hydronium ions equals the concentration of the hydroxide ions. Their concentrations can be measured experimentally:

$$[H_3O^+] = [OH^-] = 1.0 \times 10^{-7} \text{ M}$$

The product of these two concentrations is a constant and it is called the *ion product constant for water*. The equation for the ion product constant for water is:

$$K_w = [H_3O^+] \times [OH^-] = (1.0 \times 10^{-7})(1.0 \times 10^{-7}) = 1.0 \times 10^{-14}$$

The term H_3O^+ is frequently abbreviated as H^+. Thus, the equation above also may be written as $K_w = [H^+] \times [OH^-]$.

Since the numerical value for K_w must always equal 1.0×10^{-14}, any addition of hydronium ions to the water (in the form of an acid) will increase the $[H_3O^+]$ and decrease the $[OH^-]$. The addition of a base will cause a decrease in the $[H_3O^+]$ and an increase in the $[OH^-]$. If the concentration of either the $[H_3O^+]$ or the $[OH^-]$ is known, the other concentration can be calculated by rearranging the K_w equation.

Example 1:

If a solution has $[OH^-] = 1.0 \times 10^{-5}$ M, what is the $[H_3O^+]$?

$$[H_3O^+][OH^-] = 1.0 \times 10^{-14}$$ Rearrange this equation…

$$[H_3O^+] = \frac{1.0\times10^{-14}}{[OH^-]} = \frac{1.0\times10^{-14}}{1.0\times10^{-5}} = 1.0\times10^{-9}\ M$$

Example 2:

Sufficient acidic solute is added to a quantity of water to produce a solution with $[H_3O^+] = 4.0 \times 10^{-8}$ M. What is the $[OH^-]$ in this solution?

$$[H_3O^+][OH^-] = 1.0\times10^{-14}$$ Rearrange this equation…

$$[OH^-] = \frac{1.0\times10^{-14}}{[H_3O^+]} = \frac{1.0\times10^{-14}}{4.0\times10^{-8}} = 2.5\times10^{-7}\ M$$

pH of solutions

The pH of a solution is a logarithmic measure of its hydronium ion concentration (see Figure 1). The equation to calculate pH is:

$pH = -\log[H_3O^+]$ and may also be written as $pH = -\log[H^+]$

The rule for the number of significant figures in a logarithm is: the number of digits to the right of the decimal point in a logarithm is equal to the number of significant figures in the original number.

$$[H_3O^+] = 4.5\times10^{-3}\ M$$

Two significant figures

$$pH = 2.35$$

Two digits

Figure 1 ***Graphical representation of the pH scale***

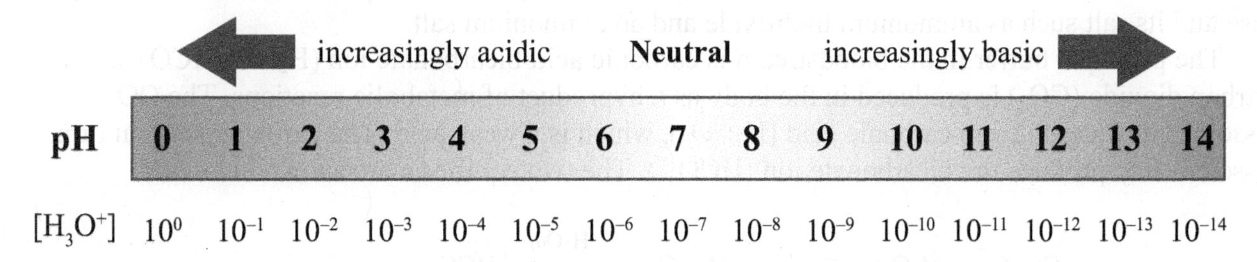

It is important to realize that a difference of 1 pH unit represents a ten-fold difference in the hydronium ion concentration.

Measuring pH

For accurate numerical values of pH, an electronic pH meter should be used. A pH meter should also be used to test colored solutions, since it may be difficult to see the color change of an indicator compound. A pH meter is an electronic device with a special electrode that generates a small voltage proportional to the $[H_3O^+]$ of the solution. The electrical signal is converted to pH and displayed on a meter or a digital readout. Be sure to use the correct number of significant figures when reading a pH meter. The electrode is very fragile and expensive. It must be cleaned after each use. When not in use, it is submerged in a buffer solution to keep the tip moist.

Another common way to measure pH is to use an *indicator*. An indicator is a complex organic compound that changes color at a certain pH. Many natural substances contain compounds that act as indicators. For example, the hydrangea bush flower is blue in acidic soil and pink/red in basic soils. Red cabbage changes colors at several different pH's and may be used as a wide range indicator.

pH indicator papers such as litmus paper or pHydrion® paper have been impregnated with indicators. These papers are easy to store and simple to use. They change color and the color shown by the paper is then compared to a color chart provided by the manufacturer. Dipping the indicator paper into a solution will contaminate the solution. It is preferable to use a clean, dry stirring rod to transfer a drop of the solution to the paper strip.

Buffers

There are a number of different systems in the body and they all have specific pH ranges. For example, the stomach is acidic, due to the production of hydrochloric acid. The small intestine is usually slightly basic. Urine is slightly acidic in the morning (pH = 6.5–7.0), generally becoming more alkaline (basic) by the evening (pH = 7.5–8.0).

The bloodstream is the most sensitive to pH changes and it must be kept in a narrow range from 7.35 to 7.45. Small departures from this range can cause serious illness, and death can result from pH changes of only a few tenths of a pH unit. To protect against pH changes, the bloodstream contains three different buffer systems to keep the pH of blood in the proper range.

Buffers minimize changes to the pH of a solution when small amounts of excess acid or base are added to the solution. This does not mean that the pH is neutral. Rather, the buffer stabilizes the solution at a certain pH. That pH depends upon the buffer system chosen. A buffer may consist of a weak acid and its salt such as acetic acid and an acetate salt. A buffer may also consist of a weak base and its salt such as ammonium hydroxide and an ammonium salt.

The principal buffer in the bloodstream is carbonic acid/bicarbonate ion (H_2CO_3/HCO_3^-). Carbon dioxide (CO_2) is produced in the body as a byproduct of metabolic reactions. The CO_2 dissolves in water to form carbonic acid (H_2CO_3), which is a weak acid. The carbonic acid, in turn, dissociates to produce the bicarbonate ion (HCO_3^-). The two equilibria are shown below:

$$CO_2(g) + H_2O(l) \rightleftharpoons \underset{\text{carbonic acid}}{H_2CO_3(aq)} \overset{H_2O(l)}{\rightleftharpoons} \underset{\text{bicarbonate ion}}{HCO_3^-(aq)} + H_3O^+(aq)$$

principal buffer in blood

When small amounts of acids or bases enter the bloodstream, the buffer reacts with them to counter their impact on blood pH. The following equations show the reaction when excess acid or base is added to the buffer solution:

$$\textit{Excess base:} \quad H_2CO_3(aq) + OH^-(aq) \rightleftharpoons HCO_3^-(aq) + H_2O(l)$$

$$\textit{Excess acid:} \quad HCO_3^-(aq) + H_3O^+(aq) \rightleftharpoons H_2CO_3(aq) + H_2O(l)$$

Excess base reacts with the carbonic acid component of the buffer and excess acid reacts with the bicarbonate ion component of the buffer. Notice that no new reactants or products are formed. All of the substances, H_2CO_3, HCO_3^-, and H_2O, were present in the original buffer solution. But the excess acid or base has been "removed" from the solution, keeping the pH relatively unchanged.

Buffer solutions do not have the ability to neutralize unlimited quantities of additional acid or base. The amount of $[H_3O^+]$ or $[OH^-]$ that can be neutralized is called the *buffering capacity*.

If the pH of the bloodstream is too low (more acidic), the condition is called *acidosis*. If the pH of the bloodstream is too high (more basic), the condition is called *alkalosis*. There are two general types of acidosis and alkalosis, one type resulting from changes in metabolic processes and the other type resulting from changes in respiratory processes. Both types can be understood in terms of Le Châtelier's principle and equilibrium shifts.

Experimental Procedures

Eye protection and appropriate clothing must be worn at all times.

Many of these chemicals are corrosive to the skin and toxic if ingested. Wash any contacted skin area immediately with running water and alert the instructor.

Discard all wastes properly as directed by the instructor. Dispose of indicator papers and cabbage in the trash containers, NOT in the sinks.

A. Red cabbage indicator

Optional: ***The pH reference sets may be prepared by the instructor before the lab meeting or students may prepare them.***

If prepared by the instructor, do not do Procedure A. Instead, continue to Procedure B. Two or more "pH reference sets" should be available for the class.

If prepared by the students, the instructor will determine how many students should work together to prepare their "pH reference set."

1. Chop several leaves of red cabbage and place in a 400 mL beaker. Fill the beaker half full with deionized water to cover the leaves. Heat gently on a hot plate until the solution turns dark purple. Turn off the burner. Use a beaker clamp to set the beaker on a wire gauze to cool.
2. Place 13 test tubes in a test tube rack. Label the test tubes #1 through #13.
3. Pour 3–4 mL of the buffer solution labeled "pH = 1" into the first test tube. Repeat this step with the remaining buffer solutions and their corresponding numbered test tubes.
   Add 2–3 mL of the *cooled* red cabbage indicator to each test tube. Stir each solution with a clean, dry stirring rod.

B. Colors of the pH reference set

1. Record the colors of the pH reference set (use words and color the wedges). It is helpful to hold a piece of white paper behind the test tubes to observe the colors.
2. Do not dispose of the reference set until the entire lab is completed.

C. Measuring pH

Universal pH paper (pHydrion®)

1. Pour 3–4 mL of each substance to be tested into clean, dry test tubes. List these substances in **alphabetical order** in Table B of the report sheet.
2. Obtain a pH paper color chart and approximately 15 cm of the universal pH paper. Place this strip of paper on a clean dry paper towel.
3. Dip a clean, dry stirring rod into the first test tube and place a drop of the substance close to one end of the strip of pH paper. Compare the resulting color with the color on the color chart. Record the pH. Save the test tube and solution for the next section of the experiment.
4. Continue to use the same strip of pH paper. Repeat the procedure for each of the household substances provided, spacing each drop approximately 1 cm apart. Save all of the test tubes and solutions for the next section of the experiment.

Natural indicator

5. To each of the test tubes prepared in the universal pH paper section, add 2–3 mL of the red cabbage solution.
6. Mix each of the solutions with a clean dry, stirring rod. Compare the resulting color of each of the solutions with the reference set prepared in Procedure A. Record the color and pH for each solution.

pH meter

7. The lab instructor will demonstrate how to calibrate and use the pH meter.
8. Select four substances tested in the previous sections. Choose the substances with the highest pH, the lowest pH, the pH closest to neutral, and one other substance of your choice.
9. Obtain new samples of each substance **without** the cabbage juice, approximately 10 mL of each one. Test the pH of each substance by pouring it into a small beaker or vial that will accommodate the electrode of the pH meter.
10. Measure the pH of the four substances using the pH meter. Use a squeeze bottle and a waste beaker to rinse the electrode with deionized water between each measurement.

D. Buffer solutions

Addition of an acid to a buffer solution

1. Obtain four test tubes and label them #1– #4.
2. Using a 10 mL graduated cylinder, measure approximately 10 mL of each of the following solutions into a different test tube. Clean and dry the graduated cylinder between solutions.

   #1 deionized water
   #2 0.1 M NaCl
   #3 a high pH buffer solution
   #4 a low pH buffer solution

3. Add 2–3 mL of cabbage indicator to **each** test tube. Mix well.
4. For each test tube, determine the pH of the solution by comparing the color with the reference set prepared in Procedure A. Record the color and pH for each test tube.
5. Add 5 drops of 0.1 M HCl to **each** test tube. Stir and determine the pH. Record.
6. Add another 5 drops of 0.1 M HCl to **each** test tube. Stir and determine the pH. Record.
7. Determine the pH change for each test tube (pH after 10 drops minus original pH). Record.
8. Indicate whether the substance in each test tube is a buffer.

Addition of a base to a buffer solution

9. Repeat steps 1–8 in Procedure D. Be sure to use clean, dry test tubes for step #1. In steps #5 and #6, use 0.1 M NaOH instead of 0.1 M HCl.

Name ______________________ Date ______________________

Partner(s) ______________________ Section ______________________

______________________ Instructor ______________________

1. Describe the correct procedure to test solutions with indicator paper.

2. What is the purpose of a buffer?

3. What should you do if you accidentally spill an acid or a base on your skin?

4. Complete the following table:

$[H_3O^+]$	$[OH^-]$	pH	Acidic, basic, or neutral?
		8.00	
2.5×10^{-2} M			
			neutral
	1.0×10^{-9} M		

Name ______________________ Date ______________________

Partner(s) ______________________ Section ______________________

______________________ Instructor ______________________

A. Red cabbage indicator

There is no data to record for this part.

B. Colors of the Cabbage pH reference set

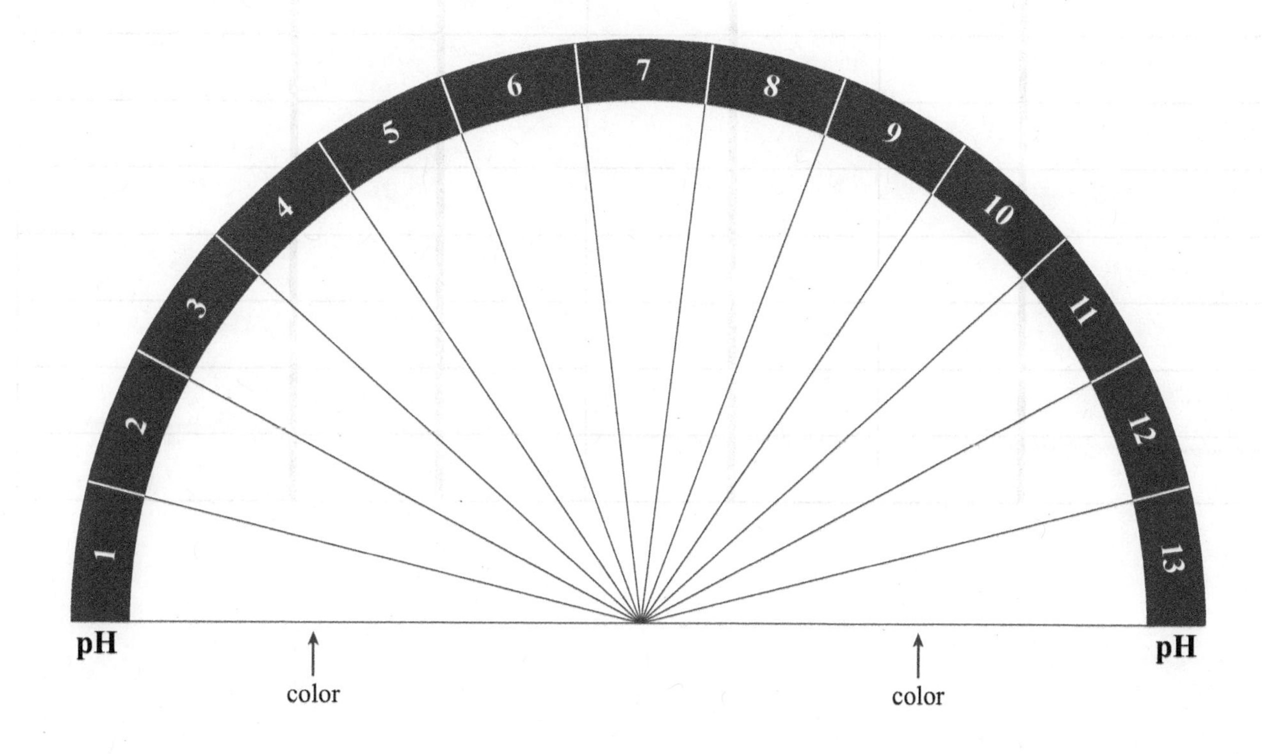

C. Measuring pH

Substance	pH paper color	pH paper pH	cabbage color	cabbage pH	pH meter pH	Acidic, basic, or neutral?
tap water						
deionized water						

D. Buffer solutions

Addition of an acid to a buffer solution

Substance	pH of original solution	pH after 5 drops of HCl	pH after 10 drops of HCl	pH change	Buffer solution?
deionized water					
0.1 M NaCl					
high pH buffer					
low pH buffer					

Addition of a base to a buffer solution

Substance	pH of original solution	pH after 5 drops of NaOH	pH after 10 drops of NaOH	pH change	Buffer solution?
deionized water					
0.1 M NaCl					
high pH buffer					
low pH buffer					

Each table above used the same four solutions.

Of the four solutions, which solutions (plural) had the greatest change in pH? Why?

Of the four solutions, which solutions (plural) had the smallest change in pH? Why?

Problems and questions

1. Compare and contrast the three methods used to test pH (red cabbage indicator, pH paper, and a pH meter). Did they give similar results? Which is the most accurate? Explain.

2. The hydroxide ion concentration of a solution is 2.5×10^{-8} M.

   a) What is the hydronium ion concentration?

   b) What is the pH?

   c) Is the solution acidic, basic, or neutral?

3. A few drops of HCl are added to two solutions, A and B. The pH of each solution is shown before and after the addition of HCl. Which solution is a buffer? Explain.

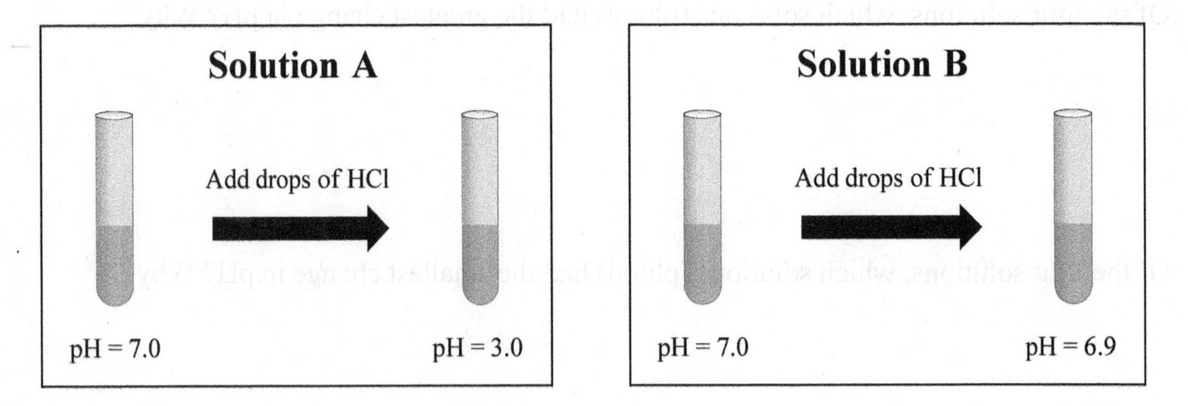

4. How many times more acidic is a solution with pH 2 relative to pH 4?

5. Circle the two substances that, when dissolved in water and mixed together, would make a buffer pair:

HF K_2CO_3 HCl Na_2CO_3 NaF $NaCl$

Goals

- Set up a buret and use proper titrating technique to reach an end point.
- Demonstrate proper technique of a pipet and buret.
- Calculate the molarity of the acid in vinegar.
- Calculate the percent by mass of acetic acid in vinegar.
- Calculate the percent error.

Materials

- Ring stand
- Buret clamp
- Buret
- 2 mL pipet
- Pipet bulb
- Phenolphthalein indicator solution
- Commercial brand of white vinegar
- Standardized NaOH labeled with actual molarity (at least 3 sig figs, approximately 0.1 M)
- Three 125 mL Erlenmeyer flasks
- Funnel
- 150 mL beaker
- 50 mL beaker
- 100 mL graduated cylinder

Discussion

Volumetric titration is a technique in which a volume of a standard solution is added gradually to a given volume of an unknown solution until the chemical reaction between the two solutions is complete. A *neutralization titration* is used for the reaction of an acid with a base. In this experiment, the base will be the "standard" solution. The concentration of the standard solution has been determined and is provided on the bottle label for use in the experiment. The acid will be the "unknown" solution.

To achieve good results, the volumes of both the standard solution and the unknown solution must be measured accurately. The unknown solution is measured and transferred to a suitable flask using a pipet (pronounced "pipe – ette") which is a glass tube calibrated to deliver an exact amount. The standard solution is usually contained in a buret (pronounced "bure – ette") which is a long glass tube calibrated to deliver precise volumes of a solution through a stopcock valve. The total volume of the standard solution used in the titration procedure is determined by measuring the initial volume in the buret before starting the procedure and then measuring the final volume after completing the procedure. The difference between the two readings is the volume of standard solution used.

$$V_{used} = V_{final} - V_{initial}$$

Unlike a graduated cylinder, where the volume readings increase from bottom to top, the numbers increase from top to bottom in a buret. To correctly read a buret, have the bottom of the meniscus at eye level and report the reading at the bottom of the meniscus. Remember to estimate one digit further than the smallest calibrated mark (see Figure 1).

Figure 1 ***Reading a buret***

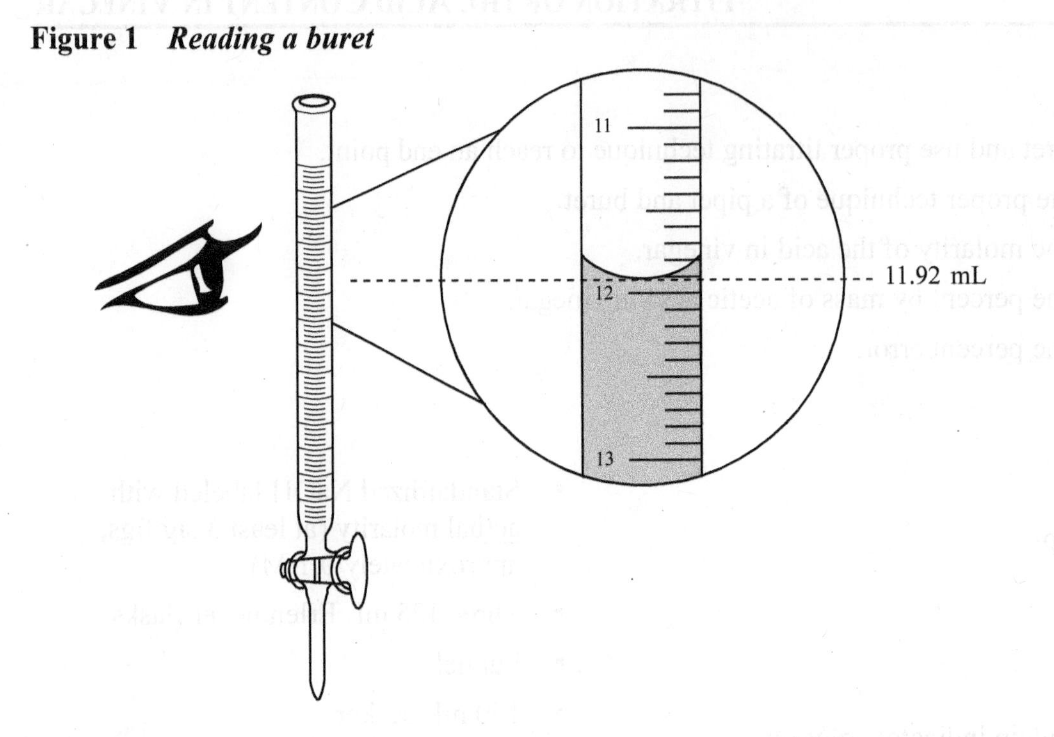

In this experiment, vinegar will be used as the acid. Regardless of brand, all household vinegars are essentially dilute solutions of acetic acid ($HC_2H_3O_2$ or CH_3COOH) in water. The amount of acetic acid in the vinegar can be determined by neutralizing the acid with a base, such as sodium hydroxide in this experiment. The balanced equation for the neutralization of acetic acid with NaOH is shown below. The products of the reaction are a salt and water.

$$\underset{\substack{\textit{Acid} \\ \textit{(acetic acid)}}}{HC_2H_3O_2(aq)} + \underset{\substack{\textit{Base} \\ \textit{(sodium hydroxide)}}}{NaOH(aq)} \longrightarrow \underset{\substack{\textit{Salt} \\ \textit{(sodium acetate)}}}{NaC_2H_3O_2(aq)} + \underset{\textit{Water}}{H_2O(l)}$$

The critical factor in a titration is the *equivalence point,* at which the moles of acid are equal to the moles of base. The equivalence point is often determined by the color change of an acid-base *indicator*, a complex organic molecule that has distinctly different colors in acidic and basic solutions. The *end point* of a titration occurs when the indicator changes color. A good indicator for a specific titration should be one that gives an end point close to the equivalence point.

Even though pH 7 is "neutral," the pH at the end point of an acid-base neutralization reaction may not be 7. It is pH 7 when a strong acid is neutralized with a strong base. The base in this experiment, NaOH, is a strong base. However, the acetic acid in vinegar is a weak acid. Thus, the pH of the solution at the equivalence point is approximately 8.72 so an indicator that changes color close to this pH is required in order for the end point (color change) to correspond to the equivalence point (reaction completion). Phenolphthalein is an appropriate indicator for the acetic acid/sodium hydroxide reaction since its color changes from colorless in acidic solution to pink in basic solutions in the pH range of 8.3 to 10.0 (see Figure 2). In this titration, the equivalence point can be assumed to be the same as the end point.

Figure 2 ***Color of phenolphthalein at different pH values***

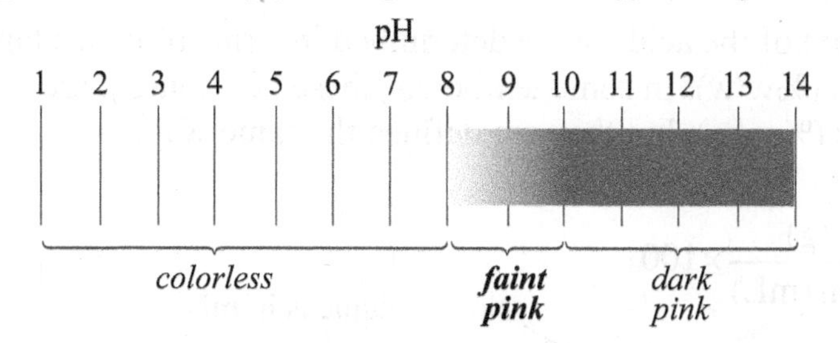

Molarity

Molarity is the measure of the concentration of a solute in a solvent. It is defined as the moles of solute in a solution divided by the liters of solution. The mathematical equation for molarity is:

$$\text{Molarity} = \frac{\text{moles of solute}}{\text{liters of solution}} \qquad \text{or simply} \qquad M = \frac{\text{mol}}{\text{L}}$$

Often in titration problems, the above equation needs to be rearranged to solve for moles or liters. The other equations are:

$$\text{mol} = M \times L \qquad L = \frac{\text{mol}}{M}$$

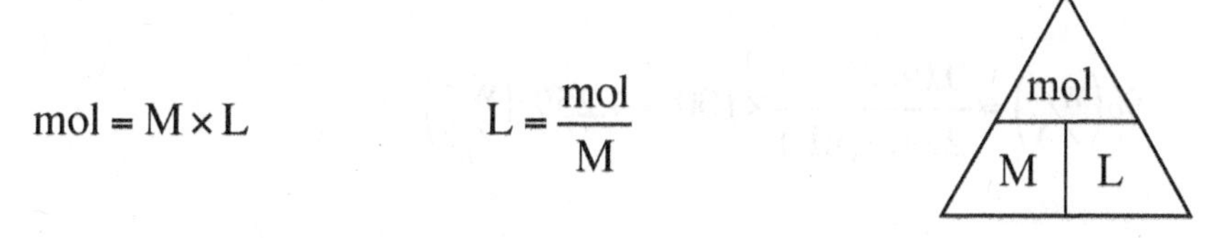

A mnemonic for remembering these equations is to refer to the triangle above. Put a finger over the quantity sought. For example, put a finger over molarity (M). It can be seen that molarity is moles over liters. Next, put a finger over moles (mol). It can be seen that moles is molarity times liters.

Solving titration problems

Refer to the balanced equation for the reaction between acetic acid and sodium hydroxide. At the equivalence point, the moles of acid and the moles of base are equal.

$\text{mol}_a = \text{mol}_b$ where "a" represents acid and "b" represents base

In the previous section, it was shown that $\text{mol} = M \times L$. Substitution gives:

$$M_a \times L_a = M_b \times L_b$$

By rearranging this equation, the concentration of the acid can be determined when all of the other factors of the equation are known. Note: volumes are usually measured in milliliters, but to use these equations, the volumes must be converted to **liters**.

$$M_a = \frac{M_b \times L_b}{L_a}$$

Percent concentration

In addition to molarity, the concentration of the acid can be determined in terms of percentage by weight/volume (%w/v) or weight/weight (%w/w). In some textbooks, these percentages are called mass/volume (%m/v) and mass/mass (%m/m), but they are defined the same way.

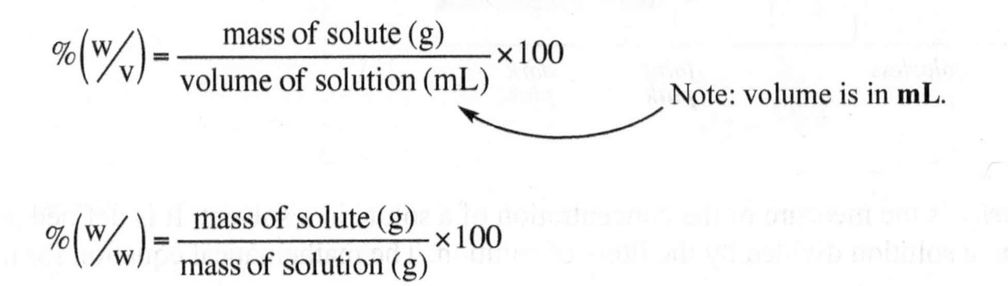

$$\%\left(\frac{w}{v}\right) = \frac{\text{mass of solute (g)}}{\text{volume of solution (mL)}} \times 100$$

Note: volume is in **mL.**

$$\%\left(\frac{w}{w}\right) = \frac{\text{mass of solute (g)}}{\text{mass of solution (g)}} \times 100$$

If the mass of solute is not known, moles can be converted to grams using the molar mass of the compound.

For example, a commercial white vinegar solution is found to contain 0.0849 g of acetic acid in a 2.00 mL sample. The %(w/v) is calculated as follows:

$$\%\left(\frac{w}{v}\right) = \frac{0.0849\ \text{(g)}}{2.00\ \text{(mL)}} \times 100 = 4.25\%\left(\frac{w}{v}\right)$$

Problems other than a stoichiometric ratio of 1:1

In the earlier example, the balanced reaction for the titration of the acid in vinegar with NaOH showed a stoichiometric ratio of 1:1, but that is not always the case. For example, suppose 25.0 mL of a 0.500 M KOH solution neutralized a 5.00 mL sample of H_2SO_4. What is the concentration of the acid?

In general, the first step in solving any titration problem is to write a balanced equation for the reaction. In all acid-base neutralization reactions, the net ionic equation (what's really happening) is the same. Hydrogen ions from the acid react with hydroxide ions from the base to produce water:

$$H^+(aq) + OH^-(aq) \longrightarrow H_2O(l)$$

from acid — *from base* — *water*

Sulfuric acid (H_2SO_4) supplies two hydrogen ions while potassium hydroxide supplies only one hydroxide ion. But the above equation says that for every one hydrogen ion, there **must** be one hydroxide ion to react with it. To make a balanced equation, the coefficient 2 must be placed before the hydroxide ion and the water:

$$\mathbf{2}H^+(aq) + \mathbf{2}OH^-(aq) \longrightarrow \mathbf{2}H_2O(l)$$

The balanced equation is nearly complete. The only formula missing is the salt.

$$H_2SO_{4(aq)} + \mathbf{2}\,KOH_{(aq)} \longrightarrow \textit{salt} + \mathbf{2}\,H_2O_{(l)}$$

supplies $2\,H^+$ *supplies* $2\,OH^-$

If the hydrogen ions from the acid react with hydroxide ions from the base to produce water, all the remaining leftover ions must comprise the salt. In the above example, there is one sulfate ion (SO_4^{2-}) and two potassium ions (K^+). The formula of the salt is K_2SO_4 (remember, the metal is written first).

$$H_2SO_{4(aq)} + 2\,KOH_{(aq)} \longrightarrow K_2SO_{4(aq)} + 2\,H_2O_{(l)}$$

The equation for the molarity of the acid must be adjusted because the stoichiometric ratio is 1:2. This can be accomplished by using a fraction. The numbers in the fraction come from the coefficients in the balanced equation. In general, it is helpful to remember that the fraction is always the coefficient of the **unknown** solution divided by the coefficient of the **standard** solution. Since the molarity of the sulfuric acid is unknown, it is the unknown solution while potassium hydroxide is the standard solution.

$$M_a = \left(\frac{M_b \times L_b}{L_a}\right)\left(\frac{1\text{ mole of acid}}{2\text{ moles of base}}\right)$$

Again, to use this equation, the volumes must be converted to liters: 25.0 mL = 0.0250 L and 5.00 mL = 0.00500 L. The molarity of the acid can be calculated as follows:

$$M_a = \left(\frac{0.500\text{ M} \times 0.0250\text{ L}}{0.00500\text{ L}}\right)\left(\frac{1\text{ mole of acid}}{2\text{ moles of base}}\right) = 6.25\text{ M}$$

Example #2:

Using the previous example as a guide, even a reaction with a complex stoichiometric ratio of 3:2 can be easily solved. What is the molarity of a 2.50 mL sample of an $Al(OH)_3$ solution that was titrated to the equivalence point using 37.6 mL of a 0.275 M H_2SO_4 solution?

Sulfuric acid supplies two hydrogen ions while aluminum hydroxide supplies three hydroxide ions. However, there must be equal numbers of hydrogen ions and hydroxide ions. The following coefficients must be used:

$$\mathbf{3}\,H_2SO_{4(aq)} + \mathbf{2}\,Al(OH)_{3(aq)} \longrightarrow Al_2(SO_4)_{3(aq)} + \mathbf{6}\,H_2O_{(l)}$$

supplies $6\,H^+$ *supplies* $6\,OH^-$ *salt*

In this problem, $Al(OH)_3$ (the base) is the unknown solution while H_2SO_4 (the acid) is the standard solution. Remember that the fraction is always the coefficient of the **unknown** solution divided by the coefficient of the **standard** solution.

$$M_b = \left(\frac{M_a \times L_a}{L_b}\right)\left(\frac{2 \text{ moles of base}}{3 \text{ moles of acid}}\right)$$

The molarity of the base can be calculated as follows:

$$M_b = \left(\frac{0.275 \text{ M} \times 0.0376 \text{ L}}{0.00250 \text{ L}}\right)\left(\frac{2 \text{ moles of base}}{3 \text{ moles of acid}}\right) = 2.76 \text{ M}$$

Note: the volumes were converted to liters.

Percent error

Accuracy is a comparison of the measured value to the accepted, or true, value. One way of assessing accuracy is to calculate a percent error. The *percent error* is the difference between the measured value and the accepted (true value) as a percentage of the accepted (true value). The equation for percent error is:

$$\% \text{ error} = \frac{|\text{Measured value} - \text{True value}|}{\text{True value}} \times 100$$

The "|" symbols mean absolute value, so negatives become positive. This insures that the percent error is always a positive number.

For example, a student measures the density of an object as 8.37 g/cm^3. The true value of the density is 8.92 g/cm^3. What is the student's percent error of the measurement?

$$\% \text{ error} = \frac{|8.37 - 8.92|}{8.92} \times 100 = \frac{0.55}{8.92} \times 100 = 6.2\%$$

Titration: real world applications

Titration is not limited to the field of chemistry. The basic titration technique learned in this experiment is used in the health sciences in a variety of ways:

- determination of the concentrations of chemicals of interest in blood and urine, such as blood glucose levels in patients with diabetes
- determination of the concentrations of vitamins
- determination of fatty acid content
- determination of correct proportion of different medicines in an intravenous drip

Experimental Procedures

Eye protection and appropriate clothing must be worn at all times.

Acids and bases are harmful to tissues. Use care when handling. Wash any contacted skin area immediately with running water and alert the instructor. Discard all wastes properly as directed by the instructor.

Record all measurements with the correct number of significant figures and units.

NEVER pipet by mouth.

Burets and pipets are fragile and should be handled with extra care.

Titration of acetic acid in vinegar with standardized NaOH

1. Set up the apparatus as shown in Figure 3.
2. Thoroughly clean and rinse three 125 mL Erlenmeyer flasks with deionized water.
3. Label the flasks as #1, #2, and #3 and pipet exactly 2.00 mL of vinegar (of unknown molarity) into each flask. The instructor will demonstrate the proper use of a pipet and pipet bulb or pipetter. Observe the type of pipet used because calibration marks differ. Also, some pipets are calibrated to retain a small amount of liquid in the tip and may not need to be completely emptied from the pipet.
4. Using a 100 mL graduated cylinder, add approximately 50 mL of deionized water to each flask. This water does not affect the reaction but merely adds sufficient volume so that the color change and swirling are easier to observe.

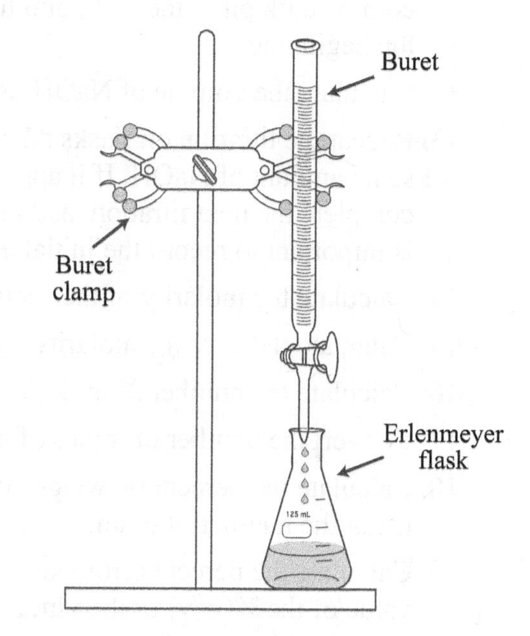

Figure 3 ***Titration apparatus***

5. Add 1–2 drops of phenolphthalein to each flask.
6. Use a 150 mL beaker and obtain approximately 75 mL of standardized NaOH solution. Record the actual molarity of the solution as shown on the bottle label.
7. Lower the buret in the buret clamp until it is at a comfortable height to work with.
   - If the NaOH solution is contained in a squirt bottle, carefully squirt it into the buret to about a centimeter above the 0.00 mL mark.
   - If the NaOH solution is contained in a beaker, use a funnel and carefully and **slowly** pour some of the NaOH solution into the buret. Fill to about a centimeter above the 0.00 mL mark.

8. Place a 50 mL beaker under the stopcock. Open the stopcock and let the NaOH drain until the bottom of the meniscus of the liquid is at or slightly below the 0.00 mL mark. The tip of the buret should be filled with NaOH solution. Remove the beaker.
9. Record the level of the meniscus as the initial reading of NaOH. Remember to estimate one place past the last calibrated mark. Notice which direction the marks are numbered on the buret. Raise the buret so the tip just fits within the Erlenmeyer flask (see Figure 3).
10. To titrate the vinegar solution in flask #1 with the standardized NaOH solution, place the flask under the tip of the buret. Add the NaOH slowly by opening the stopcock. Swirl the mixture in the flask during the titration. Optionally, a magnetic stir bar and stirrer may be used to automate the stirring. As NaOH is added, the mixture will acquire a pink tinge that will disappear while swirling. Placing a sheet of white paper or paper towel under the flask makes it easier to see the color. As the end point approaches, the pink will persist for longer periods of time before fading. When this occurs, slow the rate of addition of NaOH.
11. Continue swirling and adding NaOH drop by drop until a **faint** pink color remains for at least 30 seconds. Record the level of the meniscus as the final reading of NaOH. If the color is dark pink, the end point has been overshot and the titration must be restarted from the beginning.
12. Calculate the volume of NaOH used.
13. Repeat the titration on flasks #2 and #3. Each titration should require approximately the same amount of NaOH. If it appears that there may not be enough NaOH in the buret to complete the next titration, add additional NaOH. It does **not** need to be filled to the top. It is important to record the initial and final readings for each titration.
14. Calculate the molarity of the acetic acid for each titration.
15. Calculate the *average* molarity value for the three titrations.
16. Calculate the number of moles of acetic acid using the *average* value of the molarities.
17. Convert the number of moles of acetic acid to grams of acetic acid using the molar mass.
18. Calculate the percent by weight/volume of the acetic acid content of the vinegar solution using the number of grams of acetic acid and the volume of vinegar.
19. Calculate the percent error using the %(w/v) measurement calculated in step 18 and the true value of the %(w/v) of the vinegar.

PRE-LAB QUESTIONS: TITRATION OF THE ACID CONTENT IN VINEGAR

Name ______________________ Date ______________________

Partner(s) ______________________ Section ______________________

______________________ Instructor ______________________

1. What characteristics determine whether a compound will make a good indicator?

2. Write the correct balanced equations for the following neutralization reactions:

   a) phosphoric acid and sodium hydroxide

   b) sulfuric acid and lithium hydroxide

   b) hydrochloric acid and barium hydroxide

3. Why is it important to stop titrating when the **faint** pink color in the Erlenmeyer flask remains?

4. Why is a pipet used rather than a graduated cylinder to measure the volume of the vinegar?

Name ______________________ Date ______________________

Partner(s) ______________________ Section ______________________

______________________ Instructor ______________________

Show the correct number of significant figures and units for all measurements and calculations.

Molarity of standardized **NaOH** solution

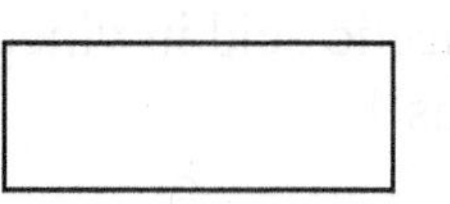

	Flask #1	Flask #2	Flask #3
Volume of ***vinegar*** in the flask			
Initial reading of ***NaOH*** in the buret			
Final reading of ***NaOH*** in the buret			
Total volume of NaOH used (show calculations)			
Molarity of vinegar solution (show calculations)			

Calculations (Show your work)

Average molarity of vinegar
(show calculations) ______________

Number of moles of acetic acid in vinegar
(show calculations) ______________

Number of grams of acetic acid in vinegar
(show calculations) ______________

Percent concentration (w/v) of vinegar
(show calculations) ______________

Accepted (true value) %(w/v) of the vinegar
(provided by the instructor) [ ]

Percent error of the measurement
(show calculations) ______________

Problems and questions (show all calculations, correct significant figures and units)

1. A solution contains 25.0 g of NaCl in 83.0 mL of solution.
   a) What is the %(w/v) concentration of the NaCl solution?

   b) What is the molarity of the NaCl solution?

2. How many moles of NaOH are required to prepare 3.00 L of a 0.325 M NaOH solution?

3. A solution of 0.312 M KOH is used to neutralize 15.0 mL of an HF solution. If 28.3 mL of KOH is required to reach the end point, what is the molarity of the HF solution?

4. How many mL of 0.15 M NaOH will be required to completely titrate 10.0 mL of 0.200 M H_2SO_4? (Hint: write a balanced equation for the reaction first.)

[*Reminder: there is one more problem on the next page*]

5. Antacids are taken to relieve heartburn or indigestion caused by excess stomach acid. The active ingredients vary depending on the brand, but all contain bases. Each Alka-Seltzer® tablet contains 1.916 g of the active ingredient sodium bicarbonate ($NaHCO_3$). When it dissolves, the sodium bicarbonate reacts with stomach acid (essentially 0.100 M HCl) according to the reaction below:

$$HCl(aq) + NaHCO_3(aq) \longrightarrow NaCl(aq) + CO_2(g) + H_2O(l)$$

How many milliliters of stomach acid can one tablet of Alka-Seltzer® neutralize?

Goals

- Label and name the monomer unit for a polymer.
- Name the polymer from the provided monomer name.
- Distinguish between chemical and physical properties of commercial polymers (plastics).
- Interpret observations of an unknown compound and determine its identity.
- Prepare the polymers Gluep and Slime®.
- Identify chemical and physical properties of lab prepared polymers (nylon, Gluep, and Slime®).

Materials

For instructor demonstration:

- Solution #1: 1,6-hexamethylenediamine/NaOH solution
- Solution #2: 4% sebacoyl chloride in hexane
- 50% ethanol
- Large diameter test tube (25×250)
- 25 mL graduated cylinder
- 50 mL beaker
- 100 mL beaker
- Forceps

For students:

- 1 cm × 1 cm samples of plastics with resin ID code numbers 1, 2, 3, 4, 5, and 6
- Plastic unknowns (labeled A, B, C)
- 70% isopropyl alcohol
- 4% polyvinyl alcohol
- Borate solution
- 6 M HCl
- 6 M NaOH
- Acetone
- Ethyl acetate
- Elmer's® School Glue
- Waste containers for plastics, acetone, ethyl acetate, and nylon wastes
- Styrofoam cups marked at 20 mL
- Bunsen burner
- Gas lighter (striker)
- 25 mL graduated cylinder
- 50 mL beaker
- 100 mL beaker
- 250 mL beaker
- Stirring rod
- Rulers
- 15×125 (or 13×100) test tubes
- Test tube rack
- Forceps
- Spatula
- Scissors
- Gloves

Discussion

Polymers are long chain molecules formed by chemically combining many smaller molecules called monomers. "Poly" comes from the Greek *polus*, meaning many and "mer" comes from the Greek *meros*, meaning part. Together, polymer means "many parts." A *monomer* (meaning "one part") is the small molecule that is the structural building block in a polymer. Synthetic and natural polymers are classified as addition polymers or condensation polymers.

In an *addition polymer*, the monomers simply "add together" with no other products formed besides the polymer. With appropriate catalysts, simple alkenes (organic molecules that have a carbon-carbon double bond) and alkynes (organic molecules that have a carbon-carbon triple bond) readily undergo polymerization.

An example of an addition polymer with an alkene monomer is polyethylene, shown below. The monomer is ethylene, which is the common name for ethene (IUPAC name). IUPAC is an acronym for the International Union of Pure and Applied Chemistry, the world authority for standardizing nomenclature (systematic rules of naming) of elements and compounds.

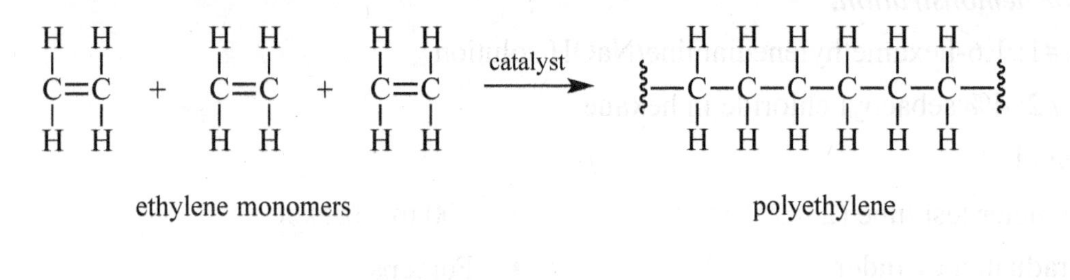

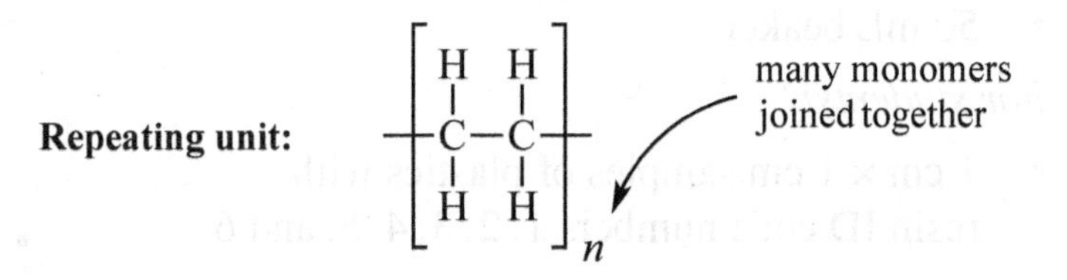

Note that the polymer formed from an alkene monomer has a single bond where the original monomer had a double bond. Because molecules of a polymer have varying lengths, an exact formula for a polymer such as polyethylene cannot be written. The wavy bonds in the above illustration indicate that the molecule continues repeating in either direction. The formula of the repeating unit is written in brackets and the subscript *n* is added after the bracket, with *n* being understood to represent a very large number.

An example of an addition polymer synthesized from an alkyne is polyacetylene. The alkyne monomer is acetylene, the common name for ethyne (IUPAC name). Polyacetylenes can be found in vegetables such as carrots and the herb fennel.

$$\mathrm{H{-}C{\equiv}C{-}H} + \mathrm{H{-}C{\equiv}C{-}H} + \mathrm{H{-}C{\equiv}C{-}H} \xrightarrow{\text{catalyst}}$$

acetylene monomers

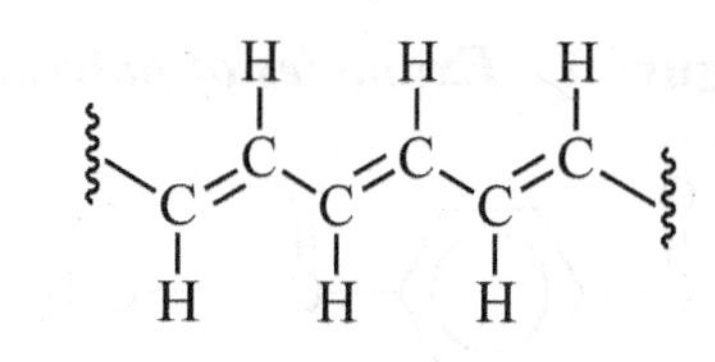

polyacetylene

Repeating unit:

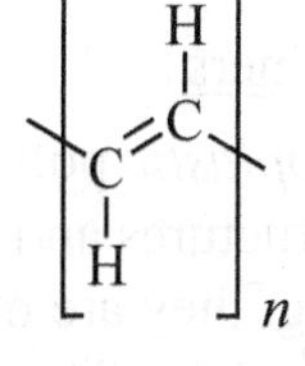

Note that the polymer formed from an alkyne monomer has a double bond where the original monomer had a triple bond.

The monomers in addition polymers need not be the same, however. A *copolymer* is a polymer in which two different monomers are present. An example of an addition copolymer is Saran Wrap™. The two monomers are vinyl chloride and 1,1-dichloroethene:

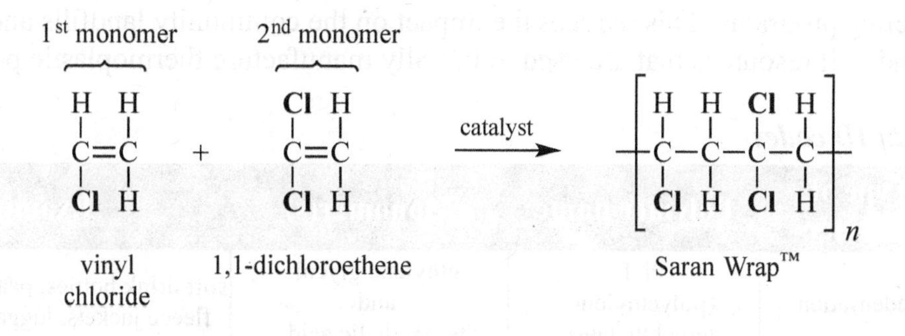

Condensation polymers form through condensation reactions. *Condensation reactions* occur when two molecules combine to form a larger molecule while liberating a small molecule such as water or HCl. The monomers do not require a double or triple bond; instead, the reaction utilizes functional groups. A *functional group* is an atom or group of atoms that has a certain structural feature. The polymer is formed by the reaction of a functional group on one monomer reacting with a functional group on another monomer. Nylon and polyesters are examples of condensation polymers. Specific examples of this type of reaction can be found in the nylon subdivision, later in this experimental discussion.

Organic compounds called carboxylic acids and alcohols react to form simple esters (and water). When the acid contains two carboxylic acid functional groups and the alcohol contains two alcohol functional groups, the polymer chain may continue growing on both ends. These condensation polymers are known as polyesters. One such polyester is dacron (see Figure 1). In the textile industry, dacron polyester is made into clothing while in the medical field, dacron is used to make arterial grafts.

Lactic acid and glycolic acid monomers can be polymerized to give a biodegradable polyester polymer (trade name Lactomer™) that is used as surgical staples. Lactomer™ staples start to dissolve (hydrolyze) after a few weeks. By the time the tissue has fully healed, the staples have fully dissolved and they do not need to be removed.

Figure 1 ***Examples of polyesters***

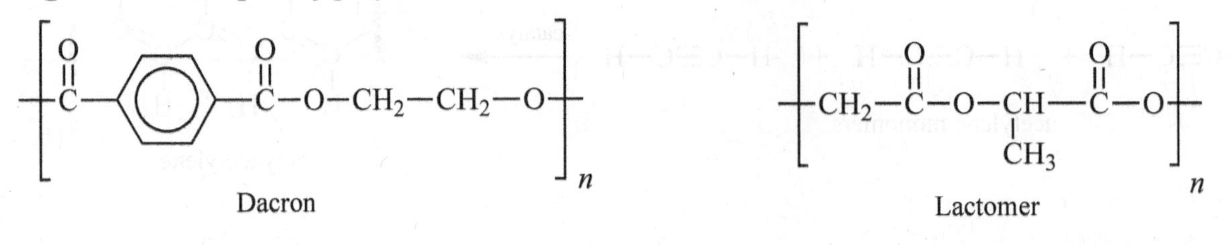

<u>Plastic polymers</u>

Thermoplastic polymers can be melted, reshaped, and reused. However, they have different chemical structures and therefore different melting points. In order to help separate these polymers for recycling, they are marked with a resin ID code (or recycling code), which identifies how easily the polymer melts. The numbers 1 to 6 correspond to specific polymers. The number 7 identifies the plastic as a miscellaneous type or a mixture of more than one polymer. The lower the number, the easier it is to recycle. A list of resin ID codes, polymer name, and examples of each type of polymer is found in Table 1.

The name of the polymer often comes from adding the prefix "poly" to the name of the monomer associated with it. Since many of the plastic polymers are based on non-reactive alkane molecules, the polymers are not biodegradable. More than 80 percent of cities in the United States now have recycling programs. This reduces the impact on the community landfills and also reduces the need for crude-oil resources that are used to initially manufacture thermoplastic polymers.

Table 1 ***Resin ID codes***

Resin ID code	Type of polymer	Polymer name	Monomer(s)	Examples
1	Condensation	PET (polyethylene terephthalate)	ethylene glycol and terephthalic acid	soft drink bottles, peanut butter jars, fleece jackets, luggage, carpeting
2	Addition	HDPE (high density polyethylene)	ethylene (ethene)	milk, juice, water, and liquid detergent bottles; squeezable ketchup and syrup bottles
3	Addition	PVC (polyvinyl chloride)	vinyl chloride	credit cards, shampoo bottles, shower curtains, plumbing pipes, garbage bags
4	Addition	LDPE (low density polyethylene)	ethylene (ethene)	six-pack rings, bread bags, dry cleaner bags, grocery bags
5	Addition	polypropylene	propylene (1-propene)	ice scrapers, straws, yogurt containers, diaper linings, artificial joints, car parts
6	Addition	polystyrene	styrene	styrofoam, toys, CD/DVD cases, insulation, cafeteria trays, disposable utensils
7	Addition	*mixture*	*mixture*	plastic lumber, safety shields and glasses, headlight lenses

The following symbols are used as generic symbols for recycling or to indicate that the material is recyclable:

Density

Density is defined as mass per unit volume. Some plastics are more dense than water. When placed in water, these plastics sink. Other plastics are less dense than water and they float on water. A large amount of human-generated trash ends up in the world's oceans. Due to ocean currents, much of the trash in the sea is carried to a number of areas where the currents meet. The collections of trash in these locations have recently been referred to as marine trash islands. They are composed mainly of plastics that are less dense than water.

A recent investigation by the Australian Research Council Center of Excellence for Climate System Science found that "even if everyone in the world stopped putting garbage in the ocean today, giant garbage patches would continue to grow for hundreds of years.[1]" Because many plastics do not biodegrade quickly, the trash islands will be around for many years.

Gluep

Gluep is a gel polymer that exhibits a number of interesting properties. The polymerization of vinyl acetate forms polyvinyl acetate, which is found in Elmer's glue. The addition reaction forming the polyvinyl acetate is shown here.

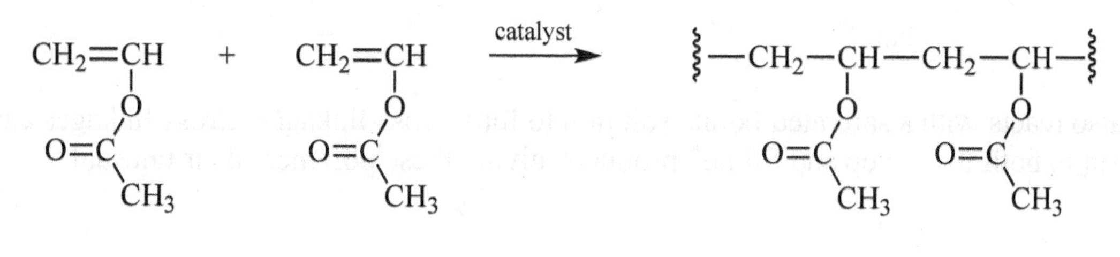

vinyl acetate monomers polyvinyl acetate

Repeating unit: $\left[-CH_2-CH(-O-C(=O)-CH_3)-\right]_n$

[1] McCauley, Lauren. "Newly Discovered 'Plastic Island' Shows Global Epidemic Worsening." *Common Dreams* 18 January 2013. Web. 21 October, 2013.

Polyvinyl acetate further reacts with a saturated borate solution to form cross-linkages that make the resulting gel. Cross-linking holds several polymer chains together in a random three-dimensional network of interconnected chains. Extensive cross-linking produces a substance that has more rigidity, hardness, and a higher melting point than the equivalent polymer without cross-linking.

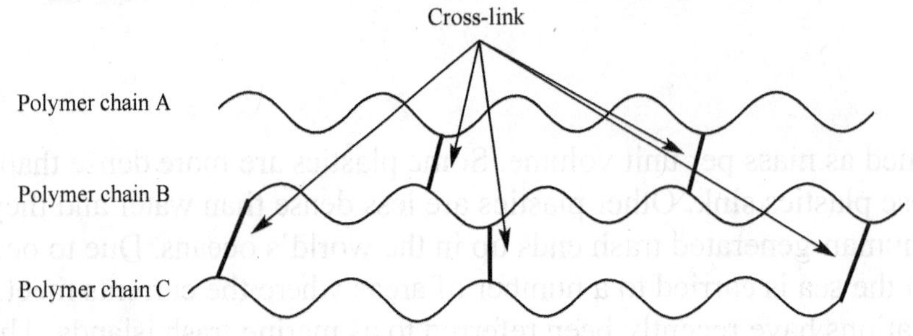

Slime®

Slime is also a gel polymer, but it differs from Gluep in that Slime® is synthesized from vinyl alcohol instead of vinyl acetate. The repeating unit of Slime® is

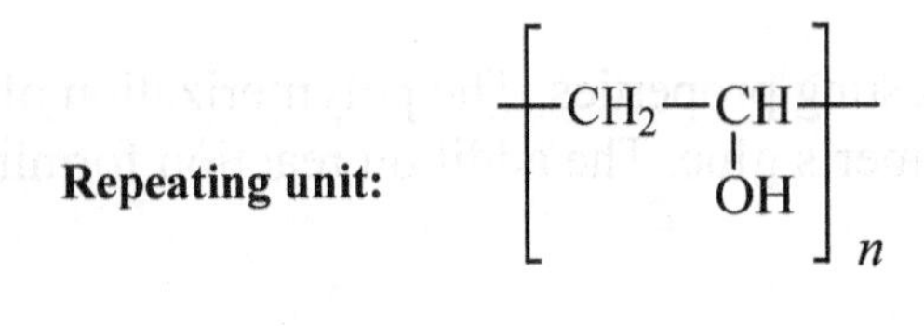

Slime®

This polymer also reacts with a saturated borate solution to form cross-linkages. Cross-linkages can break and reform in both the Gluep and Slime® products, giving these polymers their unusual properties.

Nylon

Nylon is a strong fiber used in clothing and hosiery, carpets, rope, parachutes, and surgical sutures. It was one of the first synthetic polymers and is an example of a condensation polymer. The amine ($-NH_2$) functional group on the first monomer reacts with the carboxylic acid ($-COOH$) functional group on the second monomer to form an amide bond (and water). Because the amine monomer has two amine functional groups (a diamine) and the carboxylic acid monomer has two carboxylic acid functional groups (diacid), the polymer chain may continue growing on both ends. There are many different types of nylon but all are based on diamine and diacid (or diacid chloride) monomers. The most important nylon is nylon 6,6, which is made by using 1,6-hexanediamine and hexanedioic acid. The name nylon 6,6 comes from the fact that each of the monomers has six carbon atoms.

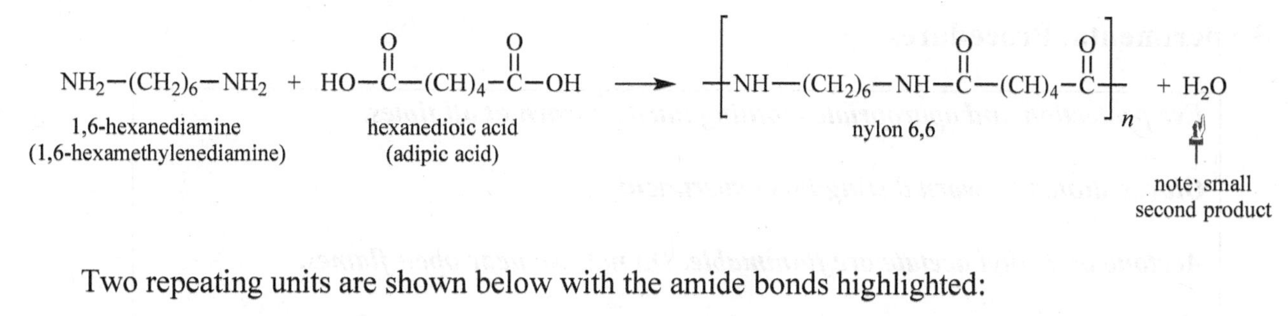

Two repeating units are shown below with the amide bonds highlighted:

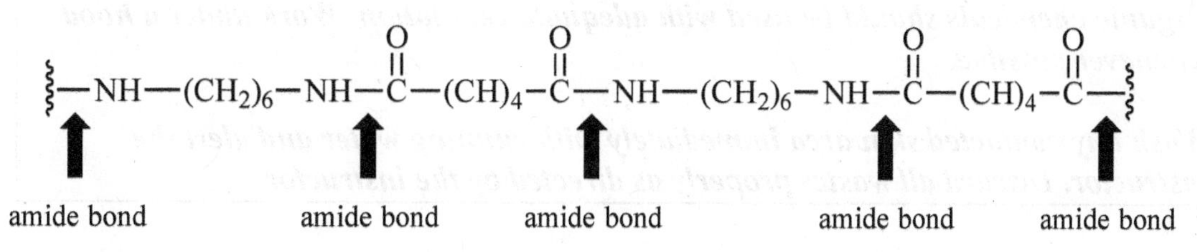

Another form of nylon, nylon 6,10, results from reacting hexamethylene diamine, which contains six carbons, and sebacoyl chloride, which contains ten carbons. The second product formed when the amide bond is made in this reaction is HCl.

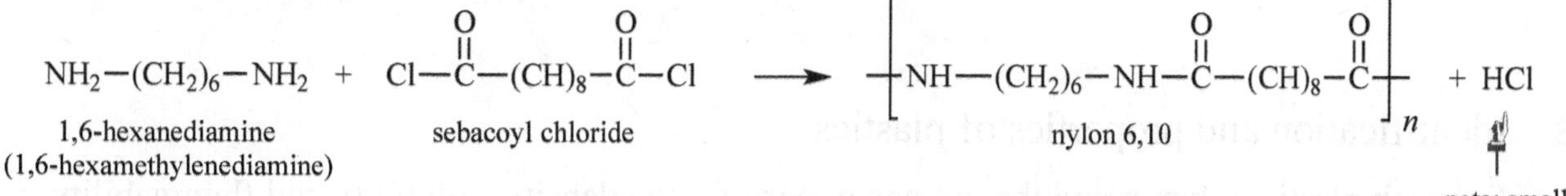

Experimental Procedures

Eye protection and appropriate clothing must be worn at all times.
Gloves should be worn during this experiment.
Acetone and ethyl acetate are flammable. Do not use near open flames.
Organic chemicals should be used with adequate ventilation. Work under a hood whenever possible.
Wash any contacted skin area immediately with running water and alert the instructor. Discard all wastes properly as directed by the instructor.

A. Recyclable plastics

1. Look at the display of plastic items. Find the resin ID code on each item.
2. Use Table 1 in the discussion to find more information.
3. Complete Table A on the report sheet.

B. Identification and properties of plastics

For each plastic polymer and the unknown, perform the density, solubility and flammability tests (steps 3 through 25). Read through all of the tests in this section before starting. Determine the most efficient way to use your time. For example, you may want to first setup the solubility tests and during the ten-minute waiting period, start the other tests. **No acetone or ethyl acetate can be in the work area when the combustion tests are being performed.**

1. Obtain five samples of each plastic polymer with resin ID codes #1, #2, #3, #4, #5, #6. Also obtain five samples of the unknown as assigned by the instructor.
2. In the table provided in the Report Sheet, identify the plastic with its resin ID code. Name the polymer and its monomer unit.
3. Measure 4 mL of water in a test tube. Use this volume as a reference level whenever 4 mL of solvent is needed. Obtain seven additional clean, dry test tubes.

Solubility in acetone

4. Pour approximately 30 mL of acetone into a 50 mL beaker. This should be enough to perform all of the solubility tests using acetone as a solvent. From the beaker, pour 4 mL of the acetone into each of the seven, clean, dry test tubes using the reference test tube as a guide.
5. Add a sample of the plastic polymer #1 to test tube #1. Add a sample of the plastic polymer #2 to test tube #2, etc. Add a sample of the unknown plastic polymer to test tube #7. Use a stirring rod to push each sample into the solvent.
6. Place the test tubes in a test tube rack and set them aside for 10 minutes.

7. After 10 minutes, use forceps to remove the samples one at a time from the test tubes and rinse with water.
8. Press the sample between your fingers and observe any changes that may have occurred. Changes may be subtle. Be sure to describe any changes in texture or appearance. For example, some organic molecules become soft, pliable, or mushy in organic solvents. Record your observations.
9. Discard the plastic samples into the "plastic waste" container. Pour the acetone from the test tubes into the "acetone/ethyl acetate waste" container as directed by the instructor.

Solubility in ethyl acetate

10. Pour approximately 30 mL of ethyl acetate into a 50 mL beaker. This should be enough to perform all of the solubility tests using ethyl acetate as a solvent. From the beaker, pour 4 mL of the ethyl acetate into each of the seven, clean, dry test tubes using the reference test tube as a guide.
11. Add a sample of the plastic polymer #1 to test tube #1. Add a sample of the plastic polymer #2 to test tube #2, etc. Add a sample of the unknown plastic polymer to test tube #7. Use a stirring rod to push each sample into the solvent.
12. Place the test tubes in a test tube rack and set them aside for 10 minutes.
13. After 10 minutes, use forceps to remove the samples one at a time from the test tubes and rinse with water.
14. Press the sample between your fingers and observe any changes that may have occurred. Changes may be subtle. Be sure to describe any changes in texture or appearance. For example, some organic molecules become soft, pliable, or mushy in organic solvents. Record your observations.
15. Discard the plastic sample into the "plastic waste" container. Pour the ethyl acetate from the test tube into the "acetone/ethyl acetate waste" container as directed by the instructor.

Density comparison test

16. Obtain two 100 mL beakers. Pour approximately 50 mL of tap water in the first beaker. In the second beaker, pour approximately 50 mL of 70% isopropyl alcohol.
17. Place one sample of the plastic polymer #1 into the beaker containing water. If the sample floats on the surface, use a stirring rod to try to push it below the surface. Observe whether the sample floats or sinks.
18. Place the second sample of the plastic polymer #1 into the beaker containing isopropyl alcohol. Use the same technique to determine whether the sample floats or sinks. Record your observations.
19. Remove the plastic samples and discard them into the "plastic waste" container as directed by the instructor.
20. Reuse the water and isopropyl alcohol solutions in the beakers to test the buoyancy for the remaining plastic polymers and the unknown.

Flammability test

Be sure that all acetone and ethyl acetate have been removed from the work area before lighting the Bunsen burner for this test.

21. Use forceps to hold the plastic polymer #1 in the flame of a Bunsen burner. Observe how quickly it melts and/or ignites. Note the color of the flame. Also, report any remaining residue. Record your observations.
22. Let the plastic residue cool and discard it in the "plastic waste" container.
23. Repeat the flammability test for the remaining plastic polymers and the unknown.

Identification of unknown

24. Compare all of the test results and determine the resin ID code **and** identity (polymer name and monomer name) of the unknown. Fill in the table in the appropriate places with this information.

C. Gluep

1. Obtain a Styrofoam cup marked with a 20 mL line. Fill this cup to the marked line with Elmer's® school glue.
2. Use a graduated cylinder to measure 20 mL of tap water. Use another graduated cylinder to measure 10 mL of the borate solution.
3. Do not start this step until you have all three solutions from C.1 and C.2 prepared. While stirring the glue in the Styrofoam cup with a stirring rod, pour the water into the cup containing the glue. **Do not add the glue to the water!**
4. Stir the mixture vigorously and pour the borate solution into the mixture.
5. Stir for an additional 10 minutes or until thoroughly mixed.
6. While wearing gloves, remove the polymer from the cup and knead (like bread dough) for 3–4 minutes.
7. Observe and record the following properties of the polymer, Gluep:
   - Odor
   - Texture
   - Slow stretchiness
   - Fast stretchiness
   - Form a ball and let it sit on the lab workbench
   - Form a ball and drop it on the lab workbench
   - Form a long roll and pull on the ends

D. Slime®

1. In a Styrofoam cup, add 20 mL of 4% polyvinyl alcohol and 5 mL of the borate solution.
2. Using a stirring rod, stir the mixture for 5 minutes or until thoroughly mixed.
3. While wearing gloves, remove the polymer from the cup and knead (like bread dough) for 3–4 minutes.
4. Observe and record the following properties of the polymer, Slime®:
   - Odor
   - Texture
   - Slow stretchiness
   - Fast stretchiness
   - Form a ball and let it sit on the lab workbench
   - Form a ball and drop it on the lab workbench
   - Form a long roll and pull on the ends

E. Nylon

Nylon will be prepared by the instructor as a demonstration. Students will obtain samples of the nylon to perform each test.

Instructor preparation of nylon: Wear gloves while performing this demonstration.

1. Prepare the rinse for step 9 by pouring approximately 25 mL of 50% ethanol into a 100 mL beaker. Set this beaker aside.
2. Pour 10 mL of Solution #1 (1,6-hexamethylenediamine and NaOH) into a 50 mL beaker.
3. Measure 10 mL of Solution #2 (4% sebacoyl chloride in hexane) in a graduated cylinder.
4. Tilt the 50 mL beaker containing Solution #1 and carefully add Solution #2 down the inside of the beaker. Avoid splashing or mixing the two solutions. They should form two distinct layers with the nylon layer forming at the interface of the two layers. Let the beaker sit undisturbed for a minute.
5. Obtain a large diameter test tube that will be used to wrap the nylon strand.
6. Use a pair of forceps to take hold of the center of the nylon layer and slowly pull the strand out of the solution beaker. Wrap the strand around the test tube and continue to roll the test tube while wrapping the strand around it. If the strand breaks, use forceps to start a new one.
7. Continue to draw out the nylon strand until one of the reactants is used up. This demonstrates the concept of "limiting reactant."
8. Rinse the nylon strand wrapped around the test tube under running tap water to remove any excess reactants.
9. Use forceps or a spatula to remove the nylon strand from the test tube and place it into the prepared 100 mL beaker containing 50% ethanol.
10. Rinse the nylon again under running tap water. Use forceps to straighten out the nylon strand and lay the nylon strand on paper towels to dry.

Student tests: Wear gloves while performing these tests.

Physical properties

1. Obtain approximately 15 cm of the nylon strand.
2. Observe the texture and appearance of the nylon. Record your observations.
3. Stretch the nylon by slowly pulling on the ends. Record your observations.

Solubility test

4. Cut five 3 cm pieces from the nylon strand.
5. Obtain and label four test tubes and place them in a test tube rack. Prepare the following:

Test tube #1	4 mL of 6 M HCl
Test tube #2	4 mL of 6 M NaOH
Test tube #3	4 mL of acetone
Test tube #4	4 mL of ethyl acetate

6. Place one piece of nylon into each test tube. Allow the nylon to sit in the solvent for 3 minutes and then observe any changes. Record your observations.
7. Dispose of the acetone and ethyl acetate in the waste bottles as directed by the instructor.

Flammability test

Be sure that all acetone and ethyl acetate have been removed from the work area before lighting the Bunsen burner for this test.

8. Light a Bunsen burner and use forceps to hold the remaining 3 cm strand of the nylon in the flame. Record your observations.
9. Dispose of the nylon strands into the "nylon waste" container.
10. Dispose of the acid and base solutions as directed by the instructor.

Name ______________________ Date ______________________

Partner(s) ______________________ Section ______________________

______________________ Instructor ______________________

1. Why do most plastics remain in landfills for such a long time?

2. What is a more familiar name for the "resin ID code?" What do the numbers indicate?

3. Describe three safety precautions that must be observed during this lab.

4. Water has a density of 1.00 g/mL and a 70% isopropyl alcohol solution has a density of 0.876 g/mL. What can you conclude about the density of a polymer if it floats on water, but sinks in the alcohol solution?

Name ______________________ Date ______________________

Partner(s) ______________________ Section ______________________

______________________ Instructor ______________________

A. Recyclable plastics

Item description	Resin ID code	Monomer	Polymer	Recyclable in your town?

B. Identification of plastics

Plastic item	Resin ID code	Polymer name	Monomer name
	1		
	2		
	3		
	4		
	5		
	6		
unknown			

B. Properties of plastics

Plastic item	Resin ID code	Acetone observations	Ethyl acetate observations	Density vs. water	Density vs. alcohol	Flammability
	1					
	2					
	3					
	4					
	5					
	6					
unknown						

C. Gluep

D. Slime®

Property	Gluep	Slime®
Odor		
Texture		
Slow stretchiness		
Fast stretchiness		
Ball—sitting		
Ball—dropping		
Thin roll—pull		

E. Nylon

Record your observations for the following properties of nylon.

Property	Observation
Texture	
Appearance	
Slow stretchiness	
HCl	
NaOH	
Acetone	
Ethyl acetate	
Flammability	

Problems and questions

1. The repeating unit of Teflon is shown below. Draw the structure of the monomer from which Teflon is made.

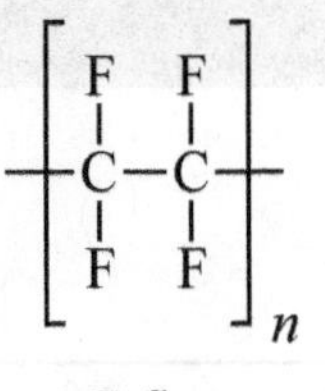

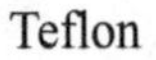

Teflon

2. Proteins and peptides are polymers in the body and their monomers are called amino acids. A general reaction for the formation of a protein is shown below. Are proteins addition polymers or condensation polymers?

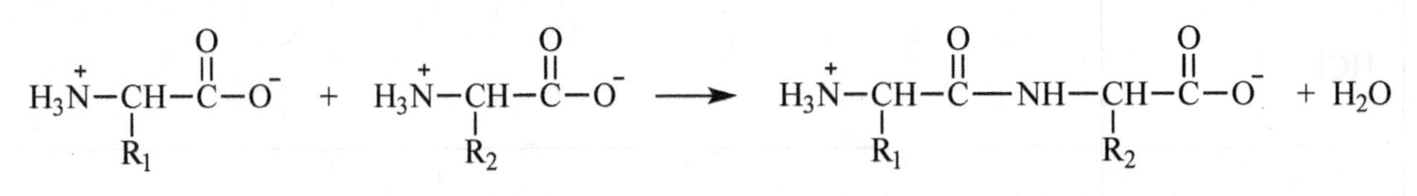

3. Which type of plastics would be best suited to make life preservers? Why?

4. Which type of plastic should <u>not</u> be used as a container for storing acetone? Why?

5. Nylon is a tough, strong, flexible material that is resistant to chemicals. However, which property tested in this lab makes nylon unsuitable for use in pajamas?

6. The natural fibers cotton and wool are fire resistant. Why do most mattress manufacturers avoid using synthetic fibers?

7. What are the similarities and differences between Gluep and Slime®?

Goals

- Identify physical properties of organic compounds.
- Identify chemical properties of organic compounds.
- Build molecular models of organic compounds.
- Identify functional groups and organic families.

Materials

- Gloves
- Molecular model kits
- *Merck Index* or *CRC Handbook of Chemistry and Physics*
- Benzoic acid
- Cyclohexane
- Toluene
- Sodium chloride
- Potassium iodide
- Evaporating dish
- Wood splint
- Matches, lighter, or Bunsen burner
- Test tubes
- Test tube rack

Discussion

In the early history of chemistry (about 200 years ago), chemists categorized compounds from living organisms as *organic*, and those obtained from minerals as *inorganic*. These terms are still used to classify compounds, but they no longer reflect their historical origins. All organic compounds contain carbon. *Hydrocarbons* are compounds that contain only carbon and hydrogen. Other organic compounds may also contain oxygen, nitrogen, sulfur, halogens, and occasionally other elements. The immense number of different organic compounds results from the ability of carbon to bond to other carbon atoms as well as to other elements.

A few compounds containing carbon generally are not studied as a part of organic chemistry. These include carbon dioxide (CO_2), and the ionic compounds containing the carbonate ion (CO_3^{2-}) and bicarbonate ion (HCO_3^-).

The physical and chemical properties of organic compounds depend upon the elements and their specific arrangement with respect to each other. The covalent bonds found in organic compounds and the ionic bonds found in inorganic compounds explain many of the differences in the physical and chemical properties of the compounds. The following table compares some of the properties of organic compounds with those of inorganic compounds.

Table 1 ***General comparison of organic and inorganic properties (*some exceptions exist)***

Organic compounds	Inorganic compounds
Covalent bonds (nonpolar or polar covalent)	Ionic, but some are polar covalent
Soluble in water only if polar groups are present	Many are soluble in water
Soluble in nonpolar solvents if few or no polar groups are present	Insoluble in nonpolar solvents
Usually low melting and boiling points Gases, liquids, or solids at 25°C	Usually high melting and boiling points Most are solids at 25°C
Often strong odors	Usually no odors
Non- or weak electrolytes (poor conductors of electricity)	Can be nonelectrolytes, weak electrolytes, or strong electrolytes (many are good conductors of electricity)
Flammable	Nonflammable

Solubility

Many organic compounds are nonpolar molecules because they contain nonpolar covalent bonds. Using the general rule "like dissolves like," nonpolar organic molecules will be soluble in nonpolar solvents such as cyclohexane. Nonpolar organic molecules are not soluble in water because water is a polar molecule containing polar covalent bonds. Organic compounds that have polar groups in their molecular structure will be soluble in water if they are small molecules. However, size is only one factor in predicting water solubility of organic molecules. There are other factors that complicate the issue from a simple yes (soluble) or no (insoluble).

Inorganic compounds contain ionic or polar covalent bonds. Many inorganic compounds are soluble in water because water is a polar molecule containing polar covalent bonds. Ionic or polar covalent bonds are not attracted to nonpolar solvents and they will not be soluble in those solvents ("like dissolves like").

Flammability

Many organic compounds react with oxygen in a reaction called combustion. The *combustion reaction* results in the formation of carbon dioxide and water as products. The following balanced equations show the combustion reactions of two common fuels. Methane is the major component of natural gas and octane is found in gasoline.

$$\underset{\text{methane}}{CH_{4(g)}} + 2\,O_{2(g)} \longrightarrow CO_{2(g)} + 2\,H_2O_{(l)} + \text{heat}$$

$$\underset{\text{octane}}{2\,C_8H_{18}(g)} + 25\,O_{2(g)} \longrightarrow 16\,CO_{2(g)} + 18\,H_2O_{(l)} + \text{heat}$$

Organic functional groups

Organic families are classified by their functional groups. A *functional group* is an atom or group of atoms that has a certain structural feature. The functional group gives similar physical and chemical properties to all compounds in that family. Many of the organic families will be studied in more detail in later labs. The ability to identify the functional group of a molecule is important, and can often be facilitated with the use of models. Molecular model kits are helpful when learning organic chemistry since they are designed to show the correct number of bonds and the correct bond angles for each element. The kits are the easiest way to observe the three-dimensional structure of the molecules.

Carbon is the principle element in organic compounds. It is located in the second row of the periodic table; its four outer shell electrons and an intermediate electronegativity make it a perfect candidate to form covalent bonds. Carbon always obeys the octet rule and it always forms four covalent bonds. There are four ways carbon can form four covalent bonds:

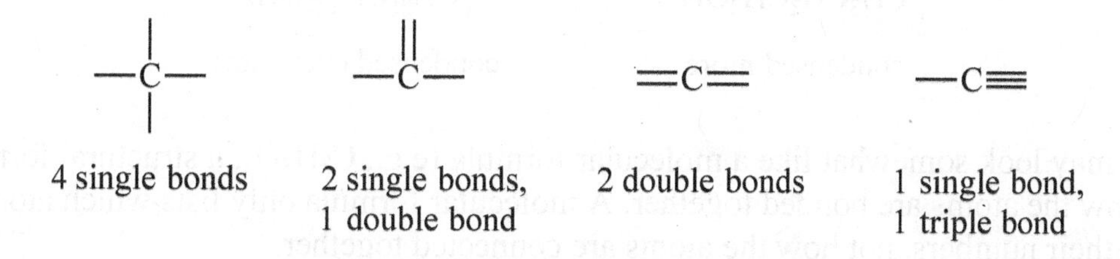

Because the other elements frequently found in organic compounds have nonbonding electrons (except hydrogen), they form a different number of covalent bonds. The acronym "HONC" refers to the typical number of covalent bonds these elements make (H makes one, O makes two, N makes three, and C makes four).

When organic chemists "draw" a molecule on a piece of paper, the drawing is a representation of the molecule. Because molecules are three-dimensional and the paper is only two-dimensional, the drawing is a simplified picture of the molecule. Typically, organic chemists use four common methods to "draw" molecules. Each type of drawing has strengths and weaknesses and organic chemists use whichever type clearly illustrates the point they want to make.

The first type is called the *complete structural formula.* In a complete structural formula, all the covalent bonds are shown. A bond (shared pair of electrons) is represented by the line connecting two atoms. Some atoms have nonbonding electrons (lone pairs). Generally, the lone pair electrons are not shown, unless needed for some reason. In the complete structural formula shown below, the oxygen's two lone pair electrons have been hidden.

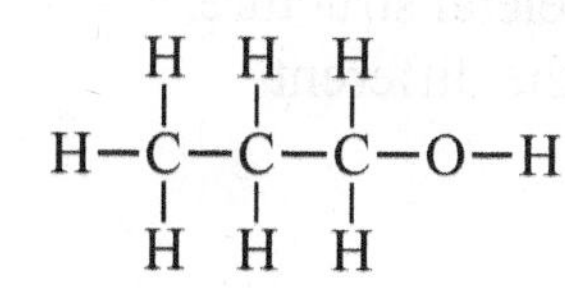

In this drawing, it is clear how the atoms are connected together. Notice that each carbon makes four bonds. However, this type of drawing uses considerable space, so organic chemists devised a method to condense the structure into a smaller space.

In a *condensed structural formula*, some, many or all of the covalent bonds are left out. Subscripts indicate the number of identical groups attached to a particular atom. Hydrogen only makes one bond, so that bond can be left out and the H placed adjacent to the atom it is bonded to. The condensed structural formula for the molecule shown above is:

$CH_3—CH_2—CH_2—OH$

The carbon on the left is bonded to three hydrogens and it has been condensed to $CH_3–$. The two middle carbons are each bonded to two hydrogens and they have been condensed to $–CH_2–$. This type of drawing uses much less space than the complete structural formula. There are successive levels of simplification and the above drawing can be condensed further by removing the bonds between the carbons and the carbon and the oxygen. One final simplification can be made because there are two CH_2's in a row.

$CH_3CH_2CH_2OH$	$CH_3(CH_2)_2OH$
condensed more	condensed even more

Although it may look somewhat like a molecular formula (e.g., C_3H_8O), a structural formula still preserves how the atoms are bonded together. A molecular formula only lists which atoms are present and their numbers, not how the atoms are connected together.

The objective is to draw the molecule in such a way that it is clear how the atoms are connected together. If a drawing is condensed too much, then it becomes hard to understand and the clarity suffers. For example, the structure shown below on the left has been condensed too much. Although it accurately represents how the atoms are bonded together, it is not very clear. The slightly less condensed structure of the same molecule on the right is much more understandable.

$(CH_3CH_2)_2C(CH_3)_2$

```
              CH3
              |
CH3—CH2—C—CH2—CH3
              |
              CH3
```

The third type of drawing is a *skeletal structure*, also known as a *line-bond drawing*. In this type of drawing, all the carbon atoms and all the hydrogen atoms that are bound to carbon atoms are omitted. It is understood that a carbon atom is present at every point where two lines meet and at the ends of the lines, unless it is labeled as some other kind of atom. Consider the following molecule, shown in its complete structural formula, condensed structural formula, and skeletal structure, respectively. The carbons have been numbered to make them easier to spot in the different representations.

```
  H  H  H  H
  |1 |2 |3 |4
H-C--C--C--C-H
  |  |  |  |
  H  H  H  H
```

```
 1      2      3      4
CH3—CH2—CH2—CH3
```

(skeletal structure: zigzag line with carbons numbered 1, 2, 3, 4)

Both ends of the zigzag line represent carbon atoms because they have not been labeled as something different. All of the internal positions where two lines intersect are carbons as well. In deciphering the skeletal structure, it can be seen that carbon 1 is bonded to carbon 2; however, carbon makes four bonds. Three bonds are "missing;" they must be to hydrogens that have been omitted. Therefore, carbon 1 is a CH_3–. Similarly, carbon 2 is bonded to carbon 1 and carbon 3. Two bonds are "missing" and so the remaining two bonds must be to hydrogens, making carbon 2 a –CH_2–. Finally, carbons 3 and 4 can be recognized as a –CH_2– and a –CH_3, respectively.

In the example below, the atom at the left end (number 3) is understood to be a carbon. But the atom at the right end it is no longer a carbon. It has been labeled as something different (an –OH). Both structures have three carbons (labeled 1–3).

$$\overset{3}{CH_3}-\overset{2}{CH_2}-\overset{1}{CH_2}-OH$$

(skeletal structure: 3, 2, 1, OH)

Because the skeletal structure provides few details, other representations will be used to develop familiarity. However, the skeletal structure may prove valuable for more complex structures.

Sometimes it's important to know where atoms point in space. *Wedge and dash notation* is similar to an artist's trick to show the 3-dimensional arrangement of the atoms in the molecule. In this notation, the solid wedge (▬) means that the atom at the end of the wedge is closer to the viewer (comes out of the paper). The atom at the end of the dash (IIIIIIII···) is farther from the viewer (goes into the paper). Bonds drawn with a line are in the same plane as the paper.

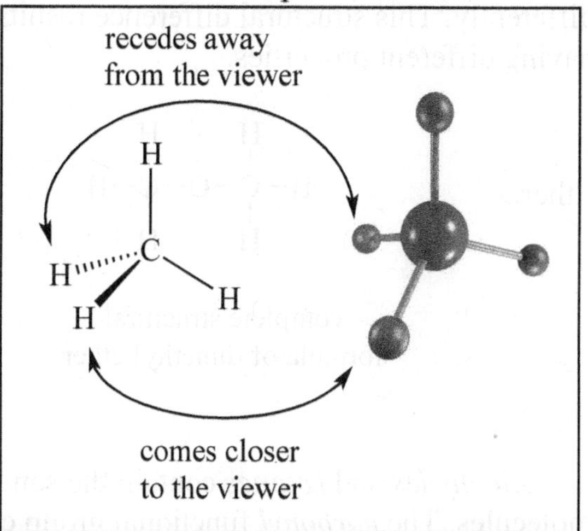

Classifying functional groups

The organic family known as *alkanes* includes saturated hydrocarbons, those containing only carbon and hydrogen atoms with all single bonds. However, if there is one or more double or triple bond between carbon atoms in the molecule, it is called *unsaturated.* The following examples show complete and condensed structural formulas for two types of unsaturated hydrocarbons: *alkenes* (containing a double bond between two carbons) and *alkynes* (containing a triple bond between two carbons).

Alkene:	$H_2C=CH_2$ (each C bonded to two H)	$CH_2=CH_2$
	complete structural formula of ethene	condensed structural formula of ethene
Alkyne:	$H-C\equiv C-H$	$CH\equiv CH$
	complete structural formula of ethyne	condensed structural formula of ethyne

Alcohols contain the functional group –OH, known as a *hydroxyl group*. The example below is the alcohol known as ethanol or ethyl alcohol and is found in alcoholic beverages. Notice that the functional group shows that the oxygen atom is bonded between a carbon atom and a hydrogen atom.

Alcohol:

```
    H  H
    |  |
H—C—C—O—H
    |  |
    H  H
```

$CH_3—CH_2—OH$

complete structural formula of ethanol

condensed structural formula of ethanol

Ethers also contain an oxygen atom, but it is bonded between two carbon atoms. The following example shows the molecule known as dimethyl ether. Notice that it has the same number of carbon, hydrogen, and oxygen atoms as the ethanol molecule shown above, but they are arranged differently. This structural difference results in the two molecules being in different families and having different properties.

Ether:

```
   H     H
   |     |
H—C—O—C—H
   |     |
   H     H
```

$CH_3—O—CH_3$

complete structural formula of dimethyl ether

condensed structural formula of dimethyl ether

Aldehydes and *ketones* contain the same functional group but in different positions in the molecules. The *carbonyl* functional group consists of a double bond between a carbon and an oxygen (C=O). *Aldehydes* have the carbonyl group on the end of a molecule. *Ketones* have the carbonyl group bonded between two other carbons. Once again, the structural difference in the placement of the functional group results in different properties between the two families. The condensed structural formulas for an aldehyde and a ketone each containing three carbons are shown below. Notice the different positions of the carbonyl groups in the molecules.

```
               O
               ||
CH3—CH2—C—H
```

Aldehyde: propanal

```
         O
         ||
CH3—C—CH3
```

Ketone: propanone

Amines contain a nitrogen atom. There are three classifications of amines, based upon the number of carbon groups attached to the nitrogen. The following examples show a primary (1°) amine, a secondary (2°) amine, and a tertiary (3°) amine. Notice that nitrogen only forms three bonds and as the number of carbon groups attached to the nitrogen changes, the number of hydrogen atoms attached to the nitrogen also changes.

$CH_3—NH_2$	$CH_3—NH(CH_3)$	$CH_3—N(CH_3)—CH_3$
primary amine (1°)	secondary amine (2°)	tertiary amine (3°)

Experimental Procedures

Eye protection and appropriate clothing must be worn at all times. Students must wear gloves.

Organic chemicals are flammable. Take precautions when using Bunsen burners or other open flames.

Organic chemicals have strong odors. Use under a hood with proper ventilation. Use the proper wafting techniques when observing the odors.

Discard all wastes properly as directed by the instructor.

A. Physical properties

Appearance and odor

1. Label six test tubes and place in a test tube rack. Put the following substances in the test tubes. If the substance is a solid, use a few crystals, about the size of a match head. If the substance is a liquid, use about 10 drops.
   - Test tube #1.......water
   - Test tube #2.......potassium iodide
   - Test tube #3.......sodium chloride
   - Test tube #4.......benzoic acid
   - Test tube #5.......cyclohexane
   - Test tube #6.......toluene
2. Record the formula for each sample on the Report Sheet.
3. Observe each sample and record its physical appearance.
4. Use the proper wafting technique to note the odor of each sample. To check for an odor, cup your hand and gently fan the odor from the test tube toward your face. Never smell or sniff a chemical directly with your nose. Record the results for each test tube.
5. Use the chemical handbook to look up the melting point for each sample. Record the information.
6. Identify and record each sample as "organic" or "inorganic."
7. Identify and record the type of bonding in each sample as nonpolar covalent, polar covalent, or ionic.
8. Dispose of the chemicals in the proper waste containers as directed by the instructor.

Solubility

9. Label five test tubes and place in a test tube rack. Prepare each test tube as follows using cyclohexane as the non-polar solvent and water as the polar solvent. Carefully observe any layers that might form. Watch the chemicals as you add them to the test tube to determine which chemical forms the top layer and which forms the bottom layer. Record your observations.

	Solvent	Solute
Test tube #1	10 drops of cyclohexane	10 drops of water
Test tube #2	15 drops of cyclohexane	a few crystals of NaCl
Test tube #3	15 drops of cyclohexane	10 drops of toluene
Test tube #4	15 drops of water	a few crystals of NaCl
Test tube #5	15 drops of water	10 drops of toluene

10. Record the solubility of the solute in each solvent. Identify each solute as organic or inorganic.
11. Dispose of the chemicals in the proper waste containers as directed by the instructor.

B. Chemical properties

Flammability

Do not panic if flames appear.
They will quickly extinguish.

1. Place a match-tip size sample of NaCl in an evaporating dish. Use a match, lighter, or Bunsen burner as directed by the instructor to ignite a wood splint and hold it on the NaCl. Observe and record the results.
2. Place 5 drops of cyclohexane in another evaporating dish. Relight the same wood splint and hold it on the cyclohexane. Observe and record the results.
3. Note any color change of the substance and/or flame color.
4. Identify each compound as organic or inorganic.

C. Bond formation

Follow the directions given by the instructor regarding the use of the molecular model kits.

1. Record the color and number of bonds for each element.
2. Record the procedure for forming single bonds and multiple bonds.

D. Molecular models

1. Make models of each compound listed on the Report Sheet table and have the models checked by the instructor.
2. Draw the complete structural formula for each compound listed.
3. Circle any functional groups on the complete structural formulas.
4. Record the name of the family for each compound.
5. Return all of the model pieces to the correct containers.

Name ______________________ Date ______________________

Partner(s) ______________________ Section ______________________

______________________ Instructor ______________________

1. Consider the differences between organic and inorganic compounds. Which type of compounds are more likely to be soluble in water? Why?

2. Convert the following formulas into **condensed** structural formulas:

a) 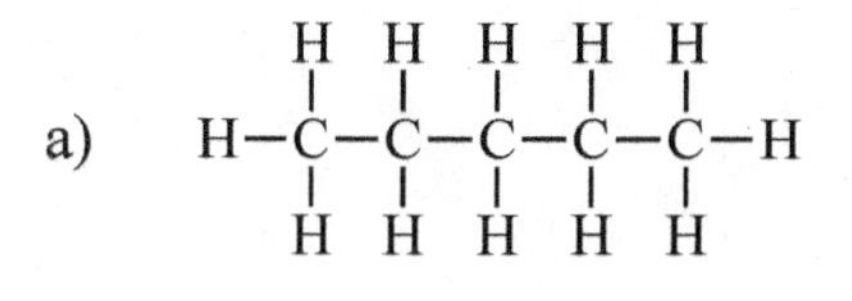

b)

3. Circle and identify the functional groups present in dihydroxyacetone, a component of several artificial tanning lotions.

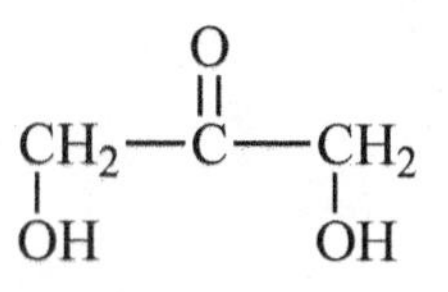

[*Reminder: there are two more problems on the next page*]

4. What structural feature is present in an unsaturated hydrocarbon that is not present in a saturated hydrocarbon?

5. Write the balanced equation for the combustion of butane (C_4H_{10}).

Name ______________________ Date ______________________

Partner(s) ______________________ Section ______________________

______________________ Instructor ______________________

A. Physical properties

Complete the following table for these physical properties.

Compound	Formula	Physical appearance (color, state of matter)	Odor	Melting point	Organic or inorganic?	Type of Bonding
Water						
Potassium iodide						
Sodium chloride						
Benzoic acid	$C_7H_6O_2$					
Cyclohexane	C_6H_{12}					
Toluene	C_7H_8					

Complete the following table for the solubility tests.

Test tube	Solvent	Solute	Did layers form? If yes, identify top and bottom layers	Is solute soluble or insoluble?	Is solute organic or inorganic?
#1					
#2					
#3					
#4					
#5					

B. Chemical properties

Complete the following table for the flammability tests.

Compound	Observations	Organic or inorganic?
NaCl		
Cyclohexane		

C. Bond formation

Complete the following table.

Element	Model color	Number of bonds formed
Carbon		
Hydrogen		
Oxygen		
Nitrogen		

Describe how to use the molecular model kit pieces to distinguish a single bond from a double or triple bond.

D. Molecular models

Complete the following table for the molecular models.

Condensed structural formula	Complete structural formula (circle any functional groups)	Organic family (functional group)	Instructor's approval of model
CH_4			
$CH_3{-}CH_3$			
$CH_2{=}CH_2$			
$CH_3{-}OH$			
$CH_3{-}CH_2{-}O{-}CH_3$			
$CH_3{-}NH_2$			
$CH_3{-}NH{-}CH_3$			

Problems and questions

1. Would you expect a solid that is odorless and soluble in water but not flammable to be an organic or inorganic compound? Why?

2. Circle the following that do not meet the "bonding requirement" for carbon atoms.

   a) two single bonds and a double bond

   b) a single bond and two double bonds

   c) three single bonds and a triple bond

   d) two double bonds

3. Write the balanced equation for the combustion of propane (C_3H_8).

4. Circle the following in which the two representations are **not** the same molecule.

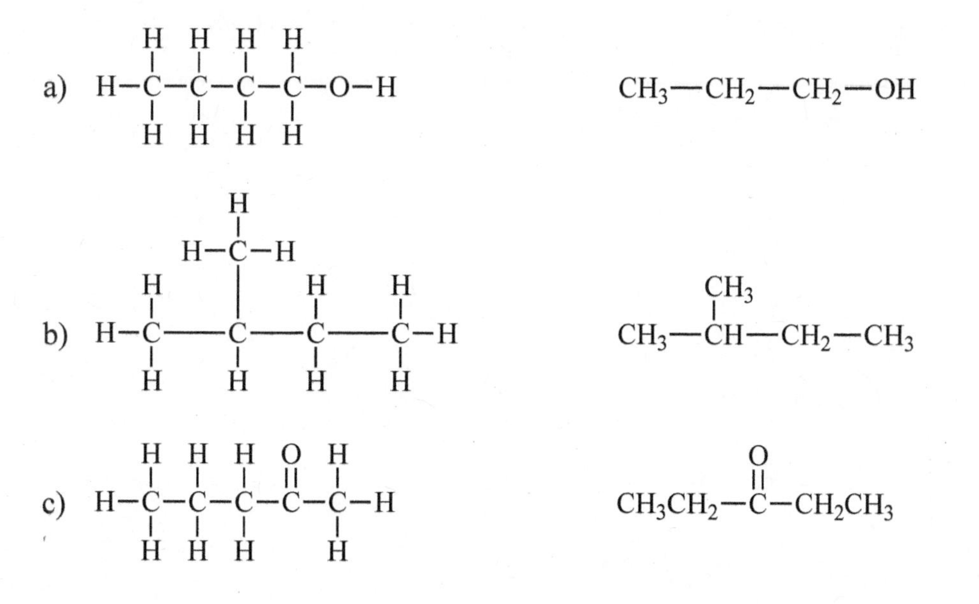

5. Predict the solubility of the following substances in water by writing *soluble* or *insoluble*:

vegetable oil	__________	ammonium nitrate	__________
potassium iodide	__________	polyethylene	__________

Goals

- Build molecular models of alkanes, cycloalkanes, and haloalkanes.
- Draw three-dimensional, complete, and condensed structural formulas for alkanes, cycloalkanes, and haloalkanes.
- Write IUPAC names for alkanes, cycloalkanes, and haloalkanes.
- Identify isomers of compounds with the same molecular formulas.
- Identify physical properties of alkanes using a reference book.

Materials

- Molecular model kits
- *Merck Index* or *CRC Handbook of Chemistry and Physics*

Discussion

The basic structure of all organic compounds centers around the electron arrangement of the carbon atom. Carbon has four valence electrons and it always obeys the octet rule. Thus, it always forms four covalent bonds. VSEPR theory indicates that when carbon makes four single bonds, the molecular geometry is tetrahedral. A molecular model kit will have the carbon bonds in this tetrahedral arrangement and will allow four bonds to be formed around the carbon. Similarly, the model kit will allow hydrogen to form one bond. When the hydrogens and carbons are put together to form the desired molecule, the resulting model will show the shape by describing the spatial arrangement of the atoms.

The simplest organic family known as *alkanes* includes saturated hydrocarbons, those containing only carbon and hydrogen atoms with all single bonds. It is useful to make models of compounds to help visualize the actual three-dimensional shape of a molecule from its written two-dimensional formula. The chemical and physical properties of a compound are dependent upon the three-dimensional shape of the molecule.

Methane is the simplest alkane. The molecule contains one carbon atom and four hydrogen atoms. The following formulas show three different ways to represent methane.

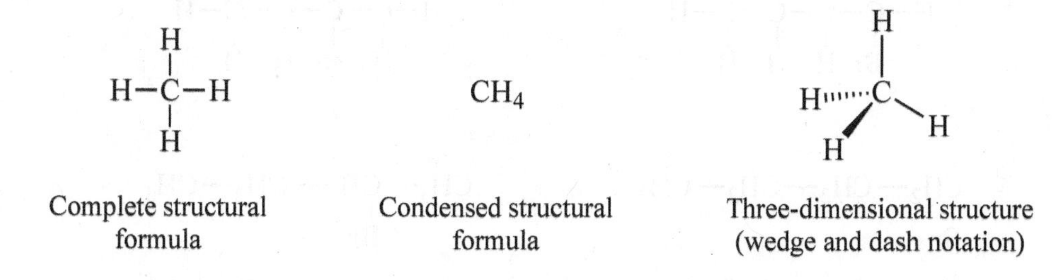

A more complex molecule is butane. Butane is an alkane with the molecular formula C_4H_{10}. Its condensed structural formula is $CH_3CH_2CH_2CH_3$. Although the condensed structural formula makes it appear as if the molecule is linear, building a molecular model shows it has a more complex shape. Because atoms may rotate around a single bond, a variety of three-dimensional shapes result when the atoms are rotated; three of the shapes are shown below. In the wedge and dash notation, the solid wedge (▬) means that the atom at the end of the wedge is closer to the viewer (comes out of the paper). The atom at the end of the dash (ıııııı···) is farther from the viewer (goes into the paper). Bonds drawn with a line are in the same plane as the paper.

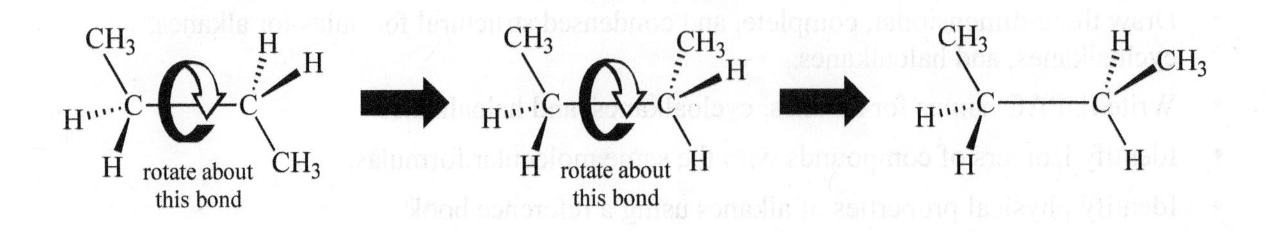

Another compound that has four carbons and ten hydrogens is named 2-methylpropane. It is a *constitutional isomer* of butane since it has the same molecular formula (C_4H_{10}) but a different arrangement of atoms. Each isomer has its own set of physical and chemical properties. No amount of bending or twisting will result in butane and 2-methylpropane having the same spatial three-dimensional arrangement. As the number of carbons increases in organic molecules, the number of possible isomers also increases.

$$CH_3-CH_2-CH_2-CH_3$$

butane

$$CH_3-\underset{}{\overset{CH_3}{\overset{|}{CH}}}-CH_3$$

2-methylpropane

Haloalkanes, also known as *alkyl halides*, are alkanes that have one or more hydrogens substituted with a halogen atom. There are two naming systems used for many of these compounds. While all compounds can be named with the IUPAC system, common names are frequently used as well. For example, the IUPAC name of $CHCl_3$ is trichloromethane but its common name is chloroform.

The placement of the halogen atom(s) gives different constitutional isomers. The following examples show the difference between 1-bromobutane and 2-bromobutane. Both isomers are written in complete and condensed structural formulas.

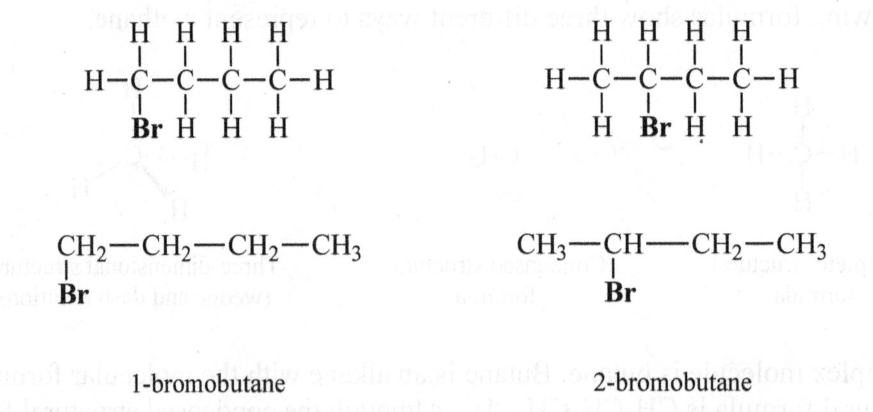

1-bromobutane 2-bromobutane

Cycloalkanes have a ring structure in the molecule. The formation of the ring requires the loss of two hydrogens. For example, the molecular formula for butane is C_4H_{10} but the molecular formula for cyclobutane is C_4H_8. The complete structural formula, condensed structural formula, and skeletal structure for cyclobutane are shown in the following example. It is important to remember that in a skeletal structure, all the carbon atoms and all the hydrogen atoms that are bound to carbon atoms are omitted. It is understood that a carbon atom is present at every point where two lines meet. Thus, each corner represents carbon with enough hydrogens attached to provide the carbon atom with four bonds.

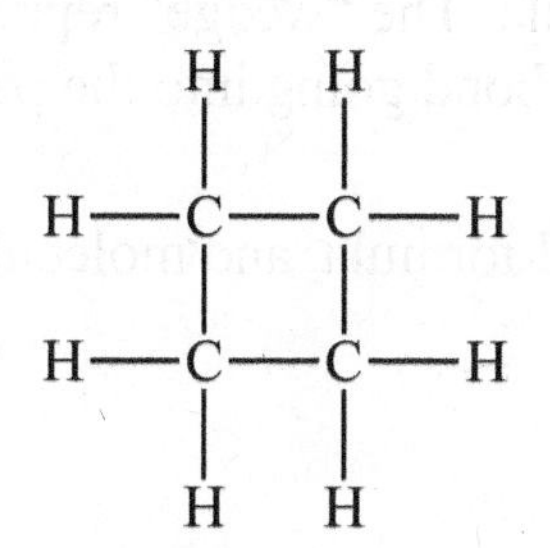

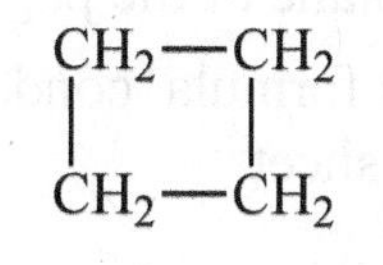

complete structural formula   condensed structural formula   skeletal structure

Experimental Procedures

Goggles are not required for this lab.

A. Alkanes

1. Use molecular model kits to build the models for methane, ethane and propane. Have the instructor approve the models.
2. For methane, draw the three-dimensional tetrahedral formula. The "wedge" represents a bond coming out of the page and the "dash" represents the bond going into the page. Solid lines represent bonds in the plane of the page.
3. Draw the complete structural formula, condensed structural formula, and molecular formula for each model on the report sheet.

B. Constitutional isomers

1. Build the models for butane and 2-methyl propane. Notice that the models cannot be twisted or turned to become the same molecule. Demonstrate to the instructor how the butane model must be changed in order to make a 2-methylpropane model. Have the instructor approve the models.
2. Make the three constitutional isomers with the molecular formula C_5H_{12}. Draw their condensed structural formulas and write the IUPAC name for each structure. Have the instructor approve the models and check the isomers.
3. Make the four constitutional isomers with the molecular formula $C_3H_6Cl_2$. Draw their condensed structural formulas and write the IUPAC name for each structure. Have the instructor approve the models and check the isomers.

C. Cycloalkanes

1. Make the five constitutional isomers with the molecular formula C_5H_{10}. Note: because these are cycloalkanes, each one must contain a ring of carbon atoms. Draw their condensed structural formulas and write the IUPAC name for each structure. Have the instructor approve the models and check the isomers.
2. Make both of the constitutional isomers with the molecular formula $C_3H_4Cl_2$. Note: because these are cycloalkanes, each one must contain a ring of carbon atoms. Draw their condensed structural formulas and write the IUPAC name for each structure. Have the instructor approve the models and check the isomers.
3. Return all of the model pieces to the correct containers.

D. Physical properties of constitutional isomers

1. Draw the condensed structural formulas for the compounds listed on the report sheet.
2. Following the instructor's directions, use a chemistry handbook to look up the physical properties for each compound and record the information on the report sheet. Include the appropriate units.

PRE-LAB QUESTIONS:

Name ______________________ Date ______________________

Partner(s) ______________________ Section ______________________

______________________ Instructor ______________________

1. What are the characteristics that identify a molecule as an alkane (elements present, type of bonds)?

2. Can there be constitutional isomers for methane? Why or why not?

3. If two alkane constitutional isomers have the same number of carbons and hydrogens, how do they differ?

4. When drawing cycloalkanes, what is the difference between a condensed structural formula and a skeletal structure?

Name ______________________ Date ______________________

Partner(s) ______________________ Section ______________________

______________________ Instructor ______________________

A. Alkanes

Compound	Molecular formula	Complete structural formula	Condensed structural formula	Instructor's approval of models
methane (draw the three-dimensional structure here)				
ethane				
propane				

B. Constitutional isomers

Compound	Condensed structural formulas	Instructor's approval of models
butane		
2-methylpropane		
C_5H_{12}	Draw all the constitutional isomers and label each with the correct IUPAC name.	
$C_3H_6Cl_2$	Draw all the constitutional isomers and label each with the correct IUPAC name.	

C. Cycloalkanes

Compound	Condensed structural formulas	Instructor's approval of models
C_5H_{10}	Draw all the constitutional isomers and label each with the correct IUPAC name.	
$C_3H_4Cl_2$	Draw all the constitutional isomers and label each with the correct IUPAC name.	

D. Physical properties of constitutional isomers

Compound	Condensed structural formula	Molar Mass	Boiling point	Density
butane				
2-methylpropane (also known as *isobutane*)				
pentane				
2-methylbutane (also known as *isopentane*)				
2,2-dimethylpropane (also known as *neopentane*)				

Compare the last three compounds in the table above. Which property is identical and which properties are different? Why?

Problems and questions

1. Draw the condensed structural formulas and write the IUPAC names for all of the constitutional isomers of C_6H_{14}.

2. Draw the condensed structural formulas and write the IUPAC names for the five cycloalkane constitutional isomers of C_5H_{10}.

3. Circle the two compounds below that have the same molar mass. Will they have the same boiling point? Explain.

   2-methylpentane 3-methylhexane cyclohexane 3,3-dimethylpentane

Goals

- Build molecular models of alkanes, alkenes, alkynes, and aromatics.
- Sketch the 3D structures of simple alkanes, alkenes, alkynes, and aromatics.
- Differentiate the molecular geometry of cyclohexane, cyclohexene, and benzene.
- Distinguish among saturated, unsaturated, and aromatic compounds by observing chemical properties.
- Interpret observations to identify an unknown.
- Write chemical equations for combustion, addition, and substitution reactions involving hydrocarbons.

Materials

- Gloves
- Molecular model kits
- Evaporating dish
- Matches, lighter, or Bunsen burner
- Wood splint
- Test tubes
- Test tube rack
- Methanol
- Cyclohexane
- Cyclohexene
- Toluene
- Unknowns (#1, #2, #3)
- 1% Bromine solution in dichloromethane
- 1% $KMnO_4$ solution
- UV light

Discussion

Hydrocarbons are organic compounds that consist of only carbon and hydrogen atoms. Many hydrocarbons are important fuels. For example, methane is the principal component of natural gas, propane is used in gas grills, and gasoline is a mixture of hydrocarbons. In the presence of oxygen gas, hydrocarbons burn quite efficiently and release a great deal of energy.

Hydrocarbons are further classified as *aliphatic* or *aromatic* hydrocarbons. Aliphatic hydrocarbons do not contain the benzene ring, whereas aromatic hydrocarbons contain the benzene ring as their building block. The aliphatic hydrocarbons are further subdivided into three families: alkanes, alkenes, and alkynes. The various families of hydrocarbons differ in the types of carbon-carbon bonds that are present, which also causes differences in their chemical reactivity. The types of carbon-carbon bonds present in the various hydrocarbons and examples of each are:

Alkanes	Alkenes	Alkynes	Aromatics
C−C All single bonds (saturated)	C=C Double bonds (unsaturated)	C≡C Triple bonds (unsaturated)	A resonance structure of alternating single and double bonds in a six-sided ring (unsaturated)
H H H−C−C−H H H ethane	H H C=C H H ethene (ethylene)	H−C≡C−H ethyne (acetylene)	benzene

Hydrocarbons are also classified as saturated or unsaturated. In saturated hydrocarbons, all of the carbon-carbon bonds are single bonds. Alkanes are saturated hydrocarbons and each carbon atom has four single bonds. Unsaturated hydrocarbons have at least one double or triple bond between two carbons. Alkenes, alkynes, and aromatic hydrocarbons are unsaturated.

Benzene (C_6H_6) is an aromatic hydrocarbon with six carbon atoms in a ring. Each carbon has one hydrogen atom covalently bonded to it. Often benzene is depicted as a ring of six carbon atoms with alternating double and single bonds. In reality, the electrons are distributed equally around the carbon atoms. The following structures show different representations of benzene. Though the third structure provides the most realistic representation of benzene, all three structures are used interchangeably.

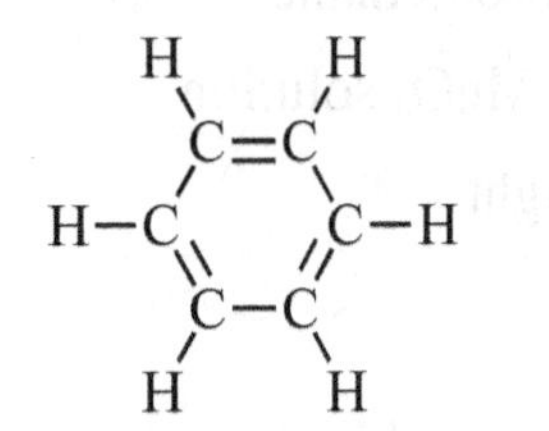

complete structural formula of benzene

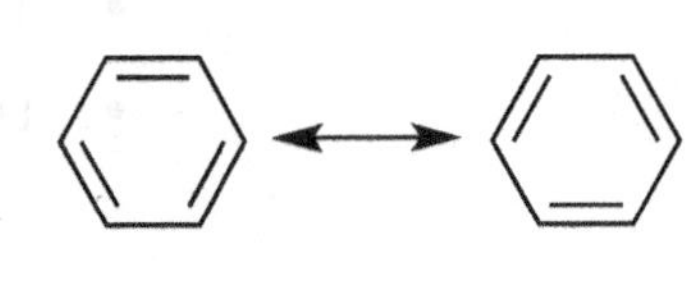

two sketetal structures (resonance structures) of benzene

alternative notation for benzene

Combustion

All hydrocarbons undergo combustion reactions. *Combustion* refers to the burning of a compound in the presence of oxygen. The products of hydrocarbon combustion are carbon dioxide (CO_2) and water (H_2O).

The efficiency with which an individual hydrocarbon burns will vary. Generally, alkanes burn with a clean, bluish flame. Substances with higher levels of unsaturation produce smokier flames. Aromatic hydrocarbons tend to burn less rapidly, with a yellow flame, and form soot (elemental carbon), in addition to CO_2 and H_2O. The following balanced equations show the combustion reactions for propane (a common fuel) and butane (lighter fluid).

$$C_3H_8(g) + 5\,O_2(g) \longrightarrow 3\,CO_2(g) + 4\,H_2O(g) + \text{heat}$$
propane

$$2\,C_4H_{10}(g) + 13\,O_2(g) \longrightarrow 8\,CO_2(g) + 10\,H_2O(g) + \text{heat}$$
butane

Halogenation

A halogenation reaction occurs when a halogen molecule, usually chlorine or bromine, reacts with a hydrocarbon. There are two types of halogenation reactions: substitution and addition. The various families of hydrocarbons differ in the types of carbon-carbon bonds that are present and this, in turn, makes them different in their chemical reactivity. Alkenes and alkynes are much more reactive than alkanes and aromatic compounds. This experiment focuses on the reactions of alkanes, alkenes, and aromatic compounds (alkynes react very similarly to alkenes). The difference in reactivity is most evident in their chemical reactivity towards the bromine molecule, Br_2.

Alkanes and benzene rings undergo *substitution* reactions because one atom of the halogen substitutes (replaces) one hydrogen atom on the alkane or benzene. An additional product of the reaction is a hydrogen halide molecule. Hydrocarbon substitution reactions are **slow** and require heat or ultraviolet (UV) light such as sunlight. The benzene substitution reaction also requires a catalyst such as Fe^{3+}. The following examples show the bromination of ethane and benzene.

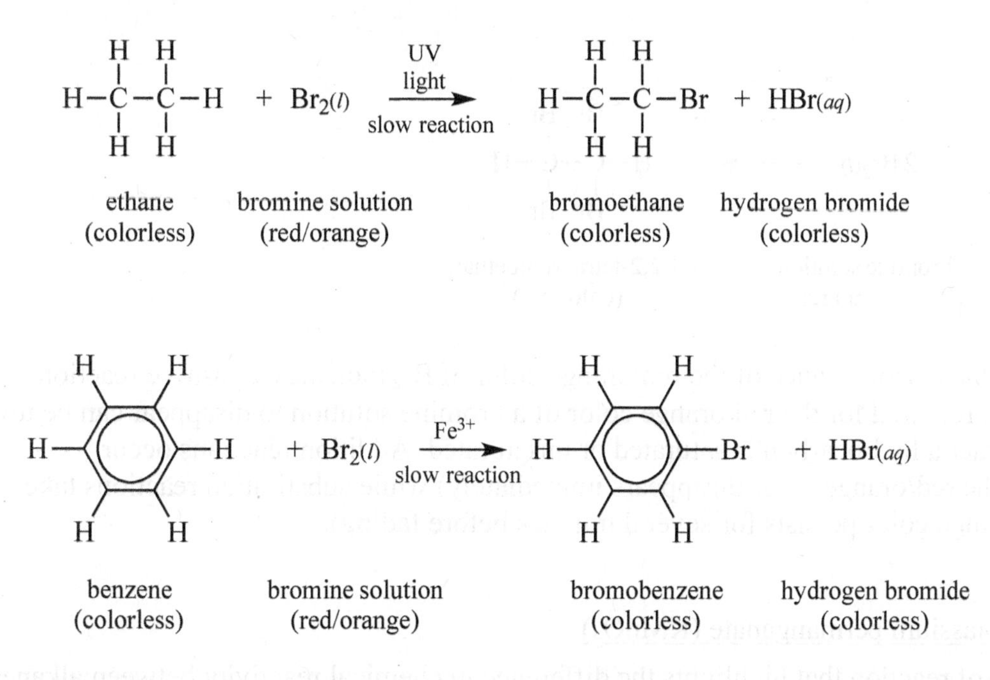

The above examples showed a monobromination reaction; that is, one bromine atom replaced one hydrogen atom. However, experimentally, halogenation reactions are uncontrolled. With excess bromine, eventually all hydrogen atoms are replaced and a mixture of products results. The bromination of methane is shown below. Note that four organic products result.

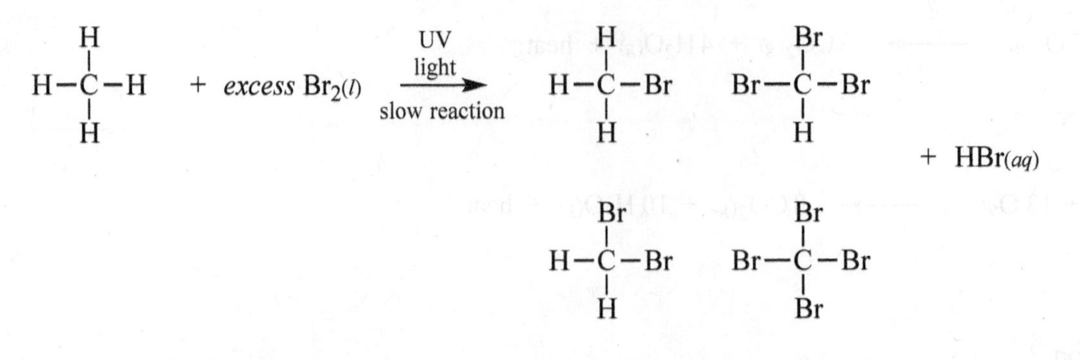

More complex molecules with more carbons (and more hydrogens) result in many more products. For example, ethane (C_2H_6) gives nine products.

Conversely, alkenes and alkynes undergo *addition* reactions. The unsaturated double or triple bond rapidly reacts with the halogen molecule. Both halogen atoms are added to the molecule when one of the multiple bonds breaks. The following examples show the bromination of ethene and ethyne. Notice that there is only one product.

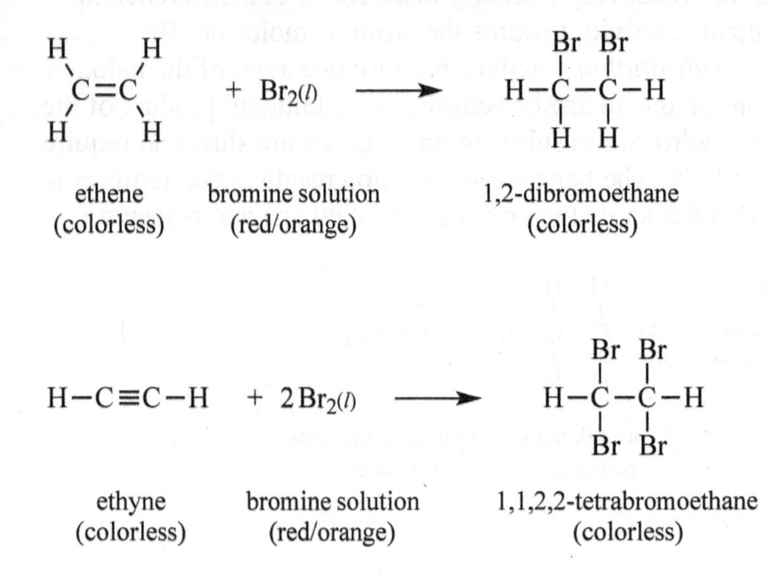

In summary, the disappearance of the red/orange color of Br_2 indicates a positive reaction. However, the time required for the red/orange color of a bromine solution to disappear can be used to determine whether a hydrocarbon is saturated or unsaturated. Addition reactions occur instantaneously (the red/orange color disappears immediately) while substitution reactions take longer (the red/orange color persists for several minutes before fading).

Oxidation with potassium permanganate ($KMnO_4$)

Another type of reaction that highlights the difference in chemical reactivity between alkanes, alkenes, alkynes, and aromatics is oxidation with $KMnO_4$. Alkenes and alkynes react with $KMnO_4$ while alkanes and aromatic compounds do not. In this reaction, oxidation involves the gain of oxygen atoms by the organic molecule. When the alkene is oxidized to form a diol (an organic compound containing two alcohol groups), the Mn^{7+} ion is reduced to Mn^{4+} and forms a brown-black precipitate, MnO_2. Alkenes react instantaneously with $KMnO_4$ (a purple solution). The disappearance of the purple color and the appearance of a solid, brown MnO_2 precipitate indicate a positive reaction. The reaction below only shows the structures of the important reactants and products and their colors.

$$H_2C{=}CH_2 + KMnO_{4(aq)} \longrightarrow HOCH_2{-}CH_2OH + MnO_{2(s)}$$

ethene	potassium permanganate	1,2-ethanediol	manganese(IV) oxide
(colorless)	(purple)	(colorless)	(brown-black)

Experimental Procedures

Eye protection and appropriate clothing must be worn at all times.

All hydrocarbons are highly flammable. Keep chemicals away from open flames. Exercise caution when performing combustion tests.

Bromine is caustic. Avoid breathing the fumes. Work under the hood. Wear gloves. Wash any contacted skin immediately with running water and alert the instructor.

$KMnO_4$ stains the skin. Handle with care.

Discard all wastes properly as directed by the instructor.

A. Molecular models

1. Use molecular model kits to build models for ethane, ethyne, cyclopropane, cyclopropene, butane, and benzene. Have the instructor approve the models.
2. Write the molecular formula for each molecule beneath its name in the table A. The first one has been done for you.
3. Using the molecular model kit, build a bromine molecule.
4. Use the model structures to demonstrate the reaction of **one** bromine molecule with each molecule in the table. Have the instructor approve each demonstration.
5. Using condensed structures, write the complete chemical reaction for the bromination of each molecule in the table. For each reaction, include any catalysts and proper reaction conditions. The first one has been done for you. In addition, classify each reaction as an addition or substitution reaction.

B. Combustion

Parts B, C, and D require an unknown compound. Use the same unknown for all three parts.

1. Work in the hood. Test only one sample at a time. Keep all other samples away from the open flame.
2. Place 5 drops of methanol in a clean, dry evaporating dish. Use a match, lighter, or Bunsen burner as directed by the instructor to light a wood splint. Keep the Bunsen burner away from all chemicals. Carefully ignite the methanol sample with a burning wood splint. Observe how quickly the sample ignites and burns, the color of the flame, and whether there is smoke or soot present. Methanol is not a hydrocarbon since it contains an oxygen atom, but it gives a good contrast to the hydrocarbon samples.
3. Record your observations.
4. Clean and dry the evaporating dish between each sample and repeat the combustion test with cyclohexane, cyclohexene, toluene, and the unknown.

C. Bromine test

1. Work in the hood and wear gloves.
2. Label five clean, dry test tubes and place them in a test tube rack.
3. Put 15 drops of the following solutions into the labeled test tubes:
   Test tube #1 cyclohexane
   Test tube #2 cyclohexene
   Test tube #3 toluene
   Test tube #4 unknown (same unknown as part B)
   Test tube #5 empty (control for color sample)
4. Add 3 drops of bromine solution to each test tube. Carefully tap the outside of each test tube with your finger to mix.
5. Hold each test tube in front of a piece of white paper. The white background makes it easier to observe any color change. Record the time it takes for the red/orange bromine color to disappear.
6. If the reaction mixture remains red, place the test tube near a UV light source or up to a window for several minutes. Observe any color changes compared to the control sample bromine solution in test tube #5.

D. Potassium Permanganate Test

1. Label five clean, dry test tubes and place them in a test tube rack.
2. Put 15 drops of the following solutions into the labeled test tubes:
   Test tube #1 cyclohexane
   Test tube #2 cyclohexene
   Test tube #3 toluene
   Test tube #4 unknown (same unknown as part B and C)
   Test tube #5 empty (control for color sample).
3. Add 10 drops of 1% $KMnO_4$ solution to each test tube. Carefully tap each test tube with your finger to mix.
4. Let each test tube set for one minute. Note any color change or formation of a precipitate. Record your observations.

E. Identification of the unknown

1. For your unknown, summarize the results from your tests.
2. Identify whether the hydrocarbon unknown is an alkane, an alkene, or an aromatic compound. The unknowns will contain one of these functional groups but they will not be the same as the known samples.

Name ____________________ Date ____________________

Partner(s) ____________________ Section ____________________

____________________ Instructor ____________________

1. Explain the difference between saturated and unsaturated hydrocarbons.

2. Draw the distinguishing structural feature for the following:

   alkanes

   alkenes

   alkynes

   aromatics

3. Toluene is the common name of methylbenzene. Draw the structure of toluene (C_7H_8).

4. Toluene is classified as an aromatic compound. It has the potential to react with bromine, Br_2. Will this be an addition or substitution reaction? What reaction conditions are required?

[*Reminder: there are more problems on the next page*]

5. Draw a condensed structure to replace the italicized reactant **and** complete the chemical equation for each of the following reactions:

a) *ethene* + $Br_{2(l)}$ $\longrightarrow$

b) *propane* + $Br_{2(l)}$ $\xrightarrow[\text{dark}]{\text{cool and}}$

c) *propane* + $Br_{2(l)}$ $\xrightarrow{\text{sunlight}}$

6. A student has two test tubes labeled A and B, each containing a colorless liquid. The student adds several drops of a potassium permanganate solution to each one. The solution in test tube A turned purple, while the solution in test tube B turned brown. Which test tube, A or B, contains an alkane?

7. Why must all open flames be extinguished before performing the bromine and potassium permanganate tests?

Name ______________________ Date ______________________

Partner(s) ______________________ Section ______________________

______________________ Instructor ______________________

A. Molecular models

Compound	Balanced equation for reaction with bromine
ethane C_2H_6 ____ Instructor's approval	$CH_3{-}CH_3 + Br_{2(l)} \xrightarrow[\text{slow reaction}]{\text{UV light}} CH_3{-}CH_2{-}Br + HBr(aq)$ ____ Instructor's approval for reaction demonstration Addition or substitution? **Substitution reaction**
ethyne ____ Instructor's approval	____ Instructor's approval for reaction demonstration Addition or substitution?____________
cyclopropane ____ Instructor's approval	____ Instructor's approval for reaction demonstration Addition or substitution?____________
cyclopropene ____ Instructor's approval	____ Instructor's approval for reaction demonstration Addition or substitution?____________
butane ____ Instructor's approval	____ Instructor's approval for reaction demonstration Addition or substitution?____________
benzene ____ Instructor's approval	____ Instructor's approval for reaction demonstration Addition or substitution?____________

B. Combustion

Compound	Combustion observations
methanol (CH_3OH)	
cyclohexane (C_6H_{12})	
cyclohexene (C_6H_{10})	
toluene (C_7H_8)	
Unknown # ________	

C. Bromine test

Compound	Observations without UV light	Observations with UV light
cyclohexane (C_6H_{12})		
cyclohexene (C_6H_{10})		
toluene (C_7H_8)		
Unknown # ________		

D. Potassium permanganate test

Compound	Observations
cyclohexane (C_6H_{12})	
cyclohexene (C_6H_{10})	
toluene (C_7H_8)	
Unknown # ________	

E. Identification of the unknown

For your unknown, summarize the results from your tests.

Combustion:

Bromine:

Potassium permanganate:

By comparing the results of the above tests on the unknown with the results of the known samples, identify whether the hydrocarbon unknown is an alkane, an alkene, or an aromatic compound. The unknowns will contain one of these functional groups but they might not be the same as the known samples.

Problems and questions

1. Explain why methanol is an organic compound but is not considered a hydrocarbon.

2. Examine your observations from part B. Which of the hydrocarbon samples (methanol is not a hydrocarbon) had the cleanest burning flame? Which had the most smoke or soot? Why did these hydrocarbons burn differently?

3. Write the complete balanced equation for the combustion of the following compounds:

   a) methanol

   b) cyclohexane

   c) cyclohexene

   d) toluene

4. Classify each reaction in part C as an addition reaction or substitution reaction. Use the observations from the "Bromine test table" in part C to explain your reasoning. Did the observation for toluene's reaction with bromine support your prediction in prelab question #4? Why or why not?

[*Reminder: there is one more problem on the next page*]

5. Complete the following reactions:

a)  [cyclobutene structure] $+ Br_2 \longrightarrow$

b) $CH_3{-}CH_2{-}CH{=}CH{-}CH_3 + Cl_2 \longrightarrow$

c) 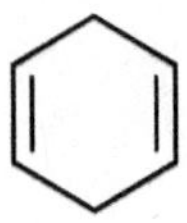 [1,4-cyclohexadiene structure] $+ 2Br_2 \longrightarrow$

Goals

- Build molecular models of alcohols and phenol.
- Identify physical properties of alcohols and phenol.
- Use chemical tests to identify primary, secondary, and tertiary alcohols and phenol.
- Use physical properties and chemical tests to identify an unknown.
- Write balanced equations for oxidation reactions of alcohols.

Materials

- Gloves
- Molecular model kits
- Test tube rack
- Test tubes
- Stirring rod
- pH paper and color charts
- Ethanol
- 1-Butanol
- 2-Butanol
- 2-Methyl-2-propanol
- 1-Octanol
- Phenol
- Unknowns (#1, #2, #3)
- Chromate solution
- 1% $FeCl_3$

Discussion

Organic families are classified by their functional groups. A *functional group* is an atom or group of atoms that has a certain structural feature. The functional group gives similar physical and chemical properties to all compounds in that family. One such functional group is a *hydroxyl* group, –OH. *Alcohols* are organic compounds that contain a hydroxyl group attached to a hydrocarbon, while *phenols* are organic compounds that have the hydroxyl group attached to a benzene ring.

Alcohols

There are three classes of alcohols: primary (1°), secondary (2°), and tertiary (3°). Each class is distinguished by the number of "R" groups attached to the carbon bearing the hydroxyl group. This carbon is often referred to as the alcohol carbon. "R" indicates either an alkyl group or an aromatic group. The following examples show the general formula for each alcohol class.

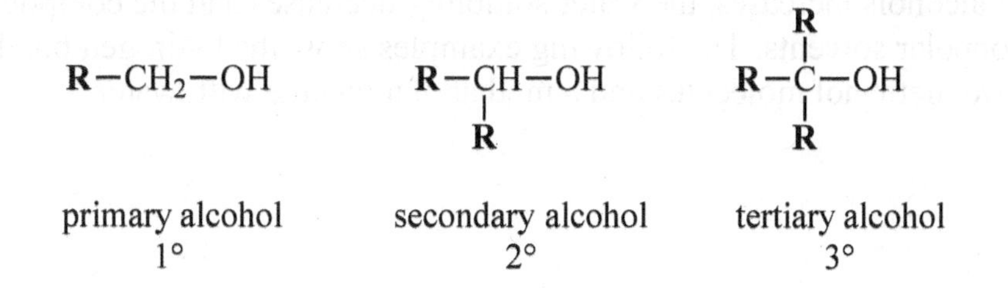

Alcohols are widely used in solvents, drugs, preservatives, and disinfectants. The following table shows the complete structural formulas, condensed structural formulas, classifications, IUPAC names and common names for some alcohols.

Table 1 ***Alcohol examples***

Complete structural formula	Condensed structural formula	Classification	IUPAC name	Common name
H–C(H)(H)–C(H)(H)–OH	CH_3-CH_2-OH or CH_3CH_2OH	1°	ethanol	ethyl alcohol or "grain alcohol"
H–C(H)(H)–C(H)(O–H)–C(H)(H)–H	$CH_3-CH(OH)-CH_3$ or $CH_3CH(OH)CH_3$	2°	2-propanol	isopropyl alcohol or "rubbing alcohol"
H–C(H)(H)–C(CH₃: H–C–H with H)(O–H)–C(H)(H)–H	$CH_3-C(CH_3)(OH)-CH_3$ or $CH_3C(CH_3)(OH)CH_3$	3°	2-methyl-2-propanol	*tert*-butyl alcohol or *t*-butyl alcohol

Physical properties of alcohols

An alcohol molecule is polar and can form hydrogen bonds with other alcohol molecules. Therefore, the boiling points of alcohols are higher than the boiling points of alkanes of similar molecular masses. The polarity of alcohols also allows the formation of hydrogen bonds with water molecules. Therefore alcohols containing five or less carbons are generally soluble in water. As the molecular mass of alcohols increases, the water solubility decreases and the compounds become more soluble in nonpolar solvents. The following examples show the hydrogen bond (dashed line) formed between two methanol molecules and a methanol molecule with water.

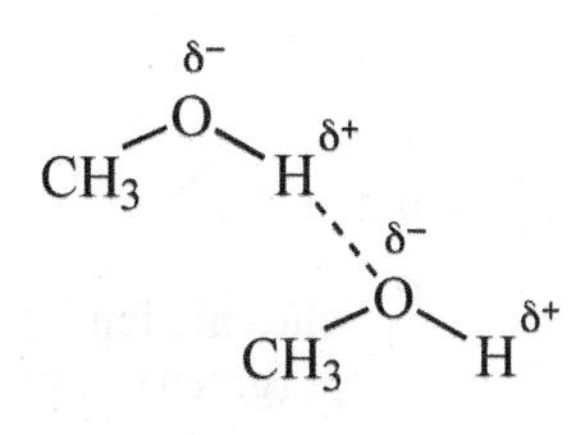

hydrogen bonding between two alcohol molecules

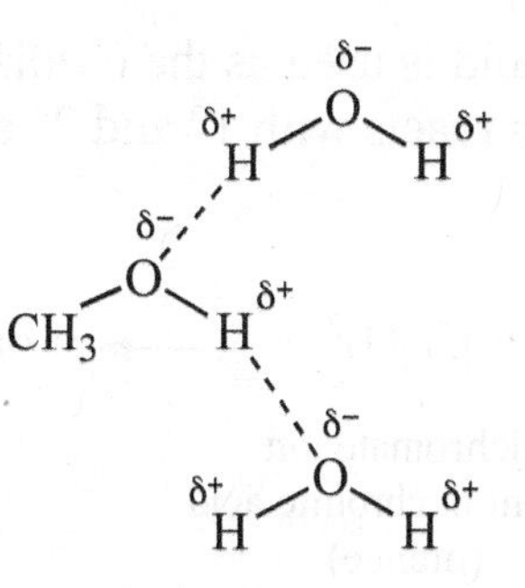

hydrogen bonding between an alcohol molecule and water

Chemical properties of alcohols

The oxidation reactions of alcohols can be used to distinguish among primary, secondary, and tertiary compounds. There are a number of different reagents available so the symbol [O] written over the arrow is frequently used to indicate an oxidizing reagent. The following reactions show the oxidation products of alcohols. The net effect is that two hydrogens are removed from the alcohol. One comes from the –OH group and the other comes from the carbon atom bonded to the –OH group. Tertiary alcohols do not undergo oxidation because they do not have a hydrogen on the carbon atom bonded to the –OH group.

$$R{-}CH_2{-}OH \xrightarrow{[O]} R{-}\overset{\overset{\displaystyle O}{\|}}{C}{-}H \xrightarrow{[O]} R{-}\overset{\overset{\displaystyle O}{\|}}{C}{-}OH$$

primary alcohol — aldehyde — carboxylic acid

$$R{-}\overset{\overset{\displaystyle R}{|}}{C}H{-}OH \xrightarrow{[O]} R{-}\overset{\overset{\displaystyle O}{\|}}{C}{-}R$$

secondary alcohol — ketone

$$R{-}\underset{\underset{\displaystyle R}{|}}{\overset{\overset{\displaystyle R}{|}}{C}}{-}OH \xrightarrow{[O]} \text{no reaction}$$

tertiary alcohol

When chromic acid is used as the oxidizing agent, a distinct color change is observed. The chromic acid solution reacts with 1° and 2° alcohols to form the green chromic ion as shown in the following reaction:

1° or 2° alcohol + $Cr_2O_7^{2-}$ ⟶ oxidized alcohol product + Cr^{3+}

dichromate ion present in chromic acid (orange) — chromic ion (green)

Notes:

- If a primary alcohol is allowed to completely oxidize to a carboxylic acid, a greenish-brown final product may be observed.
- Tertiary alcohols cannot be oxidized. This lack of color change can be used to identify an unknown alcohol as a tertiary alcohol.
- When the chromic acid test is performed on phenol compounds, a brownish-black tar-like product may be observed.

Phenols

While phenols and alcohols both have the hydroxyl group (–OH) as their functional group, the attachment of the –OH onto a benzene ring results in different properties for phenols. In some ways, phenols are similar to alcohols and in others they are so different that they seem to belong to a different class of compounds. Phenols are more acidic than alcohols because the hydrogen atom on the hydroxyl group is slightly ionized in water. Even though phenol has six carbons, it is soluble in water.

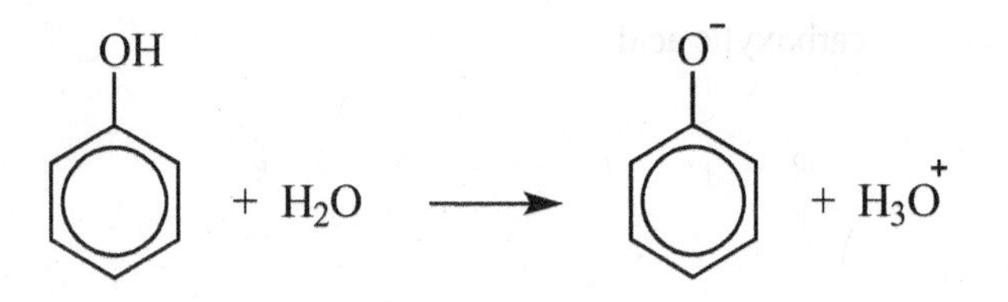

Concentrated solutions of phenols are toxic and can cause severe skin burns. However, phenol derivatives, such as thymol and resorcinol, are less dangerous and are used as antiseptics and are sometimes found in cough drops.

Phenols can be distinguished from alcohols by observing the color change that occurs when phenol reacts with the Fe^{3+} ion in the ferric chloride ($FeCl_3$) aqueous solution according to the following reaction:

Phenol + Fe^{3+} ⟶ iron(III) • phenol complex

colorless — yellow — purple

Experimental Procedures

Eye protection and appropriate clothing must be worn at all times.

Wear gloves when handling chromate solution because it contains concentrated sulfuric acid. Wash any contacted skin immediately with running water and alert the instructor.

Discard all wastes properly as directed by the instructor.

A. Molecular models

1. Use molecular model kits to build models for methanol, 2-butanol, 2-methyl-2-butanol, cyclohexanol, and phenol. Have the instructor approve the models. Write the condensed structural formula for each model. Classify each model as 1°, 2°, or 3° if applicable.

B. Properties of alcohols and phenols

All three tests in part B (odor, pH, and solubility) require an unknown compound. Use the same unknown for all three tests.

Odor

1. Label seven test tubes and place in a test tube rack. Prepare each test tube by adding 1 mL (20 drops) of the solution listed below:
   - Test tube #1 ......ethanol
   - Test tube #2 ......1-butanol
   - Test tube #3 ......2-butanol
   - Test tube #4 ......2-methyl-2-propanol
   - Test tube #5 ......1-octanol
   - Test tube #6 ......phenol
   - Test tube #7 ......unknown (as assigned by the instructor)
2. Use the proper wafting technique to note the odor of each sample. To check for an odor, cup your hand and gently fan the odor from the test tube toward your face. Never smell or sniff a chemical directly with your nose. Record the results for each test tube. **Save these solutions for the next step.**

pH

3. Dip a clean, dry stirring rod into the first test tube and touch it to a strip of pH paper. Compare the color of the paper to the color chart on the container and record the pH of the solution on the report sheet. Repeat this procedure for each test tube. **Save these solutions for the next step.**

Solubility

4. Add 2 mL of water to each test tube.
5. Use a clean, dry stirring rod to mix the solution and water in each test tube. If the solution is soluble in water the solution will be clear with no layers. If the solution is insoluble in water, the solution will appear cloudy or two layers will form. Notice if some, but not all, of the alcohol dissolved. Record your observations for each test tube.
6. **Dispose of all solutions as directed by the instructor.**

C. Oxidation of alcohols

1. Label seven clean, dry test tubes and place them in a test tube rack.
2. Put 10 drops of the following solutions into the labeled test tubes:
   - Test tube #1.......ethanol
   - Test tube #2.......1-butanol
   - Test tube #3.......2-butanol
   - Test tube #4.......2-methyl-2-propanol
   - Test tube #5.......phenol
   - Test tube #6.......unknown (**use the same unknown as part B**)
   - Test tube #7.......water (control—used to compare original color)
3. Add 3 drops of the chromate solution to each test tube. Carefully tap the outside of each test tube with your finger to mix.
4. Hold each test tube in front of a piece of white paper. The white background makes it easier to observe any color change. A color change from orange to bluish-green within two minutes is a positive test for the oxidation of an alcohol. Record your observations.
5. Write the balanced equations for any oxidation reactions that occurred. Write "NR" if no reaction occurred.
6. **Dispose of all solutions as directed by the instructor.**

D. Ferric chloride test

1. Label seven clean, dry test tubes and place them in a test tube rack.
2. Put 5 drops of the following solutions into the labeled test tubes:
   - Test tube #1.......ethanol
   - Test tube #2.......1-butanol
   - Test tube #3.......2-butanol
   - Test tube #4.......2-methyl-2-propanol
   - Test tube #5.......phenol
   - Test tube #6.......unknown (**use the same unknown as parts B and C**)
   - Test tube #7.......water (control—used to compare original color)

3. Add 5 drops of 1% $FeCl_3$ solution to each test tube. Carefully tap the outside of each test tube with your finger to mix.
4. Note any color change. Record your observations.

Name ______________________ Date ______________________

Partner(s) ______________________ Section ______________________

______________________ Instructor ______________________

1. Define the following terms:

    a) Primary alcohol

    b) Secondary alcohol

    c) Tertiary alcohol

    d) Phenol

2. Explain why the water solubilities of alcohols are much higher than those of alkanes with similar molecular masses.

3. Thymol is used to kill fungi, vanillin is a flavoring agent, and tetrahydrocannabinol (THC) is the most active ingredient in marijuana. Circle the phenol group on each structure.

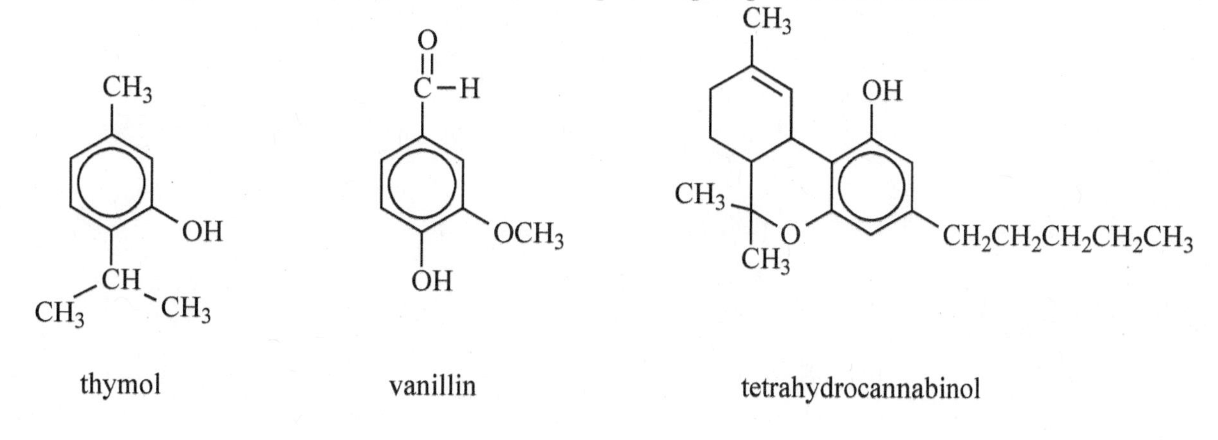

4. Draw the hydrogen bond that forms between two ethanol molecules.

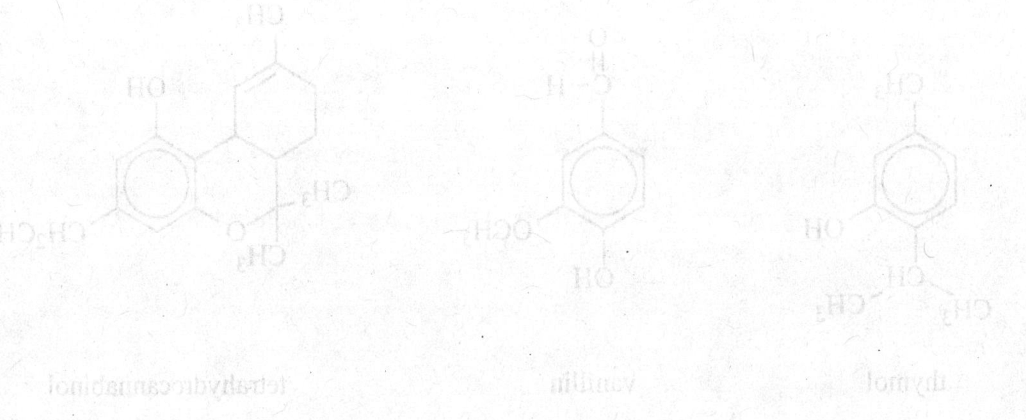

Name ______________________ Date ______________________

Partner(s) ______________________ Section ______________________

______________________ Instructor ______________________

A. Molecular models

Compound	Condensed structural Formula	1°, 2°, or 3°	Instructor's approval for models
Methanol			
2-Butanol			
2-Methyl-2-butanol			
Cyclohexanol			
Phenol		N/A	

B. Properties

Compound	Odor	pH	Solubility observations	Soluble in H_2O?
Ethanol				
1-Butanol				
2-Butanol				
2-Methyl-2-propanol				
1-Octanol				
Phenol				
Unknown # _______				

Which of these compounds has the lowest pH? Why?

Which of these compounds is the least soluble in water? Why?

C. Oxidation of alcohols

Compound	Observations	Oxidation reaction Write NR for products if there was no reaction
Ethanol		
1-Butanol		
2-Butanol		
2-Methyl-2-propanol		
Phenol		C_6H_5–OH $\xrightarrow{[O]}$ O=C_6H_4=O
Unknown # ______		

D. Ferric chloride test

Compound	Observations	Answer after testing all compounds
Ethanol		Based on your observations, the ferric chloride test is used to distinguish ______________________ from ______________________.
1-Butanol		
2-Butanol		
2-Methyl-2-propanol		
Phenol		
Unknown # ______		

Compare the results of the properties, oxidation tests, and ferric chloride tests on your unknown with the known sample tests. Identify your unknown and support your choice.

Problems and questions

1. Explain why tertiary alcohols do not undergo oxidation.

2. Use condensed structural formulas to write the equations for the oxidation reactions of the following compounds:

    a) 1-propanol (first step only)

    b) methanol (first step only)

    c) 2-hexanol

3. Circle the more water-soluble compound in each pair.

    a) propane or 1-propanol

    b) phenol or 1-hexanol

    c) methanol or 1-heptanol

Goals

- Compare different paper chromatography techniques.
- Use chromatography to separate and identify components of dyes.
- Learn how to calculate R_f values.

Materials

- Gloves
- 5.5 cm diameter filter paper
- Scissors
- Ruler
- Toothpicks
- Food colors (red, blue, green)
- Ethanol
- Glacial acetic acid
- 10 cm test tube
- Crucible
- 250 mL beaker
- "Vis-à-Vis® Wet Erase" black felt-tip pen
- Three 15 cm test tubes and stoppers
- Acetone
- Three 1 cm × 15 cm chromatography papers
- 7 cm × 8 cm chromatography paper
- Four concentrated Kool-Aid® solutions (purple, red, blue, orange)
- Watch glass
- Ethanol/acid waste container
- Test tube rack

Discussion

Chromatography is a method of separating the components of a mixture based on their differential affinity for two chemicals, one of which is immobilized (the *stationary phase*) and the other mobile (the *mobile phase*). Various types of chromatography include: size exclusion chromatography, ion exchange chromatography, gas chromatography (GC), high-performance liquid chromatography (HPLC), paper chromatography, and thin layer chromatography (TLC).

In this lab the paper chromatography technique will be used. It consists of a paper stationary phase and a liquid solvent mobile phase. The substance to be identified will be applied as a small dot to the paper. When the piece of absorbent paper is put into contact with a solvent, the solvent is drawn up the paper at a uniform rate, like a wick, creating a "wet" line across the paper. This is called the *solvent front*. As the solvent front advances, dissolved substances are carried along, each at its own rate, and can be qualitatively separated and then identified.

Those components that interact well with the solvent but poorly with the paper will travel with the solvent. However, they usually will not move up the paper as fast as the solvent. Those that interact poorly with the solvent but strongly with the paper will not travel as quickly. The interactions of the substances with the solvent and the paper are related to the polarity (or nonpolarity) of each. In solution formation, the phrase "like dissolves like" is useful in predicting solution formation. An analogous phrase for chromatography is "like interacts well with like."

When the process is completed, different components will have traveled different distances. The separations are particularly easy to follow if the substances are colored. The paper with the separated components is called a *chromatogram*.

Chromatography is often used not only to separate mixture components but also to identify them (based on their separation behavior under controlled circumstances): finding drugs in blood or urine samples, finding poisons, detecting flame accelerants in arson investigations, separating components of unknown substances found in crime scenes, and other instances where it might be desirable to separate a complex substance.

Dyes

A dye is a colored substance with the ability to stain materials such as paper, leather, plastics, and natural or synthetic fibers. Until the mid-1800s, virtually all dyes were obtained from flowers, berries, roots, bark, lichen, insects, and animals. Humans have been using dyes for thousands of years. One of the earliest written records of dye use comes from a 5,000-year-old Chinese text, listing dye recipes for coloring silks red, yellow, and black. One of the most highly-prized colors in the ancient world was purple. The Phoenicians used the hypobranchial gland mucus of a mollusk (*murex trunculus*) to make a purple-colored dye (purpura). Purple-dyed textiles were status symbols in the ancient world and purple cloth was literally worth its weight in gold.

Today, most dyes are composed of synthetic organic chemicals because they cost less and are available in a wide range of colors. More than 1000 different dyes are commercially produced in the United States. Food colorings and inks are examples of dyes. For safety, the Food and Drug Administration (FDA) regulates the use of dyes in foods, drugs, cosmetics, and medical devices. Although there are thousands of dyes available, only nine dyes are currently approved by the FDA for use in food products (as of January 2014). Food labels list the dyes in the product with names such as FD&C Blue No. 1, Citrus Red No. 2, etc. Dyes are often mixtures of compounds with complex structures and very long names. The exact chemical composition of the nine dyes can be found by referencing the Electronic Code of Federal Regulations, Title 21, Chapter 1, Subchapter A, Part 74, Subpart A.

White light (the visible region of the electromagnetic spectrum) contains all the colors of the rainbow. The color of a dye results from the selective absorption of some colors in the white light and the transmission of others. The ability to absorb light is usually attributed to a group of atoms in a molecule called a chromophore. Chromophores contain double bonds and include the nitro (NO_2), nitroso (N=O), vinyl (CH_2=CH_2), carbonyl (C=O) and azo (N=N) groups.

Some chromophores can absorb light we cannot see (ultraviolet light) and can retransmit it in the visible part of the spectrum. One use of these types of molecules is in laundry detergents. One manufacturer claimed it made the clothes "whiter than white." Because of the dye, the clothes emit more visible light than was being shone on it. This is why whites appear so mesmerizing under black light (UV light).

Experimental Procedures

Eye protection and appropriate clothing must be worn at all times.

Ethanol is flammable. Make certain there are no open flames nearby when performing this experiment.

These chemicals are corrosive to the skin. Use gloves to avoid skin contact. Wash any contacted skin immediately with running water and alert the instructor.

Touch the chromatography paper only on the edges to avoid getting fingerprints in the solvent area.

Take care not to splash solvents on chromatograms while setting up and do not disturb the apparatus while allowing the solvent front to move.

Discard all wastes properly as directed by the instructor.

A. Separation of food colorings

1. On a paper towel, use a toothpick to practice spotting samples until you can consistently form spots that are approximately 2 mm in diameter. You should put 2 to 3 drops of the same food color on top of each other to form an intense spot of color. This technique will be used in Part A and Part C.
2. Cut a piece of 5.5 cm diameter filter paper as shown in Figure 1.
3. Fold the cut portion down to form a wick.
4. At the three locations indicated by the dots as shown in Figure 1, use toothpicks to place small spots of three different food colors. The spots should be no larger than 2 mm in diameter and less than 0.5 cm from the wick.
5. While the spots are drying, put 3 mL of ethanol, 1 mL of distilled water, and 2 **drops** of glacial acetic acid into a clean 10 cm test tube.
6. Clean a crucible with soap and water. Dry the crucible and pour the mixture prepared in step 5 into it.
7. After the spots have dried, lay the filter paper across the top rim of the crucible. The wick should hang down inside the crucible and the liquid in the crucible should reach about 0.5 cm up the wick.
8. Invert a large beaker over the crucible and filter paper to minimize solvent evaporation as shown in Figure 2.
9. When the first of the colored components nears the outer edge of the filter paper, remove the filter paper, place it on a paper towel, and allow it to dry.
10. Record the color of each component from the food colorings and attach the chromatogram to your report.

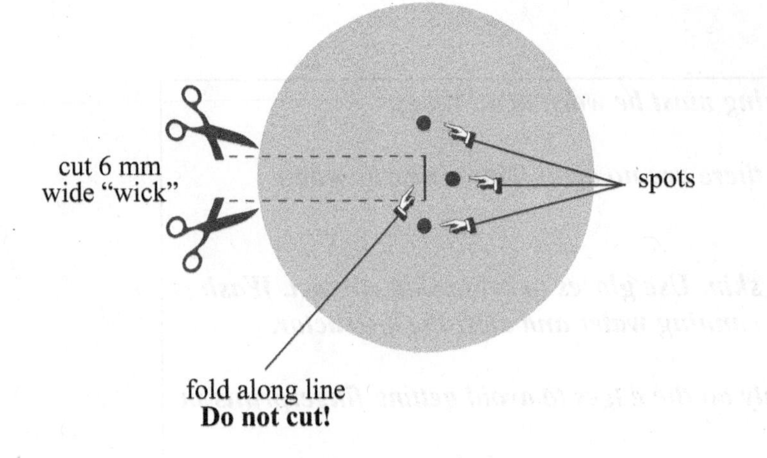

Figure 1 ***How to cut and fold the filter paper***

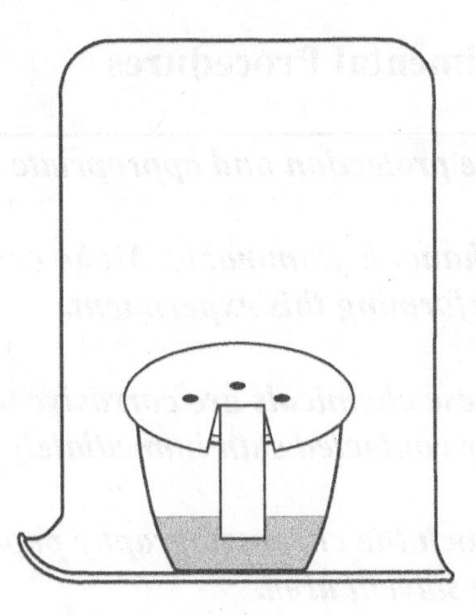

Figure 2 ***Chromatography apparatus***

B. Separation of black ink components

Most inks are complex mixtures containing several dye components. Ascending paper chromatography is used to separate the ink dye components of a black felt tip pen. Three different solvents will be used to determine the effect of solvent variation on the quality of separation.

1. Prepare three 15 cm test tubes as follows:
   - Test tube #1.......30 drops of distilled water
   - Test tube #2.......a mixture of 15 drops of distilled water and 15 drops of acetone
   - Test tube #3.......30 drops of pure acetone
2. Carry out steps 3–9 for each test tube.
3. Obtain a strip of chromatography paper, 1 cm × 15 cm.
4. Use a pencil to label the top of the strip with the solvent used.
5. Use the Vis-à-Vis® Wet Erase black felt tip pen to put a small spot of black ink on the paper strip 3 cm from the bottom end.
6. Put the paper strip into the test tube. Adjust the stopper to hold the paper strip between the stopper and the test tube so that the paper strip hangs straight and the bottom end just dips into the solvent (Figure 3). The spot of ink should be approximately 1.5 cm above the solvent surface.

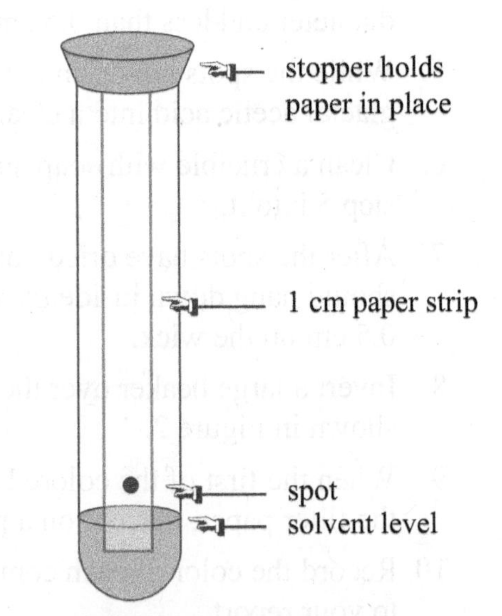

Figure 3 ***Setup for part B***

7. Allow the solvent to move up the paper strip until it reaches a height approximately 2 cm from the bottom of the stopper.
8. Remove the paper strip and allow it to dry.
9. Record the different colors detectable in the ink.
10. Record the solvent used on each chromatogram and attach the chromatograms to the report sheet.

C. Separation of Kool-Aid® food colorings

1. Obtain a 7 cm × 8 cm sheet of chromatography paper.
2. Obtain a ruler and trim the paper so it matches the dimensions given in Figure 4. Note that the paper is wider at the top than the bottom.
3. Use a pencil to draw a line across the sheet of paper 1.5 cm from the bottom, as shown in Figure 4, and mark four small dots on this line, equidistant from each other and the edges of the paper.
4. Use the toothpicks to put spots of the samples from the Kool-Aid® on the chromatography paper pencil dots. Use a different toothpick for each sample. Label under each spot with a pencil to identify the Kool-Aid®. Allow the spots to dry.

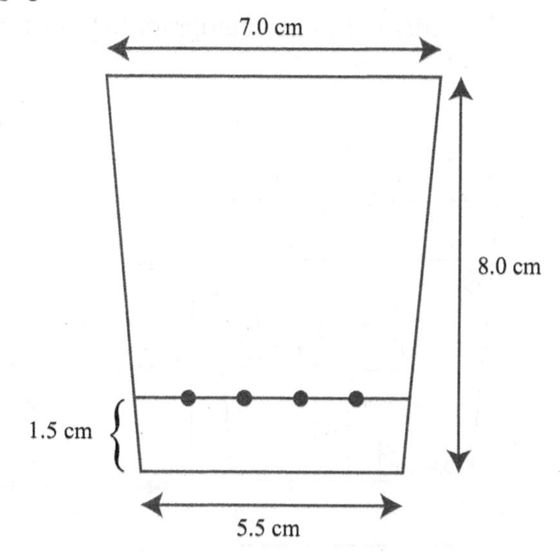

Figure 4 ***Chromatogram dimensions***

5. While the chromatogram is drying, prepare a chromatography chamber as follows:
   - Measure 3 mL of ethanol. Put the ethanol into a clean dry 250 mL beaker and cover the top of the beaker with a watch glass.
   - Add 2 drops of glacial acetic acid to the ethanol in the beaker and swirl the beaker to mix the contents.
6. Carefully put the spotted chromatogram into the beaker with the spotted end down. The paper should rest evenly on the bottom of the beaker. Take care not to splash the liquid. The liquid level in the beaker should be below the pencil line on the chromatogram. Replace the watch glass.
7. Allow the chromatogram to develop **undisturbed** until the solvent front rises to within 2 cm of the top of the chromatogram. Remove the chromatogram from the beaker and immediately mark the solvent front with a pencil line. Set the paper aside to dry. Dispose of the solvent in the waste container.
8. Spots may not be circular; they may be elliptical or teardrop shaped. Draw a line around the outside edge of each spot. Mark the center of each spot with a pencil.

9. R_f values are characteristic of a given substance. The R_f value for a given spot is the ratio of the distance the substance moved to the distance the solvent front moved. Measure the distance (cm) from the starting line to the tip of the solvent front and then measure the distance (cm) from the starting line to the center of each spot. Calculate the R_f for each spot according to this equation:

$$R_f = \frac{\text{distance traveled by spot (cm)}}{\text{distance traveled by solvent (cm)}}$$

A sample calculation is shown in Figure 5.

10. Express the R_f values as decimals (no units).
11. Attach the chromatogram to your report.

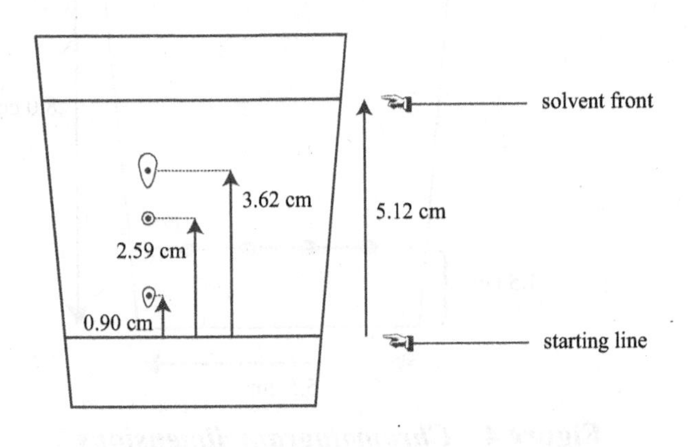

R_f calculations:

$$R_f\ (\text{spot 1}) = \frac{3.62}{5.12} = 0.707$$

$$R_f\ (\text{spot 2}) = \frac{2.59}{5.12} = 0.506$$

$$R_f\ (\text{spot 3}) = \frac{0.90}{5.12} = 0.18$$

Figure 5 ***Sample chromatogram with calculations***

Name ____________________ Date ____________________

Partner(s) ____________________ Section ____________________

____________________ Instructor ____________________

1. Identify the stationary phase and the mobile phase in Procedure A.

2. Calculate the R_f value for a spot that moved 3.25 cm with a solvent front that moved 4.83 cm.

3. Why is it important to avoid getting fingerprints on the chromatography paper?

4. Why should the filter paper and crucible be covered with a beaker while performing the chromatography separation of the food colorings?

Name ______________________ Date ______________________

Partner(s) ______________________ Section ______________________

______________________ Instructor ______________________

A. Food colors

Describe the components of each of the food colorings.

Food color #1

Food color #2

Food color #3

B. Ink

Describe the colors detected from the ink spots in each of the different solvents.

Water:

Water and acetone:

Acetone:

C. Kool-Aid®

Distance the solvent front moved: ______________

Kool-Aid® color	Distance center of spot(s) moved	R_f value(s)
Sample #1		
Sample #2		
Sample #3		
Sample #4		

Problems and questions

1. Describe an application where chromatography might be useful. Use a different example than those mentioned in the discussion section.

2. Why must you use a pencil and not a pen to write and draw on the chromatogram?

3. In Part B, compare how quickly the different solvents moved and how the different solvents affected the quality of separation?

Goals

- Draw structures of aldehydes and ketones and identify their functional groups.
- Build molecular models of various aldehydes and ketones.
- Identify physical properties of aldehydes and ketones.
- Use chemical tests to differentiate between aldehydes and ketones.
- Use physical properties and chemical tests to identify an unknown.

Materials

- Gloves
- Molecular model kits
- *Merck Index* or *CRC Handbook of Chemistry and Physics*
- Chemistry textbook
- 15 cm test tubes
- Grease pencil
- Test tube rack
- Camphor
- Cinnamaldehyde
- Benzaldehyde
- Acetone
- Vanillin
- Heptanal
- 3-Pentanone
- Unknowns (#1, #2, #3)
- Stirring rod
- 600 mL beaker
- Hot plate
- Thermometer
- 10% NaOH solution
- Iodine reagent
- Chromic acid reagent

Discussion

Aldehydes and ketones both contain the carbonyl functional group. The two families differ in the placement of the functional group within the molecule. An aldehyde has the functional group on the end of the molecule. There is at least one hydrogen attached to the carbonyl carbon. Ketones have the carbonyl carbon attached to two other carbons.

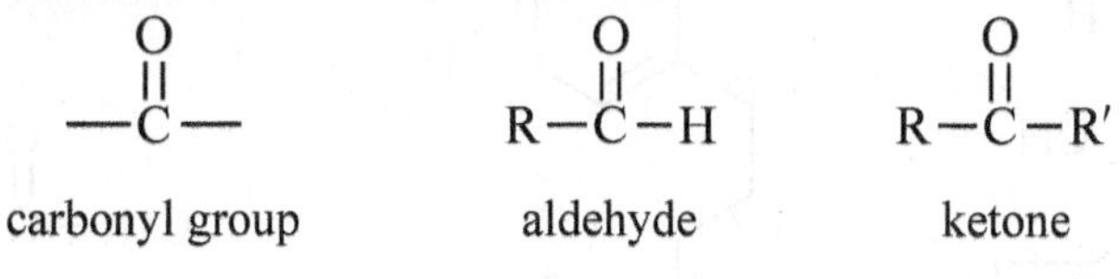

The "R" in these formulas can be an alkyl or an aromatic group. The R and R′ in the ketone formula may be the same or different groups.

The simplest of the aldehydes has two hydrogens attached to the carbonyl carbon and no R group. Its IUPAC name is methanal, but it is more widely known by its common name, formaldehyde. In the simplest ketone, both R and R′ are methyl groups. Its IUPAC name is propanone, but it is more widely known by the common names, acetone or dimethyl ketone.

$$\underset{\substack{\text{methanal}\\ \text{(formaldehyde)}}}{H-\overset{\overset{\displaystyle O}{\|}}{C}-H} \qquad \underset{\substack{\text{propanone}\\ \text{(acetone)}}}{CH_3-\overset{\overset{\displaystyle O}{\|}}{C}-CH_3}$$

Often the double bond of the carbonyl group is not shown in the formula for an aldehyde and the functional group will be written –CHO. The single line shown as a bond indicates that the oxygen must be double bonded in order for the carbon to have four bonds. This notation differs from the formulas for alcohols where the functional group is written –OH.

$$\text{Aldehyde:}\quad CH_3-CH_2-\overset{\overset{\displaystyle O}{\|}}{C}-H \Longrightarrow CH_3-CH_2-CHO$$

$$\text{Alcohol:}\quad CH_3-CH_2-CH_2-OH \Longrightarrow CH_3-CH_2-CH_2-OH$$

<u>Physical properties</u>

Because of differences in electronegativity, the carbonyl group is polar, with the oxygen end of the double bond having a partial negative charge and the carbon end of the double bond having a partial positive charge. This polarity allows low molecular weight aldehydes and ketones to be soluble in water as well as in nonpolar solvents. As the carbon chain length increases, the solubility of the compound in water decreases.

Aldehydes and ketones are often recognizable by their odors. Formaldehyde and acetone are examples of compounds frequently encountered in science labs. Other aldehydes and ketones such as camphor (recognizable odor in medical inhalants such as Vicks® VapoRub®), citronellal (odor of lemons and oranges used in perfumes and insect repellents) and butanedione (butter flavor and odor) are used as fragrances and food additives.

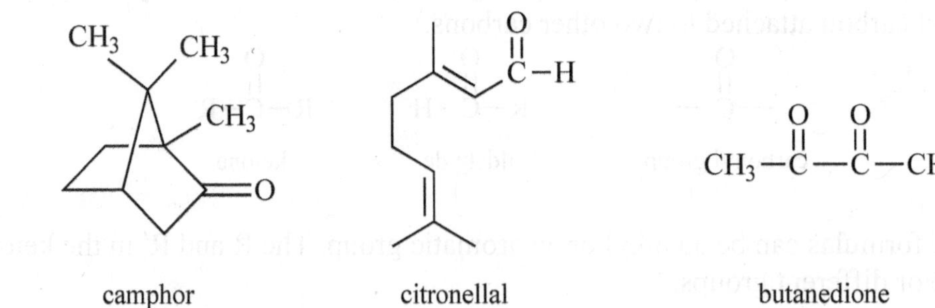

camphor citronellal butanedione

Chemical properties

Aldehydes are easily oxidized by Tollens' reagent, which is an alkaline solution containing a silver ion combined with two ammonia molecules. In the redox reaction with an aldehyde, the silver ion in the ion complex $[Ag(NH_3)_2]^+$ is reduced from a +1 oxidation state to a 0 oxidation state (metallic silver). It deposits onto the inside of the test tube as a beautiful mirror. However, due to the high cost of silver and the difficulty in cleaning the mirror residue, this test will not be used for this lab.

Benedict's reagent is a blue alkaline solution containing Cu^{2+} ions. When these ions are reduced to Cu^+ ions they form a brick red, brown, green, or yellow Cu_2O precipitate. The Benedict's reagent oxidizes aldehydes, but *simple* aldehydes do not give a positive test as clearly as aldehydes with an –OH bound to a carbon next to the carbonyl carbon, known as α-hydroxy aldehydes. In general, ketones give a negative Benedict's test, but one type of easily oxidized ketone gives a positive test: α-hydroxy ketones. Alpha-hydroxy aldehydes and ketones are important in the study of carbohydrates.

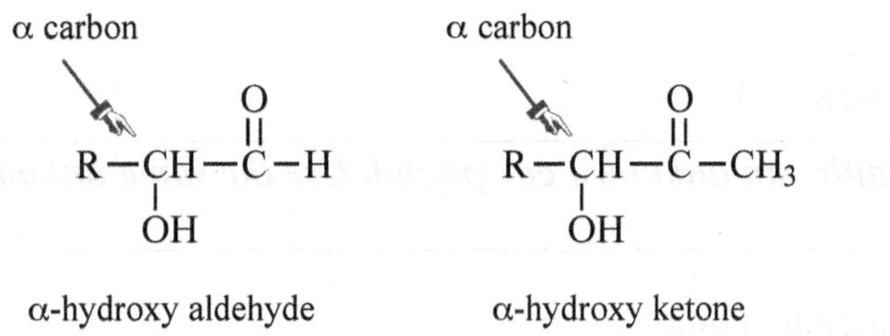

Chromic acid, H_2CrO_4, is an orange-colored solution and a strong oxidizing agent. When an aldehyde is oxidized to a carboxylic acid by the chromic acid, the acid is reduced and forms the green chromic ion (Cr^{3+}) in solution. The appearance of the blue-to-green ions indicates that an aldehyde has been oxidized. Aliphatic aldehydes react almost immediately, while aromatic aldehydes take longer. Ketones do not undergo oxidization with chromic acid.

The iodoform reagent is a reddish (or yellowish), basic (NaOH) solution of I_2. Ketones containing a methyl group as one of their R groups are frequently referred to as *methyl ketones*. These compounds react with the iodine to form a yellow precipitate of iodoform, CHI_3. This is not to be confused with a yellow oily-looking liquid. Aldehydes do not react with iodoform reagent.

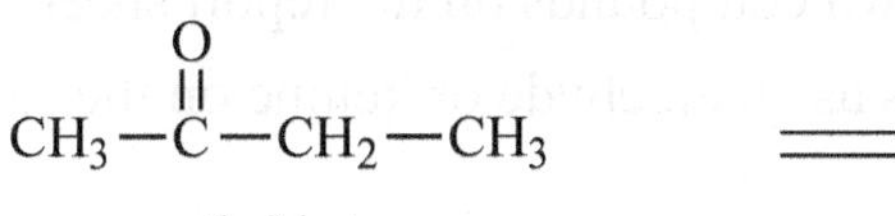

$\Longrightarrow$ positive iodoform test

a methyl ketone

$CH_3-CH_2-C(=O)-CH_2-CH_3$ $\Longrightarrow$ negative iodoform test

a ketone

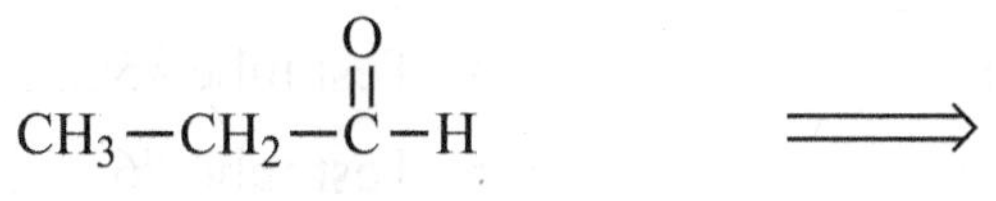

$\Longrightarrow$ negative iodoform test

an aldehyde

Experimental Procedures

Eye protection and appropriate clothing must be worn at all times.

Chromic acid is toxic and corrosive. Wear gloves to avoid skin contact. Wash any contacted skin immediately with running water and contact the instructor.

Discard all wastes properly as directed by the instructor.

A. Structures and models

1. Make models of formaldehyde, acetaldehyde, acetone, 2-pentanone, and cyclopentanone. Complete each model and have it checked by the instructor.
2. Draw the condensed structural formulas on the report sheet. Write the IUPAC names and the common names as indicated.

B. Physical properties

Parts B and C require an unknown compound. Use the same unknown for all parts.

Odor, solubility, melting point

1. Check the odor of each of the following six compounds by removing the lid from one bottle at a time and *waft* the fumes toward your nose. Record the odors on the report sheet.
   - vanillin
   - cinnamaldehyde
   - benzaldehyde
   - acetone
   - heptanal
   - camphor
2. Record the number of your unknown and its odor on the report sheet.
3. Use a chemistry handbook to look up the solubility and melting point for each of the known compounds. Record the values on the report sheet.
4. Draw the structures of each of the known compounds on the report sheet.
5. Identify each of the known compounds as an aldehyde or ketone on the report sheet.

Solubility

6. Label six 15 cm test tubes. Put 2 mL of deionized water into each test tube and place in a test tube rack.
7. Put 5 drops (a match tip sized sample for camphor) of the following solutions into the test tubes:
   - Test tube #1........ heptanal
   - Test tube #2........ benzaldehyde
   - Test tube #3........ acetone
   - Test tube #4 ....3-pentanone
   - Test tube #5 ....camphor
   - Test tube #6 ....unknown

8. Stir each solution and record your observations regarding solubility. Be sure to clean the stirring rod each time.
9. **SAVE** these samples for Part C2.

C. Chemical properties

Iodoform test

1. Set up a warm (50–60°C) water bath using a hot plate and a 600 mL beaker. Fill the beaker with approximately 200 mL of tap water.
2. To each test tube prepared in B.7, add 10 drops of 10% NaOH.
3. Place the test tubes in the warm water bath for 5 minutes.
4. While the test tubes are still in the warm water bath, add 20 drops of the iodine test reagent to each test tube. Then remove the test tubes from the warm water bath and place them in the test tube rack. Place the test tubes aside and record your observations at the end of the lab period since the formation of a yellow precipitate may be slow.

Chromic acid test

5. Label six 15 cm test tubes.
6. Put 5 drops (a match tip sized sample for camphor) of the following solutions into the test tubes:
   - Test tube #1 .... heptanal
   - Test tube #2 .... benzaldehyde
   - Test tube #3 .... acetone
   - Test tube #4 .... 3-pentanone
   - Test tube #5 .... camphor
   - Test tube #6 .... unknown
7. Add 20 drops of acetone to each test tube.
8. The chromic acid reagent is added to the test tube **one drop at a time**. After each drop, mix the solution by sharply tapping the test tube with your finger. Carefully add a total of 4 drops of chromic acid reagent to each test tube using this technique.
9. Let stand for 10 minutes.
10. Aliphatic aldehydes should show a change within a minute. Aromatic aldehydes take longer. Note the approximate time for any change in color or formation of a precipitate. Record your observations on the report sheet.

Name ______________________ Date ______________________

Partner(s) ______________________ Section ______________________

______________________ Instructor ______________________

1. Write the general structure for an aldehyde.

2. Write the general structure for a ketone.

3. a) What observation indicates a positive iodoform test?

   b) What type of compound is indicated by the positive iodoform test?

4. a) What observation indicates a positive chromic acid test?

   b) What type of compound is indicated by the positive chromic acid test?

   c) What type of compound takes a longer time to give a positive chromic acid test?

Name ____________________ Date ____________________

Partner(s) ____________________ Section ____________________

____________________ Instructor ____________________

A. Structures and Models

IUPAC name	Common name	Condensed structural formula	Instructor's approval of model
	formaldehyde		
	acetaldehyde		
	acetone		
2-pentanone			
cyclopentanone	X		

B. Physical properties

Compound	Aldehyde or ketone?	Odor	Condensed structural formula	Solubility in H_2O	Melting point
vanillin (4-hydroxy-3-methoxybenzaldehyde)					
cinnamaldehyde					
benzaldehyde					
acetone					
heptanal					
camphor					
Unknown # _____					

B. Solubility

Compound	Observed solubility in H_2O
heptanal	
benzaldehyde	
acetone	
3-pentanone	
camphor	
Unknown # _____	

C. Chemical properties

Compound	Iodoform test observation	Iodoform test indicates?	Chromic acid test observation	Chromic acid test indicates?
heptanal				
benzaldehyde				
acetone				
3-pentanone				
camphor				
Unknown # _____				

Using the results from this experiment, identify your unknown:

Unknown #____________________ is ____________________

How did you arrive at the identity of your unknown?

Problems and questions

1. Why are long chain aldehydes and ketones insoluble in water?

2. Draw the structure and give the three names (one IUPAC name and two common names) for the ketone with the molecular formula C_3H_6O.

3. a) Draw the structures and give the IUPAC names of the three ketones that have the molecular formula $C_5H_{10}O$.

   b) Which of the three ketones above would give a positive iodoform test?

4. Determine whether each of the following compounds will give a positive or negative test result for each of the following tests. Put a + in the table for a positive result and a – for a negative result.

Compound	Benedict's	Chromic acid	Iodoform
$CH_3-CH(OH)-C(=O)-CH_2-CH_3$			
$CH_3-(CH_2)_3-C(=O)-H$			
$CH_3-C(=O)-CH_2-CH_3$			

Goals

- Draw structures of carbohydrates and identify their functional groups.
- Identify aldoses and ketoses.
- Classify a carbohydrate structure as a monosaccharide, disaccharide, or polysaccharide.
- Build molecular models of various types of carbohydrates.

Materials

- Molecular model kits
- Instructor-prepared maltose model
- Instructor-prepared polysaccharide model
- Students' textbooks

Discussion

Carbohydrates are the most abundant class of bioorganic molecules on earth. Carbohydrates are a common source of energy in living organisms and may be found in a variety of foods such as flour products, vegetables, fruits, and milk products. Starch, for example, is a carbohydrate.

Although many people equate carbohydrates with sugar, they are much more than that. Carbohydrates supply living organisms with the carbon atoms needed to synthesize other biochemical substances. Genetic information is stored and transferred by way of specialized derivatives of carbohydrates. Paper, cotton, and wood are largely cellulose, which is a carbohydrate.

A carbohydrate is a polyhydroxy aldehyde, a polyhydroxy ketone, or a compound that yields a polyhydroxy aldehyde or polyhydroxy ketone upon hydrolysis. Polyhydroxy means that there are many –OH groups in the molecule.

The Latin word for sugar is *saccharum*, and the derived term *saccharide* is the basis of a system of carbohydrate classification. Based on their molecular size, carbohydrates are classified as monosaccharides, disaccharides, or polysaccharides.

Monosaccharides

A *monosaccharide*, also called a simple sugar, is a carbohydrate that contains a single polyhydroxy aldehyde or polyhydroxy ketone unit. It cannot be broken down into simpler carbohydrates by hydrolysis. Monosaccharides have a molecular formula $C_nH_{2n}O_n$, where n is an integer. This same formula can be written $C_n(H_2O)_n$. Written this way, the formula appears to be "hydrates of carbon" and historically the term "carbohydrates" has been used to describe these molecules.

There is no limit to the number of carbons that can be present in a monosaccharide, but only monosaccharides with three to seven carbons are commonly found in nature. A three-carbon monosaccharide is called a *triose*. The suffix *–ose* is used for monosaccharides, just as the suffix *–ol* is used for alcohols and *–ene* is used for alkenes. Monosaccharides containing four, five, and six carbons are called *tetroses*, *pentoses*, and *hexoses*, respectively.

Monosaccharides may also be classified based on the type of carbonyl group present in the molecule. An *aldose* is a monosaccharide containing an aldehyde functional group, whereas a *ketose* is a monosaccharide containing a ketone functional group.

Often, the two classifications are combined in one word. For example, an *aldohexose* is a monosaccharide that contains an aldehyde group and has six carbons. Similarly, a ketopentose is a monosaccharide that contains a ketone group and has five carbons.

Three of the most common monosaccharides are all hexoses:

- Glucose, also known as dextrose or blood sugar (an aldohexose)
- Galactose, found in the disaccharide lactose (an aldohexose)
- Fructose, found in honey and also in the disaccharide sucrose (a ketohexose)

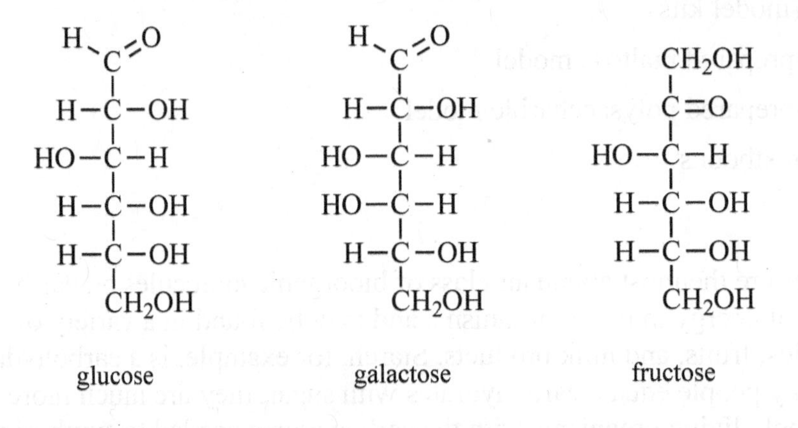

These three compounds all have the same molecular formula, $C_6H_{12}O_6$, but they differ in their structural formulas.

Fischer projections

Using the wedge and dash notation, the three-dimensional structure of a molecule can be shown. For example, in the hypothetical molecule below left, atoms *X* and *Y* point toward the viewer while atoms *W* and *Z* point away from the viewer. However, this type of drawing is somewhat awkward. The German chemist Hermann Emil Fischer developed a two-dimensional structural notation for showing three-dimensional molecules. The vertical lines in a Fischer projection represent bonds going away from the viewer (*W* and *Z*). The horizontal lines represent bonds toward the viewer (*X* and *Y*). Although the Fischer projection looks flat, it carries three-dimensional information depending on whether the bonds are vertical or horizontal lines.

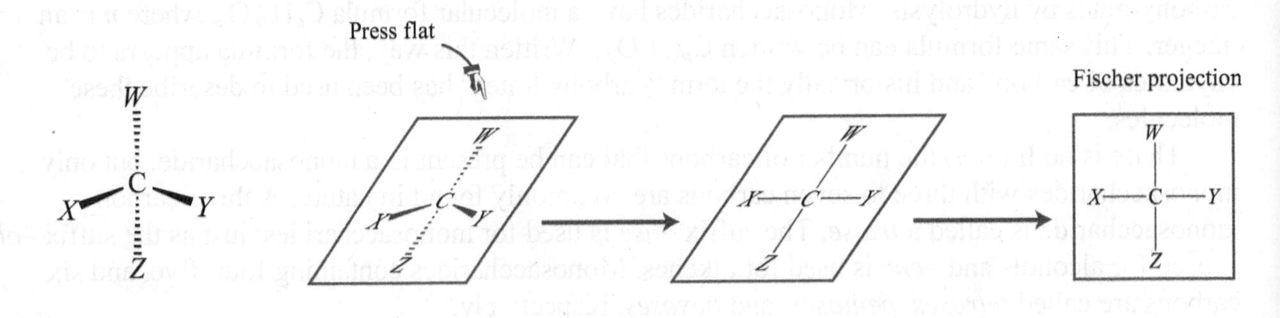

Instead of specifying the exact designation of every chiral carbon in a carbohydrate, an older system uses the small capital letters D and L to specify enantiomers (nonsuperimposable mirror image molecules) of carbohydrates. The focus is on the –OH group attached to the *chiral* carbon farthest away from the carbonyl group. Note that in Figure 1 the last carbon is not chiral. By convention, the –OH group is on the right in D compounds and it is on the left in L compounds. An example is shown in Figure 1.

Figure 1 ***Fischer projections of D- and L-glucose***

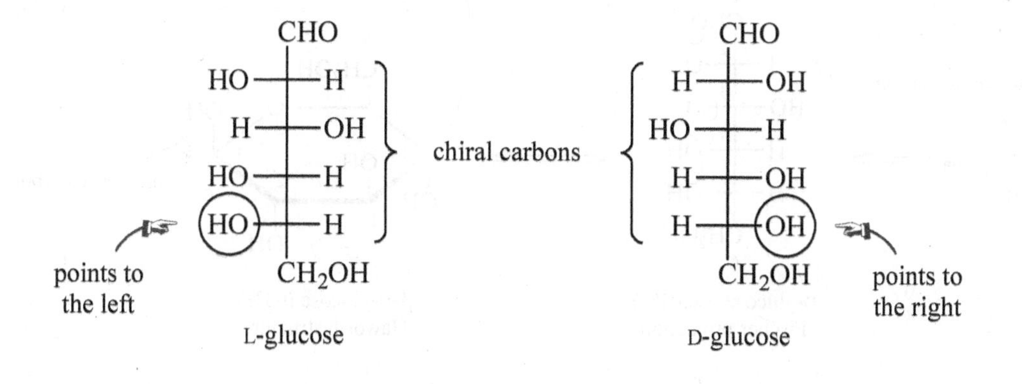

The molecule D-erythrulose is a ketotetrose. Its molecular formula is $C_4H_8O_4$. To draw the Fischer projection, draw the molecule vertically with the carbonyl group (C=O) at the top (in an aldose) or as close to the top as possible (in a ketose). Because this molecule is a ketose, the carbonyl carbon will be #2, not #1. With one chiral carbon, two different stereoisomers exist for this molecule and they are nonsuperimposable mirror images of each other (enantiomers). The mirror is shown as a dotted line in Figure 2. Note the left and right directions of –H and –OH bonded to the *chiral* carbon have been swapped in enantiomers. Achiral carbons do not have any of their groups change directions. Look at Figure 1 for another example.

Figure 2 ***Fischer projections of D- and L-erythrulose***

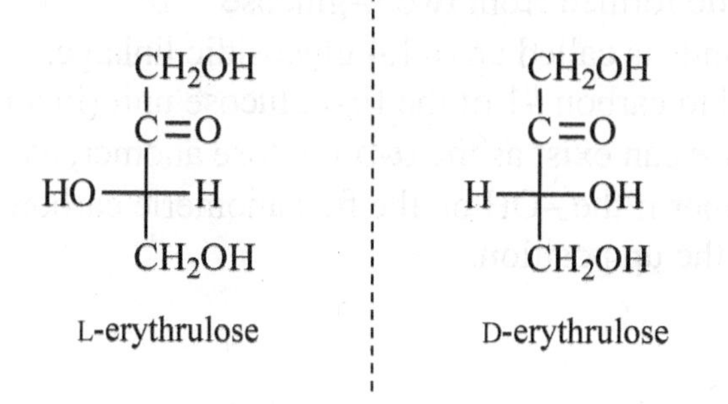

Haworth structures

An aldehyde or ketone functional group may react with an alcohol functional group to form a hemiacetal. Carbohydrates have both functional groups in the same molecule. When this reaction occurs in a carbohydrate, a cyclic hemiacetal is formed. The new –OH group on the former carbonyl carbon (now referred to as the anomeric carbon) may point up or down. Thus, the resulting ring has two possible structures called *anomers*, depending on the direction of the –OH (on carbon #1 for aldoses, on carbon #2 for ketoses). The "down" position is designated as the α anomer and the "up" position is designated as the β anomer. The two different anomers and the relative abundance of the three forms of glucose present in an aqueous solution are illustrated in Figure 3.

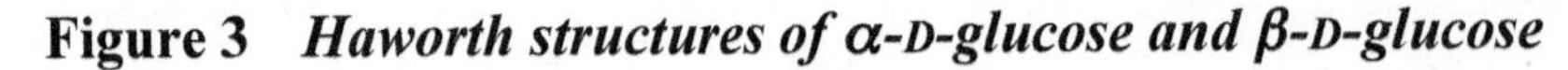

Figure 3 ***Haworth structures of α-D-glucose and β-D-glucose***

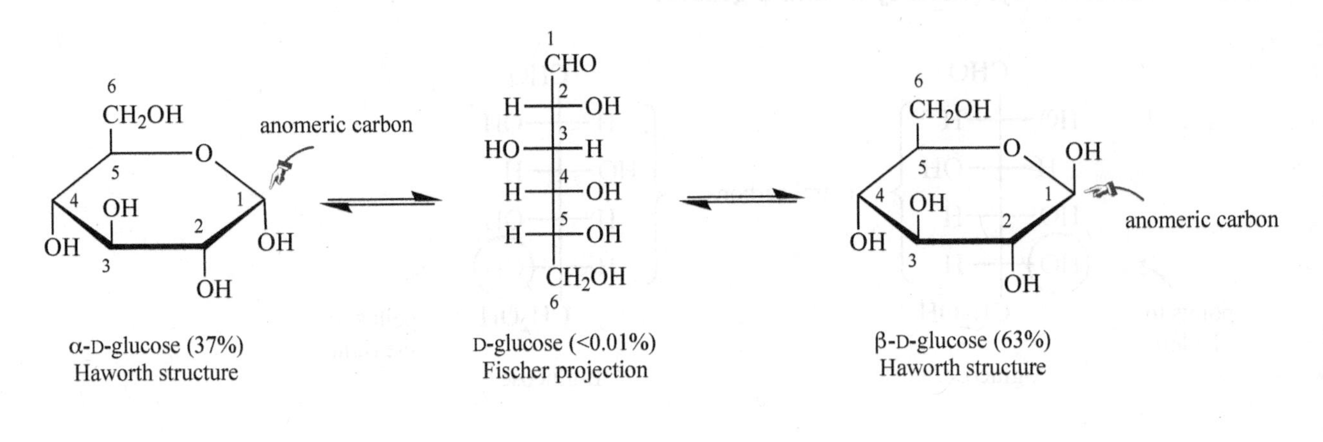

α-D-glucose (37%)
Haworth structure

D-glucose (<0.01%)
Fischer projection

β-D-glucose (63%)
Haworth structure

Disaccharides

A *disaccharide* is a carbohydrate that contains two monosaccharides bonded together. The bond that links the two monosaccharides together is called a *glycosidic linkage*. The reaction linking the two monosaccharides together occurs between the anomeric –OH group of one monosaccharide and an –OH group on a second monosaccharide. In this dehydration reaction, the product is an acetal (the disaccharide) and water.

The linkage is always a carbon–oxygen–carbon bond in a disaccharide. In naming the linkage, the α- or β- refer to the direction of the bond from the anomeric carbon (on carbon #1 for aldoses, on carbon #2 for ketoses). The numbers refer to the carbon atoms involved on each monosaccharide.

In the example in Figure 4, maltose is a disaccharide formed from two D-glucose monosaccharides. The bond between the two glucose units is called an α-1,4 glycosidic linkage. The two –OH groups that form the linkage are attached to carbon #1 of the first glucose unit (in an α configuration) and to carbon #4 of the second. Maltose can exist as the α-D-maltose anomer, as shown below. It may also exist as the β-D-maltose anomer if the –OH on the free anomeric carbon (the anomeric carbon not involved in the linkage) is in the up position.

Figure 4 ***Haworth structure of α-D-maltose***

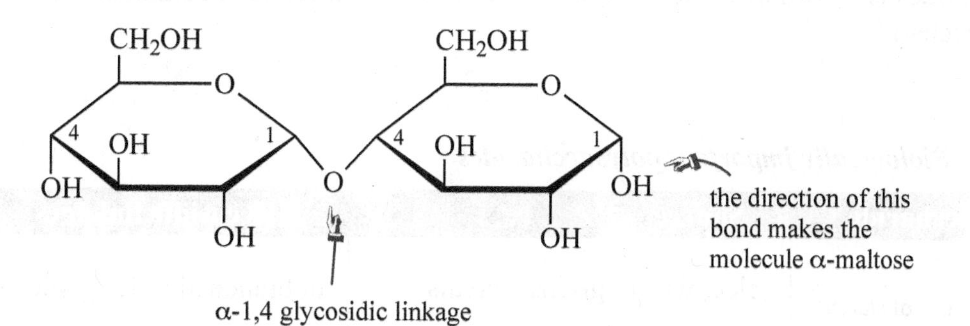

Three of the most common disaccharides are maltose, lactose, and sucrose. Their monosaccharide components and glycosidic linkages are:

glucose + glucose ⟶ maltose (found in grains) α-1,4 glycosidic linkage

galactose + glucose ⟶ lactose (found in dairy products) β-1,4 glycosidic linkage

glucose + fructose ⟶ sucrose (table sugar) α,β-1,2 glycosidic linkage

The acid hydrolysis of disaccharides results in the cleaving of the disaccharide into two monosaccharides. The following equation shows the hydrolysis of maltose to produce two glucose molecules:

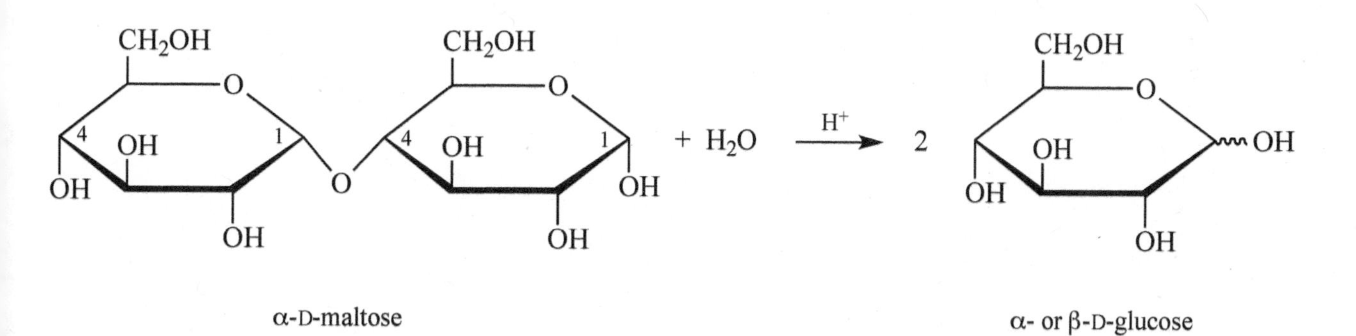

Polysaccharides

Polysaccharides are long chain polymers of monosaccharides characterized by high molecular weights. Four biologically important polysaccharides (amylose, amylopectin, glycogen, and cellulose) are all polymers of D-glucose. They differ in the type of glycosidic linkages and the amount of branching. For example, while most linkages are between carbon #1 of the first monosaccharide and carbon #4 of the next monosaccharide, *branched* linkages may occur between carbon #1 of the first monosaccharide and carbon #6 of the next monosaccharide. A summary of the polysaccharides is shown in Table 1. Figures 5–7 show the structures of small sections of each of the polysaccharides.

Table 1 ***Biologically important polysaccharides***

Polysaccharide	Source	Glycosidic linkages
amylose (makes up 20% of starch)	rice, wheat, grains, cereals	unbranched α-1,4 glycosidic linkages
amylopectin (makes up 80% of starch)	rice, wheat, grains, cereals	α-1,4 glycosidic linkages and α-1,6 glycosidic branches
glycogen	liver and muscles	α-1,4 glycosidic linkages and many α-1,6 glycosidic branches
cellulose	plant fiber, bran, beans, celery	unbranched β-1,4 glycosidic linkages

Figure 5 ***Glycosidic linkages in amylose***

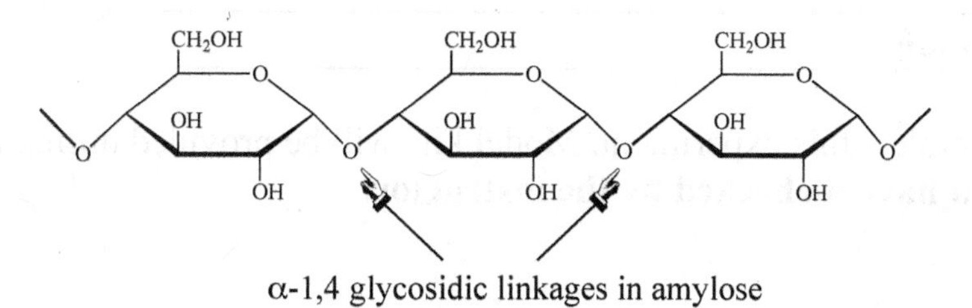

Figure 6 ***Glycosidic linkages in amylopectin and glycogen***

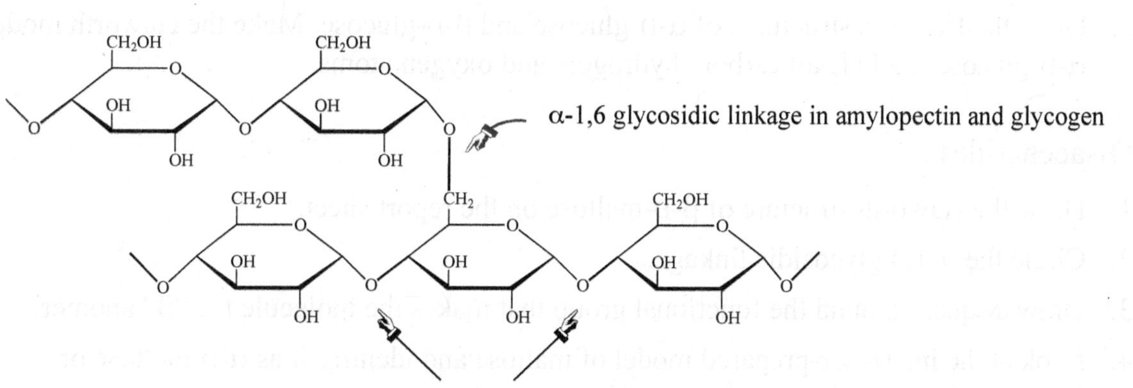

Figure 7 ***Glycosidic linkages in cellulose***

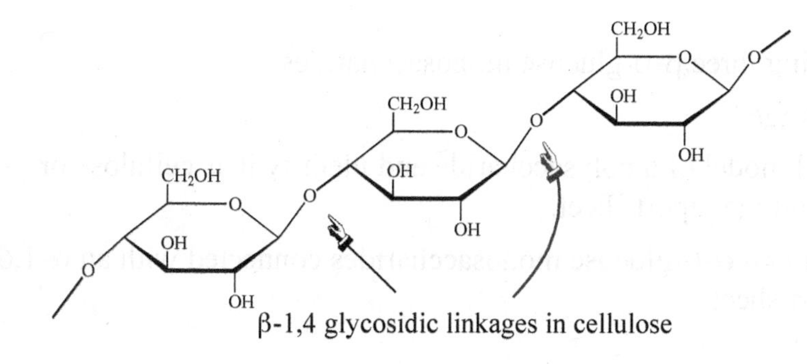

Experimental Procedures

Goggles are not required for this lab.

Use your textbook as a resource to complete this experiment. Model kits will be provided during the lab period. **Complete each model and have it checked by the instructor.**

A. Monosaccharides

1. Draw the Fischer projections of L-glyceraldehyde and D-glyceraldehyde on the report sheet.
2. Make models of L-glyceraldehyde and D-glyceraldehyde.
3. Draw the Fischer projections of L-glucose and D-glucose on the report sheet.
4. Make a model of D-glucose and have it checked by the instructor.
5. Draw the Haworth structures of α-D-glucose and β-D-glucose. Make the Haworth model of α-D-glucose. Include all carbon, hydrogen, and oxygen atoms.

B. Disaccharides

1. Draw the Haworth structure of β-D-maltose on the report sheet.
2. Circle the α-1,4 glycosidic linkage.
3. Draw a square around the functional group that makes the molecule the "β" anomer.
4. Look at the instructor-prepared model of maltose and identify it as α-D-maltose or β-D-maltose. Record your answer on the report sheet.
5. Draw the Haworth structure of α-D-lactose on the report sheet.
6. Circle the β-1,4 glycosidic linkage.
7. Draw a square around the galactose monosaccharide.

C. Polysaccharides

1. Draw a portion of cellulose using three β-D-glucose monosaccharides.
2. Circle the β-1,4 glycosidic linkages.
3. Look at the instructor-prepared model of a polysaccharide and identify it as cellulose or amylose. Record your answer on the report sheet.
4. Draw the Haworth structure of two α-D-glucose monosaccharides connected with an α-1,6 glycosidic linkage on the report sheet.

Name ______________________ Date ______________________

Partner(s) ______________________ Section ______________________

______________________ Instructor ______________________

1. Name one food source for each type of carbohydrate below.

   a) monosaccharide:

   b) disaccharide:

   c) polysaccharide:

2. What is the difference between an aldose and a ketose?

3. In the Fischer projections of monosaccharides, how do D monosaccharides differ from L monosaccharides?

4. What do the letters α and β indicate in the Haworth structure of a carbohydrate?

5. What two polysaccharides are found in starch?

Name ______________________ Date ______________________

Partner(s) ______________________ Section ______________________

______________________ Instructor ______________________

A. Monosaccharides

Drawings		Instructor's approval of models
Fischer: L-glyceraldehyde	Fischer: D-glyceraldehyde	L-glyceraldehyde D-glyceraldehyde
Fischer: L-glucose	Fischer: D-glucose	D-glucose
Haworth: α-D-glucose	Haworth: β-D-glucose	α-D-glucose

B. Disaccharides

Drawings
Haworth: β-D-maltose
Haworth: α-D-lactose

Identify the instructor-prepared maltose as α-D-maltose or β-D-maltose. Describe the feature(s) that support your answer.

C. Polysaccharides

Drawings
Haworth: cellulose using three β-D-glucose monosaccharides
Haworth: two α-D-glucose monosaccharides connected with an α-1,6 glycosidic linkage

Identify the instructor-prepared model of a polysaccharide as cellulose or amylose. Describe the feature(s) that support your answer.

Problems and questions

1. Are **all** carbohydrates soluble in water? Explain.

2. Draw the Fischer projection of an aldopentose.

3. The Haworth structure of D-talose is shown below.

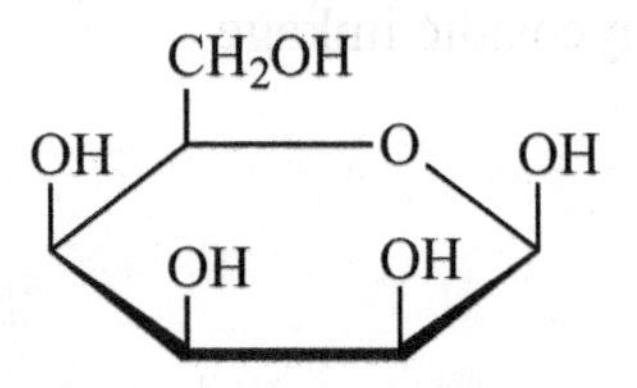

a) Is talose a monosaccharide, disaccharide, or polysaccharide?

b) Does the above structure represent α-D-talose or β-D-talose?

4. Compare amylose and cellulose. Describe one similarity and one difference.

5. Which monosaccharide is the product of the complete hydrolysis of glycogen?

Goals

- Use chemical tests to identify reducing sugars.
- Use chemical tests to differentiate between monosaccharide and disaccharide reducing sugars.
- Use chemical tests to differentiate between aldohexoses and ketohexoses.
- Use chemical tests to identify polysaccharides.
- Use chemical tests to identify the presence of starch in foods.
- Use chemical tests to identify an unknown carbohydrate.
- Use chemical tests to determine the types of carbohydrates formed by the hydrolysis of disaccharides and polysaccharides.

Materials

- Gloves
- 2% glucose
- 2% fructose
- 2% galactose
- 2% lactose
- 2% sucrose
- 2% maltose
- 2% starch
- Unknown solutions (#1, #2, #3)
- 15 cm test tubes
- Test tube rack
- Grease pencil
- 600 mL beaker
- Stirring rod
- Hot plate
- Droppers
- Benedict's reagent
- Barfoed's reagent
- Seliwanoff's reagent
- Iodine reagent
- Food samples (table sugar, pasta, bread, cereal flake, potato, apple)
- Spot plate
- 10% HCl
- 10% NaOH
- Red litmus paper

Discussion

Carbohydrates may be classified into various categories by performing a number of chemical tests. To identify an unknown carbohydrate, test results of the unknown carbohydrate are compared to those of known carbohydrates. Carbohydrate terms used in this experiment include:

- hexose: a monosaccharide with six carbon atoms
- aldose: a carbohydrate containing an aldehyde functional group
- ketose: a carbohydrate containing a ketone functional group

Benedict's test

Benedict's test distinguishes reducing sugars from nonreducing sugars. Benedict's reagent is a blue alkaline solution containing Cu^{2+} ions. When these ions are reduced to Cu^{+} ions they form a brick red, brown, green, or yellow Cu_2O precipitate. Carbohydrates that give a positive Benedict's test result are known as *reducing sugars* because they are the reducing agent in the reaction (they reduce the copper ions). A colored "solution" without a precipitate is a negative result and indicates the absence of a reducing sugar.

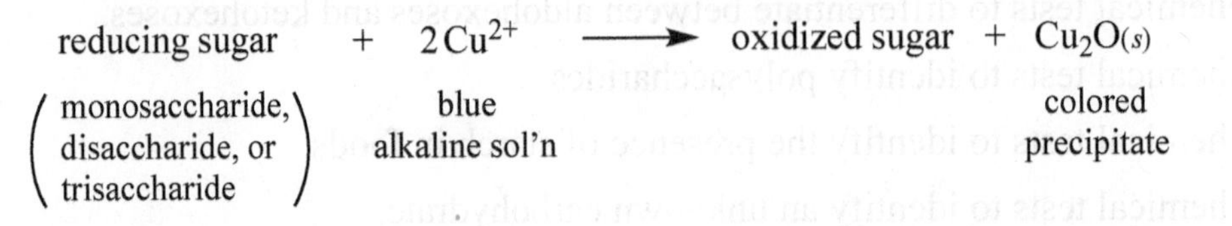

All monosaccharides are reducing sugars whether they are aldoses or ketoses. Disaccharides and trisaccharides may or may not be reducing sugars. The structural feature needed to be a reducing sugar is a free –OH on an anomeric carbon, not involved in a linkage between monosaccharides. The example below shows a reducing sugar and a nonreducing sugar.

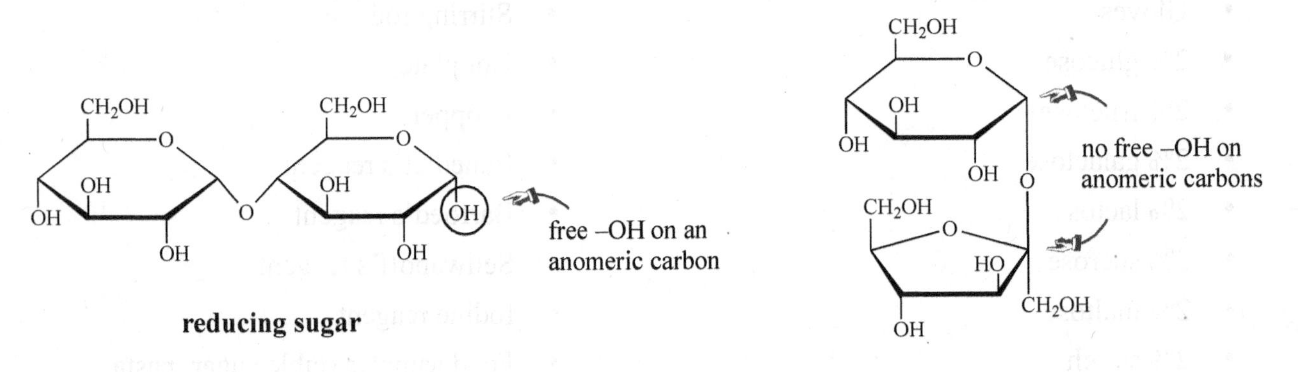

Barfoed's test

Barfoed's test distinguishes reducing sugars that are monosaccharides from reducing sugars that are disaccharides. Barfoed's reagent is a blue acidic solution containing Cu^{2+} ions. Like Benedict's test, a brick red, brown, green, or yellow Cu_2O precipitate indicates a positive result. It is a weaker oxidizing agent than Benedict's reagent, however, so it will oxidize monosaccharides within three minutes (fast color change). Disaccharides that are reducing sugars undergo the same reaction but at a much slower rate (slow color change).

$$\underset{\text{(monosaccharide only)}}{\text{reducing sugar}} + \underset{\text{blue acidic sol'n}}{2\,Cu^{2+}} \longrightarrow \text{oxidized sugar} + \underset{\text{colored precipitate}}{Cu_2O(s)}$$

Seliwanoff's test

Seliwanoff's test distinguishes between ketohexoses and aldohexoses. The Seliwanoff's reagent is a colorless solution of HCl and resorcinol. The hexoses are dehydrated when heated with the HCl in the reagent and react with the resorcinol in the reagent to form reddish solutions. The time required for the formation of the reddish color is used to distinguish ketohexoses from aldohexoses.

Ketohexoses (fructose and sucrose) usually form the reddish solution within a minute after adding the Seliwanoff's reagent. Aldohexoses may produce faintly pink solutions that take a longer time for the reaction to occur.

ketohexoses + Seliwanoff's reagent ⟶ reddish color (quickly)

aldohexoses + Seliwanoff's reagent ⟶ pinkish color (more slowly)

Iodine test

The iodine test detects the presence of polysaccharides. The iodine reagent is a yellow/orange solution of I_2 and KI dissolved in water. The I_2 molecules are trapped inside the large polysaccharide structures resulting in blue-black (amylose), reddish-purple (glycogen), or reddish-brown (amylopectin, cellulose, glycogen) colors. Monosaccharides and disaccharides are too small to trap iodine molecules, leaving the solution with its original yellow/orange color. This negative test result indicates the absence of a polysaccharide.

polysaccharides + iodine reagent ⟶ colored solution (blue-black, reddish-purple, or reddish-brown)

mono-, di-, or trisaccharides + iodine reagent ⟶ colored solution (yellow/orange from the iodine reagent)

Hydrolysis of carbohydrates

Disaccharides react with water and acid catalysts or enzymes to hydrolyze into their monosaccharides. Polysaccharides will also hydrolyze in the presence of acid catalysts or enzymes to produce smaller polysaccharides known as dextrins. The dextrins further hydrolyze to form disaccharides and eventually monosaccharides.

Tests can be used to determine the monosaccharide composition of disaccharides and polysaccharides after they have been hydrolyzed. For example, the disaccharide sucrose will give a negative Benedict's test (it is a nonreducing sugar), but after it is hydrolyzed to form glucose and fructose, the Benedict's test result will be positive because all monosaccharides are reducing sugars.

The polysaccharide amylose will give a negative Benedict's test, but after it is completely hydrolyzed to glucose, the Benedict's test result will be positive. The human body uses enzymes to catalyze these hydrolysis reactions as shown in the following reaction:

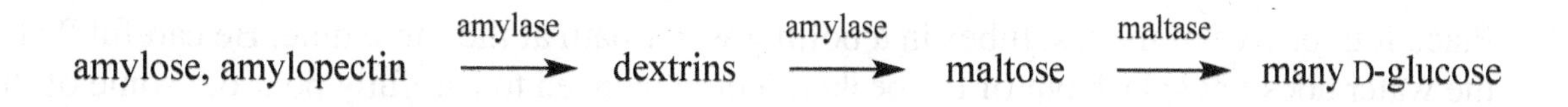

Experimental Procedures

Eye protection and appropriate clothing must be worn at all times. Students must wear gloves.

All of these chemicals are corrosive to the skin and toxic if ingested. Strong acids and bases are used in many of these procedures. Wash any contacted skin area immediately with running water and alert the instructor.

Observe proper safety precautions when handling hot objects.

For experimental procedures using a test tube, hold the test tube with a test tube clamp.

Use just one stirring rod for the many solutions that require stirring. Keep a small beaker of deionized water nearby to rinse off the stirring rod and then dry it before proceeding to the next solution to avoid contamination.

Discard all wastes properly as directed by the instructor.

A. Benedict's test

1. Set up a boiling water bath using a hot plate and a 600 mL beaker. Fill the beaker with approximately 200 mL of tap water. If enough equipment is available, set up two boiling water baths. Only heat four or five test tubes in one beaker at a time to avoid splashing.
2. Label nine 15 cm test tubes and place them in a test tube rack.
3. Put 10 drops of the following solutions into the labeled test tubes:
   - Test tube #1 ...... 2% glucose
   - Test tube #2 ...... 2% fructose
   - Test tube #3 ...... 2% galactose
   - Test tube #4 ...... 2% lactose
   - Test tube #5 ...... 2% sucrose
   - Test tube #6 ...... 2% maltose
   - Test tube #7 ...... 2% starch
   - Test tube #8 ...... unknown
   - Test tube #9 ...... deionized water (control—used to compare original color)
4. Add 2 mL of Benedict's reagent to each test tube. Stir the solution in each test tube using a stirring rod. Be sure to clean the stirring rod before stirring the next test tube.
5. Place four or five of the test tubes in a boiling water bath at the same time. Be careful that the water does not splash out of the beaker. You may need to carefully pour out some of the water and/or turn down the heat.

6. Remove the test tubes after 5 minutes, place them in the test tube rack, and let them cool. After 10 minutes, record your observations.
7. If only one boiling water bath is available, repeat steps 5 and 6 with the remaining test tubes.
8. Classify each solution as a *reducing* or a *nonreducing* sugar.

B. Barfoed's test

1. Prepare another set of nine test tubes using 10 drops of the same solutions used for Part A. Make sure that the test tubes are clean and dry.
2. To each test tube add 2 mL of the Barfoed's reagent. Stir the solution in each test tube using a stirring rod. Be sure to clean the stirring rod before stirring the next test tube.
3. Follow the same directions from Part A.1 regarding the heating of the test tubes in the boiling water bath(s).
4. Remove the test tubes after 5 minutes, place them in the test tube rack, and let them cool. After 10 minutes, record your observations.
5. Identify any solutions that are *reducing sugars*.

C. Seliwanoff's test

1. Prepare another set of nine test tubes using 10 drops of the same solutions used for Part A. Make sure that the test tubes are clean and dry.
2. To each test tube add 2 mL of the Seliwanoff's reagent. Stir the solution in each test tube using a stirring rod. Be sure to clean the stirring rod before stirring the next test tube.
3. Follow the same directions from Part A.1 regarding the heating of the test tubes in the boiling water bath(s).
4. Remove the test tubes after **1** minute and record your observations.
5. Place the test tubes into the boiling water bath for an additional **2** minutes.
6. Remove the test tubes and record your observations.
7. Place the test tubes back into the boiling water bath for another **2** minutes.
8. Remove the test tubes and record your observations for a third time.
9. Identify the solutions as ketoses, aldoses, or polysaccharides.

D. Iodine test

1. Using nine wells in a spot plate, place 5 drops of the same nine solutions used in Part A.
2. Add 1 drop of iodine solution to each well.
3. Identify polysaccharides by the color change. The iodine solution will not change color with monosaccharides and disaccharides. Record your observations.

E. Testing food samples

1. Obtain a pea-sized sample of each food item.
2. Place the samples in clean wells in a spot plate.
3. Add 1 drop of iodine solution to each sample.
4. Identify the samples containing polysaccharides by the color change.
5. Record your observations on the report sheet.

F. Hydrolysis of disaccharides and polysaccharides

1. Label four test tubes and place them in a test tube rack.
2. Put 2 mL (40 drops) of the following solutions into the test tubes:
   - Test tube #1 ...... 2% starch
   - Test tube #2 ...... 2% starch
   - Test tube #3 ...... 2% sucrose
   - Test tube #4 ...... 2% sucrose
3. To test tubes #1 and #3 add 12 drops of 10% HCl.
4. To test tubes #2 and #4 add 12 drops of deionized water.
5. Stir the solution in each test tube using a stirring rod. Be sure to clean the stirring rod before stirring the next test tube.
6. Place all four test tubes in the boiling water bath for 10 minutes.
7. Remove the test tubes and place them in the test tube rack for 10 minutes to cool.
8. To test tubes #1 and #3 add 6 drops of 10% NaOH. Use a stirring rod to mix each solution and then touch the tip of the stirring rod to a piece of red litmus paper. Continue to add more of the NaOH solution 2 drops at a time until the red litmus turns blue when touched with the stirring rod. This change indicates that the HCl has been neutralized.
9. Place 5 drops from test tube #1 into a spot plate well. Repeat with the solutions in test tubes #2, #3 and #4.
   - Add one drop of iodine solution to each well.
   - Record your observations. Indicate if any of the solutions contain polysaccharides.
10. Add 30 drops of Benedict's solution to each of the four test tubes.
11. Place the four test tubes into the boiling water bath for 5 minutes. Remove and let cool for ten minutes. Record your observations. Indicate if any of the solutions have been hydrolyzed.

Name ______________________ Date ______________________

Partner(s) ______________________ Section ______________________

______________________ Instructor ______________________

1. List the three major classes of carbohydrates and give a specific example of each.

2. Write the reaction (using words, not formulas) for the hydrolysis of sucrose.

3. Cellulose and amylose are both polysaccharides. What specific test can be used to differentiate between the two compounds?

4. What is the definition of a reducing sugar?

5. What purpose does the "control" test tube of water serve for the carbohydrate tests?

[*Reminder: there is one more problem on the next page*]

6. Complete the following table summarizing the carbohydrate tests. The completed first line is an example.

Test	Observations for a positive test	Interpretation of a positive result
Benedict's	blue solution ⟶ colored precipitate	A reducing sugar was present.
Barfoed's		
Seliwanoff's		
Iodine		

Name ______________________ Date ______________________

Partner(s) ______________________ Section ______________________

______________________ Instructor ______________________

Parts A–D. Turn to the next page to record observations and results for Parts A–D.

Using the results from the tests A–D, identify your unknown:

Unknown # ____________ is ______________________.

Note: the tests in this lab will not differentiate between glucose and galactose. If that is what you believe your unknown is, write "galactose or glucose."

How did you determine the identity of your unknown?

	TUBE #1	TUBE #2	TUBE #3	TUBE #4
Test tube contents				
Part A				
Benedict's observations				
Reducing or non-reducing sugar?				
Part B				
Barfoed's observations				
Monosaccharide or disaccharide?				
Part C				
Seliwanoff's observations after 1 minute				
3 minutes				
5 minutes				
Ketose? Aldose? Other?				
Part D				
Iodine observations				
Polysaccharide? (which one)				

TUBE #5	TUBE #6	TUBE #7	TUBE #8	TUBE #9

E. Testing food samples

Record the results from the iodine test on the food samples:

Food	Observed color after iodine test	Is a polysaccharide present?

Were the food samples that gave positive iodine tests what you would have expected? Why or why not? List three other foods that you might expect to contain polysaccharides.

F. Hydrolysis of disaccharides and polysaccharides

	Test tube #1	Test tube #2	Test tube #3	Test tube #4
Test tube contents ➡				
Iodine test observations				
Polysaccharide present? (yes/no)				
Benedict's test observations				
Hydrolyzed? (yes/no)				

What observation indicated the presence of a polysaccharide?

What observation indicated that hydrolysis had occurred?

Problems and questions

1. Is the following disaccharide a reducing sugar? Explain.

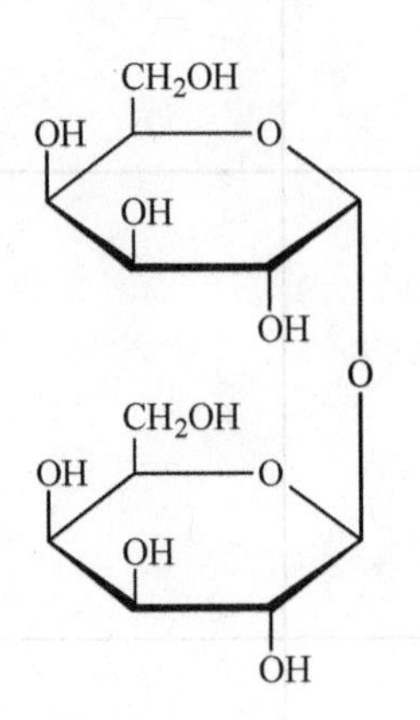

2. The following test results were observed in a lab experiment:

   - A reddish-orange solid with the Benedict' test
   - A reddish precipitate with the Barfoed's test
   - A red color within 1 minute with the Seliwanoff's test
   - A yellow color with the iodine test

   Name the carbohydrate that you would expect to show these results.

[*Reminder: there is one more problem on the next page*]

3. Determine whether each of the following compounds will give a positive or negative test result for each of the following tests. Put a + in the table for a positive result and a – for a negative result. The structures of the compounds are shown below.

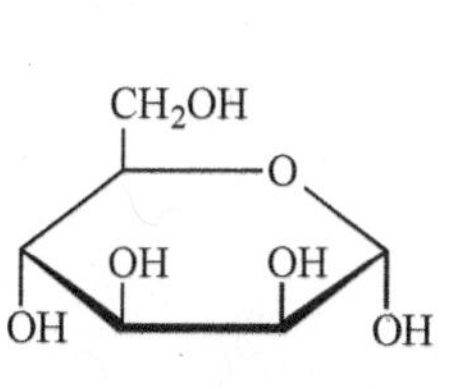

mannose

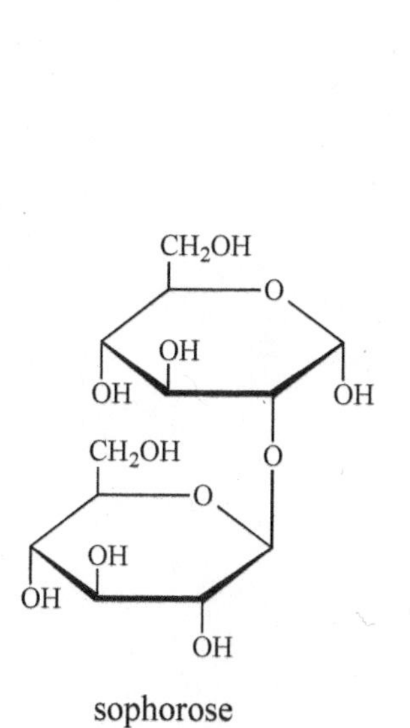

sophorose

raffinose

Compound	Benedict's	Barfoed's	Seliwanoff's	Iodine
mannose				
sophorose				
raffinose				

Goals

- Name and draw structural formulas of carboxylic acids, carboxylate salts, and esters.
- Build molecular models of carboxylic acids, carboxylate salts, and esters.
- Write the equations for the neutralization of carboxylic acids.
- Prepare salts of carboxylic acids and compare the solubility of the salts with the solubility of their acids.
- Write the equations for the esterfication of carboxylic acids.
- Prepare esters of carboxylic acids and compare the odors of the reactants with the odors of the products.
- Hydrolyze esters and compare the solubility of the reactants with the solubility of the products.

Materials

- Gloves
- Molecular model kits
- Hot plate
- Thermometer
- 600 mL beaker
- Test tubes
- Test tube rack
- Stirring rod
- Glacial acetic acid
- Benzoic acid
- Formic acid
- Salicylic acid
- pH paper
- 10% NaOH
- 10% HCl
- Methanol
- Ethanol
- 1-Pentanol
- 1-Octanol
- Isobutyl alcohol (2-methyl-1-propanol)
- Methyl salicylate
- Concentrated H_2SO_4
- Ice
- Watch glass

Discussion

Carboxylic acids contain a *carboxyl group.* The functional group is often written as –COOH but its structure is more easily seen when it is written as:

$$-\overset{\displaystyle O}{\overset{\|}{C}}-OH$$

Carboxylic acids are weak acids because they only slightly ionize in water and exist mostly as unionized molecules.

$$R-\overset{O}{\overset{\|}{C}}-OH + H_2O \rightleftharpoons R-\overset{O}{\overset{\|}{C}}-O^- + H_3O^+$$

Many common household products contain carboxylic acids. For example, folic acid, niacin (vitamin B-3), and vitamin C are all carboxylic acids. The sour acidic taste of fruits such as lemons, limes, oranges, and grapefruits is due to citric acid. Commercial vinegar is a 5% solution of acetic acid and has a distinctive taste and odor. Aspirin (acetylsalicylic acid) is also a carboxylic acid.

Neutralization reactions

Carboxylic acids react with bases such as NaOH and KOH in a neutralization reaction to form carboxylate salts and water. Carboxylic acids with low molecular weights are soluble in water because the polarity of the carboxyl group allows the formation of hydrogen bonds with water molecules. But as the length of the nonpolar carbon chain increases, carboxylic acids with five or more carbons are generally insoluble in water. However, their ionic salts are usually soluble in water due to the ionic charges. Ionic salts are commonly used as preservatives in food products. For example, benzoic acid has very limited solubility in water but its salt, sodium benzoate, is soluble and used as a preservative in soft drinks.

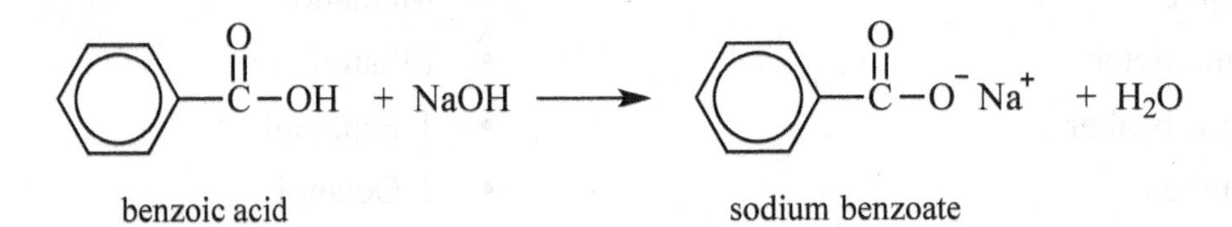

Reacting the carboxylic acid with a base gives the carboxylate salt. The reverse reaction occurs when the carboxylate salt reacts with an acid to give the original carboxylic acid. This explains why the exact form of the molecule is pH dependent.

$$\text{carboxylic acid (at low pH)} \rightleftharpoons \text{carboxylate ion (at high pH)}$$

Esterification reactions

An ester is formed by reacting a carboxylic acid with an alcohol in the presence of a strong acid catalyst. The reaction is a *dehydration* reaction since water is chemically removed from the reactants and becomes a second product. The general formula for an ester is:

$$R-\overset{O}{\overset{\|}{C}}-O-R'$$

An example of an esterification reaction is:

$$R{-}\overset{\overset{\displaystyle O}{\|}}{C}{-}OH \;+\; R'{-}OH \xrightarrow{H^+} R{-}\overset{\overset{\displaystyle O}{\|}}{C}{-}O{-}R' + H_2O$$

The **R group** comes from the carboxylic acid and the **R′ group** comes from the alcohol. The name of the ester is a combination of the alcohol and acid names. The alcohol R′ group is named first, using the "yl" ending of the alkyl group, followed by the acid group as a separate word. If the acid had an IUPAC name, the *–oic* ending is changed to *–oate*. If the acid had a common name, the *–ic* ending is changed to *–ate*. The examples in Table 1 show both the IUPAC names and common names.

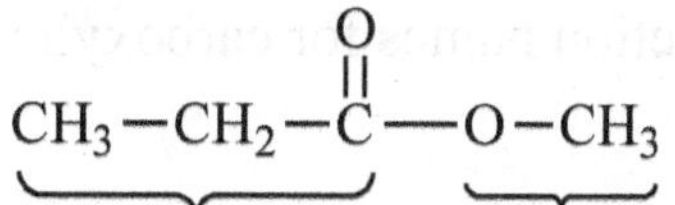

⟹ methyl propanoate

acid group (IUPAC: propanoic acid) — alcohol group (methyl)

Table 1 ***IUPAC and common names of esters***

Structural formula	IUPAC name	Common name
$H{-}\overset{\overset{O}{\|}}{C}{-}O{-}CH_3$	methyl methanoate	methyl formate
$CH_3{-}\overset{\overset{O}{\|}}{C}{-}O{-}CH_3$	methyl ethanoate	methyl acetate
$C_6H_5{-}\overset{\overset{O}{\|}}{C}{-}O{-}CH_3$	methyl benzoate	N/A

The odors of carboxylic acids and alcohols are often noticeably unpleasant. The odors of esters are frequently pleasant and fruity. Thus, esters may be used for fragrances and flavoring agents since their agreeable odors are easily detected. Esters are volatile (evaporate quickly) even when their molecular weights are high since they do not form hydrogen bonds between their molecules.

Hydrolysis of esters

The hydrolysis of an ester using a base such as NaOH or KOH is called *saponification*. The products of the reaction are the carboxylate salt and the alcohol. Even though the ester is usually insoluble in water, the hydrolysis products are soluble if the carbon chains are reasonably short.

$$R-\overset{O}{\overset{||}{C}}-O-R' + NaOH \longrightarrow R-\overset{O}{\overset{||}{C}}-O^-Na^+ + R'-OH$$

A reaction map of reactants, products, and reaction names for carboxylic acids and esters is shown in Figure 1.

Figure 1 ***Reaction map for carboxylic acids and esters***

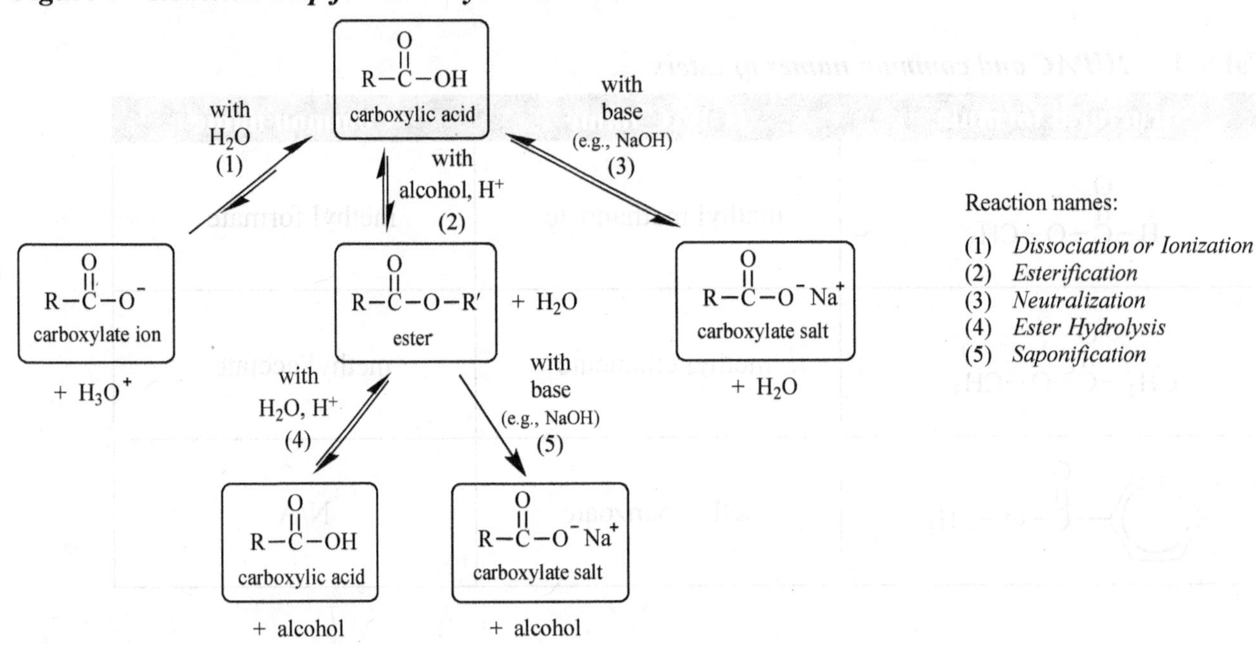

Experimental Procedures

Eye protection and appropriate clothing must be worn at all times. Students must wear gloves.

These chemicals have strong odors. Use the proper wafting technique when observing the odors.

All of these chemicals are corrosive to the skin and toxic if ingested. Strong acids and bases are used in many of these procedures. Wash any contacted skin area immediately with running water and alert the instructor.

Observe proper safety precautions when handling hot objects.

For experimental procedures using a test tube, hold the test tube with a test tube clamp.

Discard all wastes properly as directed by the instructor.

In order to make the best use of time, it may be helpful to start Part C while working on Part A or Part B.

A. Properties and reactions of carboxylic acids

1. Set up a warm (60°C) water bath using a hot plate and a 600 mL beaker. Fill the beaker with approximately 200 mL of tap water.
2. Make molecular models for acetic acid and benzoic acid. Have the instructor check the models on the report sheet.
3. Draw the condensed structural formulas on the report sheet.
4. Obtain two test tubes and put approximately 2 mL of deionized water in each.
5. In one test tube add 5 drops of acetic acid and stir with a stirring rod.
6. In the second test tube add a small amount (about the size of a match head) of benzoic acid and stir with a clean stirring rod.
7. Notice the solubility of the acids in water. Record your observations on the report sheet.
8. Dip a stirring rod into the first test tube, then touch it to a small strip of pH paper. Compare the color of the paper to the color chart on the container and record the pH of the solution on the report sheet.
9. Clean the stirring rod and then repeat procedure #8 with the second test tube.
10. Place the test tube containing benzoic acid into a warm (60°C) water bath for 5 minutes. Record your observations. Let the test tube cool before proceeding with the next procedure. Save the warm water bath for Part B.
11. Add about 10 drops of 10% NaOH to each of the test tubes. Test the pH of each test tube to ensure that the pH is slightly greater than 7. If not, add NaOH one drop at a time and retest the pH until the solutions are basic. Record your observations.

12. Write the equations for the neutralization reactions of the carboxylic acids with NaOH on the report sheet. Also, give the names of the organic products.
13. Add about 10 drops of 10% HCl to each test tube. Test the pH of each test tube to ensure that the pH is slightly less than 7. Record your observations.
14. Write the equations for the reactions of the sodium carboxylate salts with HCl on the report sheet. Also, give the names of the organic products.

B. Formation and properties of esters

1. Make molecular models of acetic acid and methanol. Use the models to demonstrate the formation of methyl acetate. Have the instructor approve the models and demonstration.
2. Write the equation for the reaction on the report sheet.
3. Label five test tubes and place them in a test tube rack. Using the table on the report sheet as a guide, prepare each ester with the appropriate carboxylic acid and alcohol.
   - Carboxylic acid — If the acid is a liquid, pour 1 mL into the numbered test tube. Otherwise, if the acid is a solid, put a match-head size piece into the numbered test tube.
   - Alcohol — Add 1 mL of the alcohol to the same numbered test tube.
4. Carefully add 3–5 drops of concentrated sulfuric acid (consider wearing gloves to protect yourself) to each test tube and heat the test tubes in a warm (60°C) water bath for 10–15 minutes.
5. Remove the test tubes and carefully place them in an ice-water bath.
6. Add 5 mL of deionized water to each test tube.
7. Using the wafting technique, carefully note the odors of the esters formed. It may be helpful to pour a few drops of the solution onto a watch glass to aid in the detection of the odor.
8. Record any odor observed on the report sheet.
9. Draw the condensed structural formula and name each ester on the report sheet.
10. Dispose of the esters as directed by the instructor.

C. Hydrolysis of esters

1. Heat the warm water bath used in parts A and B until it boils.
2. Place 3 mL of water in a test tube. Add 5 drops of methyl salicylate.
3. Record the appearance and odor.
4. Add 1 mL (20 drops) of 10% NaOH. Carefully observe whether the NaOH remains on the surface above the ester or sinks to the bottom under the ester.
5. Place the test tube in a boiling water bath for 30 minutes. Use the wafting technique to detect any odor change, note the appearance of the layers, and record your observations.
6. After the solution has cooled, add approximately 1 mL of 10% HCl. Check the pH to assure that the solution has turned acidic and add HCl dropwise if necessary.
7. Record your observations. Write the condensed formula and name the precipitate that formed.

Name ______________________ Date ______________________

Partner(s) ______________________ Section ______________________

______________________ Instructor ______________________

1. Draw the condensed structure of formic acid.

2. Write the balanced equation showing the ionization of formic acid in water.

3. Is formic acid a weak acid or a strong acid? Why?

4. Write the balanced equation showing the reaction of formic acid with NaOH.

5. Write the **balanced** equation showing the formation of an ester from the reaction of formic acid and 1-propanol. Give the IUPAC name of the ester.

6. Octyl formate has the odor of oranges. Name the carboxylic acid and alcohol needed to synthesize this ester.

Name ______________________ Date ______________________

Partner(s) ______________________ Section ______________________

______________________ Instructor ______________________

A. Properties and reactions of carboxylic acids

Condensed structural formula	Instructor's approval of models	pH	Solubility in cold water	Solubility in warm water	Observation after addition of NaOH	Observation after addition of HCl
Acetic acid				X		
Benzoic acid						

Write the balanced equation for the reaction of acetic acid with NaOH. Name the organic product.

Write the balanced equation for the reaction of benzoic acid with NaOH. Name the organic product.

Write the balanced equation for the reaction of sodium acetate with HCl. Name the organic product.

Write the balanced equation for the reaction of sodium benzoate with HCl. Name the organic product.

B. Formation and properties of esters

Use condensed structural formulas to write the balanced equation for the reaction of acetic acid and methanol to form methyl acetate.

Instructor's approval of models and reaction _______________.

Test tube	Carboxylic acid	Alcohol	Condensed structural formula of ester	Name of ester	Observed odor of ester
#1	Salicyclic acid (2-hydroxybenzoic acid)	Methanol			
#2	Formic acid	Isobutyl alcohol (2-methyl-1-propanol)			
#3	Formic acid	Ethanol			
#4	Acetic acid	1-Octanol			
#5	Acetic acid	1-Pentanol			

C. Hydrolysis of esters

Procedure	Results
Odor of ester before heating	
Appearance of layers before heating	
Odor after heating	
Appearance of layers after heating	
Observations after addition of HCl	

Write the condensed formula and name the precipitate that was formed from the hydrolysis of methyl salicylate.

Problems and questions

1. Draw the condensed structural formulas and write the IUPAC and common names for the first three carboxylic acids.

Structural formula	IUPAC name	Common name

2. Write a balanced chemical equation for the conversion of butanoic acid to potassium butanoate.

3. Draw structures of the products of the following reaction:

$$CH_3-(CH_2)_4-\overset{\overset{\displaystyle O}{||}}{C}-O^-\,Na^+ \quad + HCl \longrightarrow$$

4. Draw the structures of the "parent" acid and "parent" alcohol from which the following ester is made:

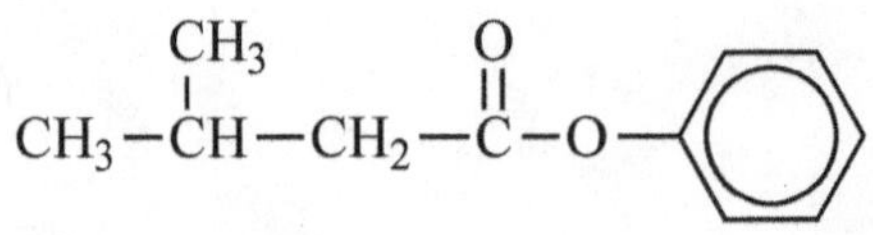

$$CH_3-\overset{\overset{\displaystyle CH_3}{|}}{CH}-CH_2-\overset{\overset{\displaystyle O}{||}}{C}-O-C_6H_5$$

5. Explain why esters are less soluble in water than carboxylic acids of similar molecular mass.

6. Draw the structure of the product that is the **same** in both reactions below.

$$CH_3-\underset{\displaystyle CH_3}{\overset{|}{CH}}-CH_2-\overset{\displaystyle O}{\overset{||}{C}}-O-CH_3 + H_2O \xrightarrow{H^+}$$

$$CH_3-\underset{\displaystyle CH_3}{\overset{|}{CH}}-CH_2-\overset{\displaystyle O}{\overset{||}{C}}-O-CH_3 + NaOH \longrightarrow$$

7. Rank the following compounds in terms of water solubility (1 = least soluble, 5 = most soluble):

acetic acid ______

ethanol ______

dimethyl ether ______

sodium formate ______

pentane ______

Goals

- Prepare a soap.
- Write the equation for a saponification reaction.
- Prepare a detergent.
- Compare properties of soaps and detergents.

Materials

- Gloves
- Hot plate
- Magnetic stir bar
- 400 mL beaker
- Beaker tongs
- Stirring rod
- Buchner funnel
- Filter paper
- Suction flask
- Vacuum apparatus
- Spatula
- Watch glass
- 50 mL beaker
- 100 mL beaker
- Test tubes
- Rubber stoppers
- pH paper with color chart
- Ice
- Solid shortening (Crisco)
- Absolute ethanol
- 20% NaOH
- Saturated NaCl solution
- 1-Dodecanol (container must be kept in a warm water bath)
- Concentrated H_2SO_4
- NaCl
- 6 M NaOH
- Mineral oil
- 1% $MgCl_2$
- 1% $FeCl_3$
- 1% $CaCl_2$

Discussion

Soaps are salts of carboxylic acids that contain a very long alkyl group. The polar (ionic salt) end is soluble in water and the nonpolar alkyl chain is soluble in nonpolar substances. The formulas for sodium palmitate (a common soap) are shown below in the long chain and condensed versions, and the line-angle formula.

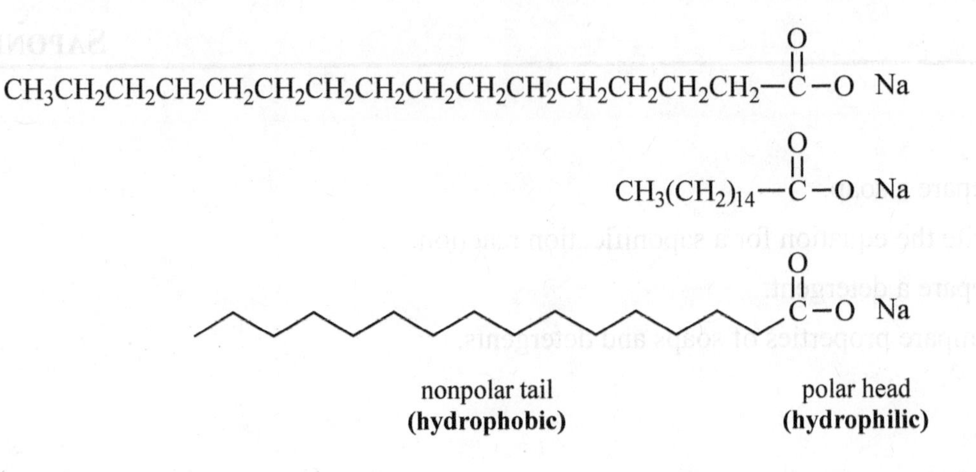

sodium palmitate formulas

Detergents are similar to soaps in their structure but differ in the water-soluble portion of the molecule. The formulas for sodium lauryl sulfate, a common type of detergent, are shown below in the long chain and condensed versions, and the line-angle formula. Compare the formulas for sodium palmitate with the formulas of sodium lauryl sulfate. Notice the difference between the polar (ionic salt) ends. The detergent includes a sulfate (SO_4^{2-}) bonded to the sodium ion rather than the carboxylate (COO^-) found in the soap molecule.

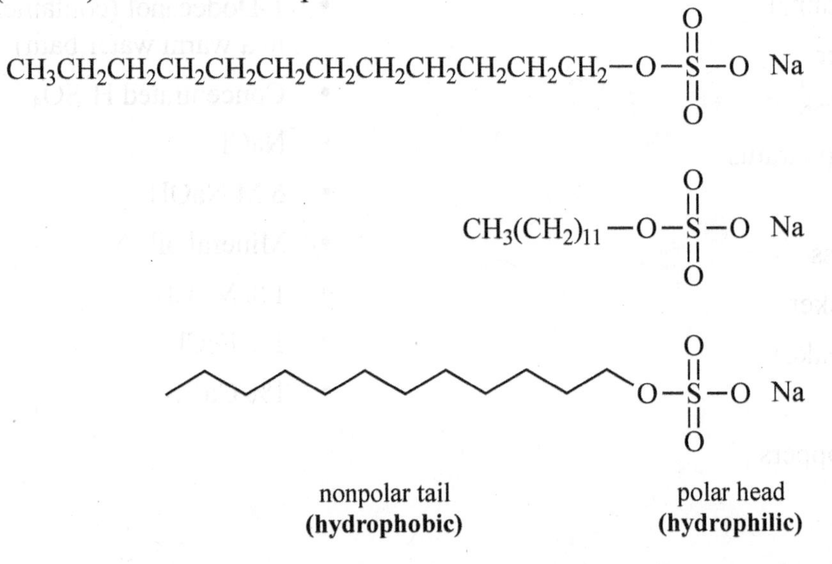

sodium laurel sulfate formulas

Soaps can be made from solid fats such as lard or liquid fats such as vegetable oils. Soaps made with NaOH are "solid" and are used for bar soaps. Soaps made with KOH form potassium salts and are used for liquid soaps. While soaps and detergents both have nonpolar and polar regions in their molecules, it is their solubility in hard water that differs. Hard water contains Fe^{2+}, Fe^{3+}, Mg^{2+}, and Ca^{2+} ions that replace the Na^+ or K^+ ions in the soap molecules. The soap molecules then become insoluble in water and form precipitates often seen as "soap scum." Detergents remain soluble in the presence of the hard water ions because the sulfate ion has a much lower attraction for the hard water ions.

The improved cleaning power of detergents over soaps led soap manufacturers to increase their production of detergents and by the 1950's most detergents included phosphates as the ionic salts. Those molecules eventually found their way into lakes and streams where the phosphates acted as fertilizers for algae just as phosphate fertilizers are used on agricultural crops, lawns, and gardens. The increase in algae decreased the oxygen supply in the waterways, resulting in the death of many fish and upsetting the ecological balance. By the late 1980's the presence of phosphates in detergents was banned or limited and has now been replaced by sulfates or other non-phosphate ionic groups.

Preparation of soaps and detergents

Soaps are prepared by the base hydrolysis of fats or oils. This reaction is known as *saponification.* Glycerol is also obtained as a product in the reaction and is used in the production of antifreeze and moisturizers. A process known as *salting out* is used to separate the soap from the glycerol and any remaining reactants. This is achieved by mixing a saturated NaCl solution to form a soap precipitate that can easily be filtered. The following balanced equation shows the saponification reaction of glyceryl tristearate (tristearin) and NaOH to form glycerol and sodium stearate:

$$\begin{array}{l} CH_2{-}O{-}\overset{\overset{\large O}{\|}}{C}{-}(CH_2)_{16}CH_3 \\ | \\ CH{-}O{-}\overset{\overset{\large O}{\|}}{C}{-}(CH_2)_{16}CH_3 \\ | \\ CH_2{-}O{-}\overset{\overset{\large O}{\|}}{C}{-}(CH_2)_{16}CH_3 \end{array} + 3\,NaOH \longrightarrow \begin{array}{l} CH_2{-}OH \\ | \\ CH{-}OH \\ | \\ CH_2{-}OH \end{array} + 3\ CH_3(CH_2)_{16}{-}\overset{\overset{\large O}{\|}}{C}{-}O\ Na$$

glyceryl tristearate   glycerol   sodium stearate

Cleaning action of soaps and detergents

Soaps and detergents form *micelles* in an aqueous solution (see Figure 1). The micelle is a spherical cluster of molecules in which the polar heads of the soap or detergent molecules face outward. The nonpolar tails are located in the interior of the micelle.

Dirt is largely composed of nonpolar molecules such as grease and oil. Dirt molecules can dissolve inside the interior of the micelles, which are also nonpolar. The polar heads of the soap or detergent molecules are attracted to the polar water molecules. Thus, the dirt on clothes can be "pulled" into the micelle and then washed down the drain with the water.

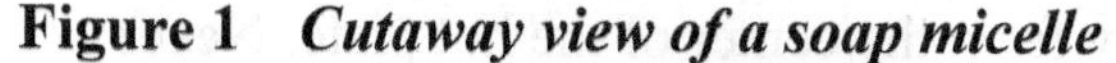
Figure 1 ***Cutaway view of a soap micelle***

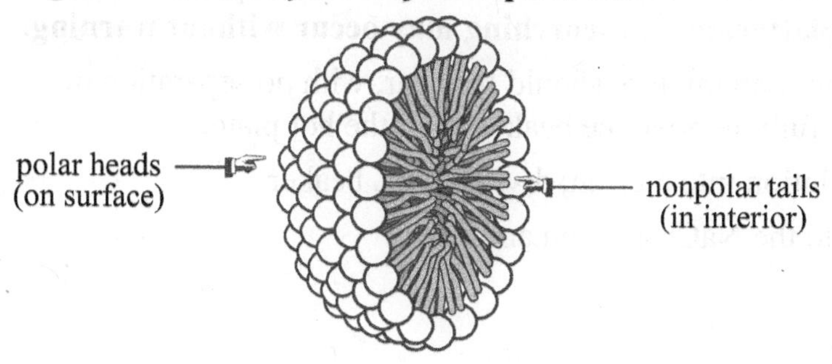

Experimental Procedures

Eye protection and appropriate clothing must be worn at all times.

Boiling the NaOH and fat mixture can be dangerous since it can splatter easily. Follow the directions carefully to avoid injury.

These chemicals are corrosive to the skin. Use gloves to avoid skin contact. Wash any contacted skin immediately with running water and alert the instructor.

Observe the correct procedure to handle the hot beaker using beaker tongs.

Discard all wastes properly as directed by the instructor.

A. Preparation of a soap

1. Set up a hot plate. If available, use a hot plate with a magnetic stirrer. Be sure to retrieve the stirring bar at the end of the experiment.
2. Weigh an empty 400 mL beaker. Record the mass on the report sheet.
3. Add approximately 3 grams of solid shortening to the beaker. Reweigh and record the mass. Add or remove solid shortening until the mass is within 2.8 to 3.2 grams. Reweigh if necessary.
4. Add approximately 10 mL of absolute ethanol to the beaker. The ethanol is a solvent for both the solid shortening and the NaOH.
5. Add approximately 10 mL of 20% NaOH to the beaker.
6. Gently put the magnetic stirring bar into the beaker. Do not drop it in.
7. Have beaker tongs nearby **before** starting the saponification reaction.
8. Place the beaker on the hot plate and turn the magnetic stirrer on. Avoid splattering by setting the stirrer on a low speed.
9. Prepare 10 mL of a diluted ethanol solution by mixing 5 mL of absolute ethanol and 5 mL of deionized water in a 20 mL beaker. Set the solution aside. It may be necessary to add this diluted solution as the mixture is being heated in step 10.
10. Use a low to medium temperature setting and heat the solid shortening and ethanol solution to a gentle boil for about 20 minutes. If the soap mixture begins to overheat or burn, add 3 mL portions of the diluted ethanol solution from step 9. Reduce the heat if necessary or use beaker tongs to move the soap mixture away from the center of the hot plate. **Never leave the reaction unattended. Splattering and scorching may occur without warning.**
11. After approximately 20 minutes, the soap mixture should be clear, with no separation of layers. Use the beaker tongs to carefully remove the beaker from the hot plate.
12. Pour 30 mL of a saturated NaCl solution into a clean, dry, 250 mL beaker.
13. Carefully pour the soap mixture into the NaCl solution and stir.

14. Set up a Buchner funnel and suction flask to the vacuum as demonstrated by the instructor. Place a piece of filter paper in the funnel. It should cover the flat surface of the funnel. Then turn the vacuum on.
15. Pour the NaCl/soap mixture from step 13 into the funnel. Rinse the collected soap in the funnel with two 10 mL portions of **cold** water.
16. Turn off the vacuum. Wear gloves to protect your hands from the NaOH still remaining on the soap. Use a spatula to remove the filter paper and soap from the funnel. Scrape the soap onto a watch glass. Use a paper towel to pat the soap dry. **Save the soap for Part C.**
17. Wash the Buchner funnel and place a clean piece of filter paper in the funnel. You will use this apparatus again in Part B.
18. Retrieve the magnetic stirring bar from the beaker. Wash and dry the stirring bar and return it to the equipment area.

B. Preparation of a detergent

1. Prepare a warm water bath by putting approximately 50 mL of hot tap water into a 100 mL beaker.
2. Pour 2 mL of 1-dodecanol into a small test tube and place the test tube into the warm water bath.
3. Stir slowly while carefully adding 1 mL (20 drops) of concentrated H_2SO_4 to the test tube.
4. Leave the test tube in the warm water bath for 10 minutes.
5. Fill a 100 mL beaker with approximately 30 mL of ice. Add approximately 5 g of solid NaCl and stir to mix thoroughly. Add enough water to bring the volume of the iced salt water to 40 mL.
6. In a 50 mL beaker, mix 2 mL of 6 M NaOH and 5 mL of water. Stir.
7. Slowly pour the 1-dodecanol/sulfuric acid mixture from step 4 into the NaOH/water solution from step 6.
8. Stir for 2 minutes. The detergent will precipitate out of the solution.
9. Pour the iced salt water from step 5 into the solution containing the detergent precipitate and stir vigorously for 1 minute.
10. Filter the detergent using the clean Buchner funnel and new piece of filter paper from Part B. Use **ice-cold** water to rinse the detergent.
11. Remove the detergent from the funnel and scrape onto a watch glass. Use a paper towel to pat the detergent dry. **Save the detergent for Part C.**

C. Properties of soaps and detergents

Prepare solutions of your soap and detergent:

- Obtain two 100 mL beakers and fill each with 50 mL of deionized water.
- Weigh out approximately 1 gram of your prepared soap and add it to the first beaker. Stir to dissolve.
- Weigh out approximately 1 gram of your prepared detergent and add it to the second beaker. Stir to dissolve.
- Perform each of the following tests on your soap solution and your detergent solution. Record the results on the report sheet.

pH test

Obtain two test tubes. Place 10 mL of your soap solution in the first test tube and 10 mL of your detergent solution in the second test tube. Dip a clean stirring rod into each test tube and then touch the stirring rod to a piece of pH paper. Compare the color to the color chart. Record the pH for each solution. Save the test tubes for the next test.

Suds test

Use the test tube solutions saved from the pH test. Insert a clean, dry rubber stopper and shake each test tube for 10 seconds. Record your observations.

Emulsification test

Emulsification is the ability of soaps and detergents to disperse water-insoluble substances. Obtain two test tubes. Place 10 mL of your soap solution in the first test tube and 10 mL of your detergent solution in the second test tube. Add 5 drops of mineral oil to each test tube. Use clean rubber stoppers for each test tube and shake for 20 seconds. Record your observations.

Hard water test

Label six test tubes. Prepare the test tubes as directed, stopper each test tube with a clean rubber stopper and shake for 10 seconds. Record your observations.

- Test tube #1 ...... 5 mL of soap solution + 2 mL of 1% $CaCl_2$
- Test tube #2 ...... 5 mL of soap solution + 2 mL of 1% $MgCl_2$
- Test tube #3 ...... 5 mL of soap solution + 2 mL of 1% $FeCl_3$
- Test tube #4 ...... 5 mL of detergent solution + 2 mL of 1% $CaCl_2$
- Test tube #5 ...... 5 mL of detergent solution + 2 mL of 1% $MgCl_2$
- Test tube #6 ...... 5 mL of detergent solution + 2 mL of 1% $FeCl_3$

Name ______________________ Date ______________________

Partner(s) ______________________ Section ______________________

______________________ Instructor ______________________

1. This lab has a number of safety precautions. Describe three of them and why they are important.

2. Why is ethanol added to the reactants in the saponification reaction?

3. The structure of glyceryl tripalmitate is shown below. Write the balanced equation for the saponification reaction of glyceryl tripalmitate, also known as tripalmitin.

$$\begin{array}{l} CH_2\text{-}O\text{-}\overset{\displaystyle O}{\overset{\|}{C}}\text{-}(CH_2)_{14}CH_3 \\ | \\ CH\text{-}O\text{-}\overset{\displaystyle O}{\overset{\|}{C}}\text{-}(CH_2)_{14}CH_3 \\ | \\ CH_2\text{-}O\text{-}\overset{\displaystyle O}{\overset{\|}{C}}\text{-}(CH_2)_{14}CH_3 \end{array}$$

glyceryl tripalmitate

4. Explain the differences between soaps and detergents.

Name ______________________ Date ______________________

Partner(s) ______________________ Section ______________________

______________________ Instructor ______________________

A. Preparation of a soap

Mass of empty beaker ______________

Mass of beaker and solid shortening ______________

Mass of solid shortening used ______________

Describe the appearance of your soap product after it has been washed and dried.

B. Preparation of a detergent

Describe the appearance of your detergent product after it has been washed and dried.

C. Properties of soaps and detergents

Test	Soap	Detergent
pH		
Suds		
Emulsification		

Complete the following table for the hard water test results:
(Complete the "Contents" column following the example for test tube #1)

Test tube	Contents	Observations
#1	soap and Ca^{2+}	
#2		
#3		
#4		
#5		
#6		

Problems and questions

1. According to Pliny the Elder (23 A.D.–79 A.D.), the Phoenicians were using soap as early as 600 B.C. Animal fat (or olive oil) was boiled with ashes from a wood fire. The ashes from the fire contained potassium carbonate (K_2CO_3) which were reacted with slaked lime [$Ca(OH)_2$] to form potassium hydroxide (KOH). Finally, the potassium hydroxide was reacted with the animal fat or oil to produce soap. Assume you are a Phoenician with "alchemical knowledge." What is the formula of your soap using the following component of olive oil and the potassium hydroxide from the wood fire ashes as starting materials?

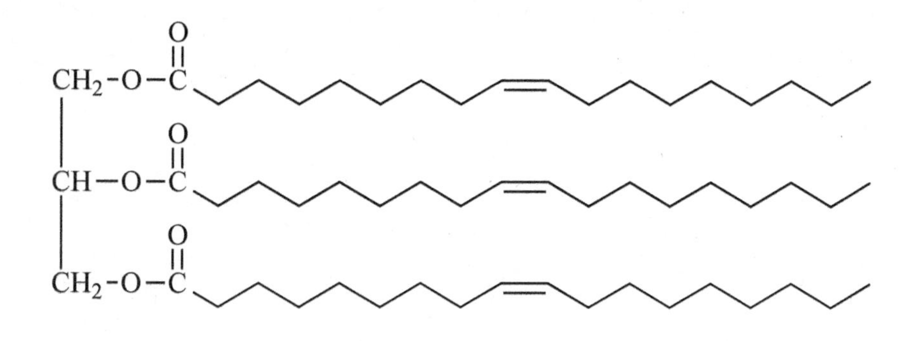

2. What properties of soaps and detergents allow them to remove oil and grease stains from clothing?

3. What is the purpose of the NaCl solution that is used in the process of preparing soaps and detergents?

Goals

- Draw structures of amines and amides.
- Name amines and amides.
- Classify amines as primary, secondary, or tertiary.
- Build molecular models of amines and amides.
- Identify physical and chemical properties of amines and amides.
- Write the equations for the neutralization of amines.
- Write the equations for the formation of amides.
- Write the equations for the hydrolysis of amides.

Materials

- Gloves
- Molecular model kits
- Test tube rack
- Test tubes
- Test tube holder
- Stirring rod
- pH paper and color charts
- Red litmus paper
- Hot plate
- 600 mL beaker
- Triethylamine
- Aniline
- *N,N*-dimethylaniline
- 6 M HCl
- 6 M NaOH
- 6 M NH_3 (NH_4OH)
- Acetamide
- Benzamide

Discussion

Amines

Amines are organic derivatives of ammonia (NH_3) and may be classified as primary (1°), secondary (2°), or tertiary (3°). The number of alkyl or aromatic groups attached to the amine nitrogen determines the classification. The following examples use R to represent an alkyl or aromatic group to illustrate these classifications:

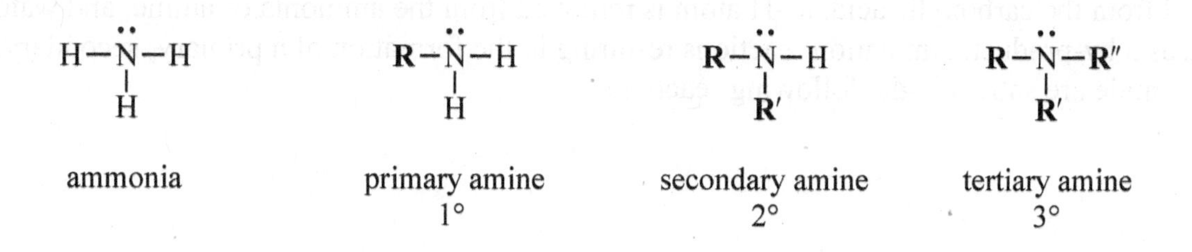

Properties of amines

Amines have distinctive odors. Lower molecular weight molecules often smell fishy, unpleasant, pungent, or similar to ammonia. For example, two amines are responsible for the odor of decaying flesh: putrescine and cadaverine.

All amines are soluble in organic solvents. Amines with four or fewer carbons are soluble in water due to extensive hydrogen bonding with water molecules. Because amines acts as weak bases, a solution containing a water-soluble amine will have a pH higher than 7. The unshared pair of electrons on the nitrogen atom accepts a proton from a water molecule, forming a positively charged alkyl ammonium ion and a negatively charged hydroxide ion as shown in the following reaction:

$$R-\ddot{N}H_2 + H_2O \rightleftharpoons R-\overset{+}{N}H_3 + OH^-$$

Since amines are bases, they can be neutralized with acids to produce ammomium salts. These salts are water soluble because of their ionic bonds even if the parent amine is not soluble in water. Many drugs containing amines are prepared as ammonium salts using HCl, H_2SO_4, or other acids to take advantage of this solubility property. This is known as the solubility switch.

Acids such as lemon juice can be used to neutralize the odor of fish because it converts the odiferous amine to a salt, which can easily be washed away. The neutralization of a primary amine is shown in the following reaction:

$$R-\ddot{N}H_2 + HCl \longrightarrow R-\overset{+}{N}H_3 + Cl^-$$

Amides

Amides contain a functional group consisting of a nitrogen atom attached to a carbonyl group.

$$-\overset{O}{\overset{\|}{C}}-\underset{|}{N}-$$

amide group

Amides are formed by the *amidation* reaction of a carboxylic acid and a primary alkyl amine, secondary alkyl amine, aromatic amine, or ammonia. In the amidation reaction, an –OH group is removed from the carboxylic acid, a –H atom is removed from the ammonia or amine, and water is formed as a by-product. Amidation reactions resulting in the formation of a primary, secondary, and tertiary amide are shown in the following reactions:

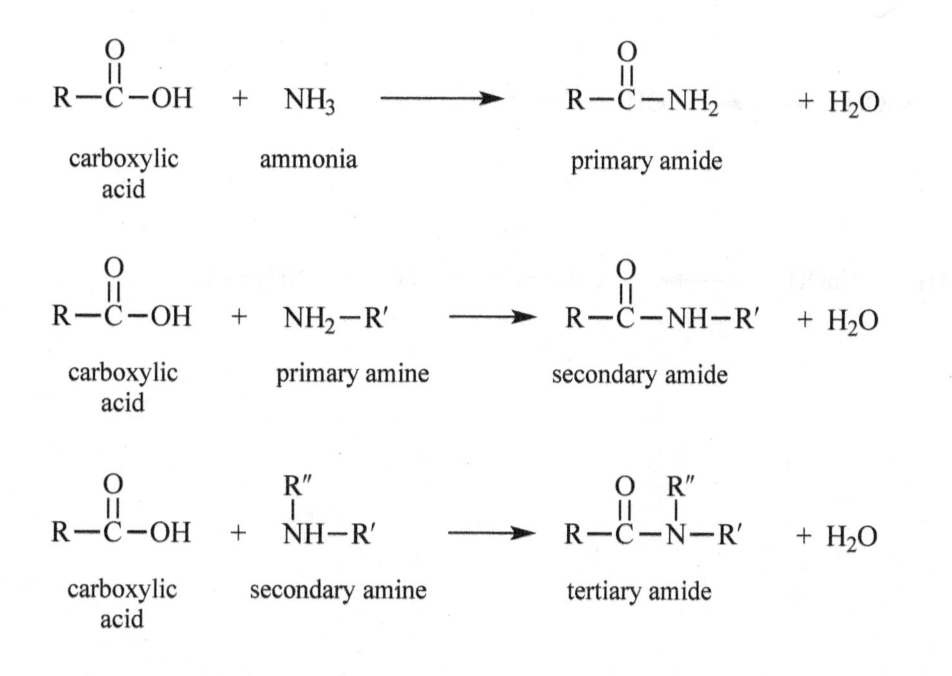

Properties of amides

Amides are neutral compounds since the electrons on the nitrogen atom are in resonance with the electrons in the carbonyl group and are not available to attract a hydrogen ion. Thus, amides show none of the acidic properties of carboxylic acids nor the basic properties of amines.

The reaction of an amide with water is a *hydrolysis* reaction. This reaction can occur in an acidic solution or a basic solution. In the reaction, the amide bond is broken and the carboxylic acid and the amine are separated. Amides are generally odorless compounds, but odors are detected during a hydrolysis reaction because the products (either an amine or a carboxylic acid) have very distinct odors.

If the reaction is carried out in an acidic solution, the products are a carboxylic acid and an amine salt (or ammonium salt if the amide is a primary amide). The following examples show an acid hydrolysis reaction:

$$CH_3-\overset{O}{\overset{\|}{C}}-NH_2 + HCl + H_2O \longrightarrow CH_3-\overset{O}{\overset{\|}{C}}-OH + NH_4Cl$$

primary amide

$$CH_3-\overset{O}{\overset{\|}{C}}-NH-CH_3 + HCl + H_2O \longrightarrow CH_3-\overset{O}{\overset{\|}{C}}-OH + CH_3-\overset{+}{N}H_3\,Cl^-$$

secondary amide

If the reaction is carried out in a basic solution, the products are a carboxylate salt and an amine (or ammonia if the amide is a primary amide). Often, the odor of ammonia or an amine will be detected as the reaction progresses. The following examples show a basic hydrolysis reaction:

$$CH_3-\overset{O}{\overset{||}{C}}-NH_2 \;+\; NaOH \longrightarrow CH_3-\overset{O}{\overset{||}{C}}-O^-\,Na^+ \;+\; NH_{3(g)}$$

primary amide

$$CH_3-\overset{O}{\overset{||}{C}}-NH-CH_3 \;+\; NaOH \longrightarrow CH_3-\overset{O}{\overset{||}{C}}-O^-\,Na^+ \;+\; NH_2-CH_3$$

secondary amide

Experimental Procedures

Eye protection and appropriate clothing must be worn at all times.

These chemicals have strong odors and are irritating. Work in the hood. Use the proper wafting technique when observing the odors.

These chemicals are corrosive to the skin. Use gloves to avoid skin contact. Wash any contacted skin immediately with running water and alert the instructor.

Discard all wastes properly as directed by the instructor.

A. Structures and models of amines

1. Make models of ammonia, ethylamine, dimethylamine, trimethylamine, and aniline. Complete each model and have it checked by the instructor.
2. Draw their condensed structural formulas and classify the amines as 1°, 2°, or 3° on the report sheet.

B. Properties of amines

1. Label four test tubes and place in a test tube rack.
2. Put 5 drops of liquid or 0.1 g (match tip sized) of solid into the test tubes:
   - Test tube #1 ......6 M NH_3
   - Test tube #2 ......triethylamine
   - Test tube #3 ......aniline
   - Test tube #4 ......*N,N*-dimethylaniline
3. Use the proper wafting technique to note the odor of each sample. To check for an odor, cup your hand and gently fan the odor from the test tube toward your face. Never smell or sniff a chemical directly with your nose. Record your observations on the report sheet.
4. Add 2 mL of deionized water to each test tube.
5. Using a clean stirring rod, carefully stir the contents of test tube #1. Note the absence or presence of layers to determine the solubility of the amine. Record your observations.
6. Touch the stirring rod to a small strip of pH paper. Compare the color of the paper to the color chart on the container and record the pH of the solution.
7. Repeat step 5 and 6 with the remaining test tubes.
8. **Save the test tubes for the Part C.**

C. Reaction of amines

1. To each test tube from B.8, add 10 drops of 6 M HCl. Using a clean stirring rod, carefully stir the contents of each test tube.
2. Cautiously check the odors of each test tube. If any odor remains in a test tube, continue to add HCl dropwise, stirring after each addition, until little or no odor remains (this should only require a few more drops). Record your observations.
3. Using the technique from B.6, check the pH of each test tube. Record your observations.
4. To each test tube, add 10 drops of 6 M NaOH. Using a clean stirring rod, carefully stir the contents of each test tube. Record your observations.

D. Structures and models of amides

1. Make models of acetic acid and ammonia. Use these models to demonstrate the formation of acetamide. Have the instructor check the models and reaction.
2. Write the balanced equation for the reaction on the report sheet.

E. Properties of amides

1. Label two test tubes and place in a test tube rack.
2. Put 0.1 g (match tip sized) of solid into the test tubes:
   - Test tube #1.......acetamide
   - Test tube #2.......benzamide
3. Use the proper wafting technique to note the odors of test tubes #1 and #2. Record your observations.
4. Add 2 mL of deionized water to each test tube. Using a clean stirring rod, carefully stir the contents of each test tube.
5. Note the presence or absence of layers to indicate solubility. Record your observations.
6. Cautiously recheck the odors. Record your observations.
7. Touch the stirring rod to a small strip of pH paper. Compare the color of the paper to the color chart on the container and record the pH of the solution.
8. **Save test tube #1 for the first part of Part F.**

F. Reactions of amides

Base hydrolysis (saponification)

1. Set up a boiling water bath using a hot plate and a 600 mL beaker. Fill the beaker with approximately 200 mL of tap water.
2. To test tube #1, containing acetamide saved from step E.8, add 20 drops of 6 M NaOH. Gently heat the test tube in the boiling water bath for approximately 3 minutes.

3. Moisten a strip of red litmus paper with deionized water. Carefully hold the strip just inside the mouth of test tube #1 without touching the glass or the liquid. Record your observations.
4. Continue heating the test tube for 10 more minutes.
5. Remove the test tube and put it in a cold water bath for 5 minutes.
6. If any solid has formed at the bottom of the test tube, carefully pour off the liquid into a clean test tube and discard the solid as directed by the instructor. It is excess reactant.
7. Add 20 drops of 6 M HCl to the clear solution in the test tube. Record your observations.
8. Write the balanced equation for the reaction on the report sheet.

<u>Acid hydrolysis</u>

9. Label two test tubes and place in a test tube rack.
10. Put 0.1 g (match tip sized) of solid into the test tubes:
    - Test tube #1 ......acetamide
    - Test tube #2 ......benzamide
11. Add 2 mL of 6 M HCl to each test tube. Gently heat the test tubes in the boiling water bath for 5 minutes.
12. Cautiously note any odors. Record your observations.
13. Write the balanced equations for the reactions on the report sheet.

Name ______________________ Date ______________________

Partner(s) ______________________ Section ______________________

______________________ Instructor ______________________

1. Identify each of the following as a primary, secondary or tertiary amine:

$CH_3-\ddot{N}(CH_2CH_3)-CH_2CH_3$ $CH_3-\ddot{N}(H)-CH_3$ $C_6H_5-CH_2-\ddot{N}H_2$

2. Aqueous solutions of amines have a pH greater than 7 because amines are weak bases. In basic solutions, $[OH^-] > [H_3O^+]$. Why does the hydroxide ion concentration increase when an amine dissolves in water?

3. What precautions must be taken during this lab regarding odors?

4. Circle the compound in each pair that you would expect to be more soluble in water:

a) dimethyl amine OR triethylamine

b) sodium benzoate OR benzoic acid

c) *N,N*-diethylacetamide OR acetamide

5. Predict the approximate pH of solutions of the following compounds. Will the pH be greater than 7, less than 7, or approximately equal to 7?

a) acetic acid ______________

b) methanamine ______________

c) acetamide ______________

Name ____________________ Date ____________________

Partner(s) ____________________ Section ____________________

____________________ Instructor ____________________

A. Structures and models of amines

Compound	Condensed structural formula	1°, 2°, or 3°	Instructor's approval for models
ammonia		X	
ethylamine			
dimethylamine			
trimethylamine			
aniline			

B. Properties of amines

Test tube	Compound	Odor	Solubility in H_2O	pH
#1	ammonia			
#2	triethylamine			
#3	aniline			
#4	*N,N*-dimethylaniline			

C. Reactions of amines

Test tube	Compound	Odor after 10 drops of HCl	Total number of drops of HCl needed to remove odor	pH	Observations after addition of NaOH
#1	ammonia + HCl				
#2	triethylamine + HCl				
#3	aniline + HCl				
#4	*N,N*-dimethylaniline + HCl				

D. Structure and models of amides

Compound	Condensed structural formula	Instructor's approval for models
acetic acid		
ammonia		
acetamide		

Write the balanced equation for the reaction of acetic acid and ammonia to form acetamide.

E. Properties of amides

Test tube	Compound	Odor? If so, describe	Solubility in H_2O	Odor with H_2O? If so, describe	pH
#1	acetamide				
#2	benzamide				

F. Reactions of amides

Base hydrolysis

Test tube	Compound	Observations of red litmus paper	Observations after addition of HCl
#1	acetamide + H_2O + NaOH		

Write the balanced equation for the reaction of acetamide and NaOH.

Acid hydrolysis

Test tube	Compound	Odor
#1	acetamide + H_2O + HCl	
#2	benzamide + H_2O + HCl	

Write the balanced equation for the reaction of acetamide and H_2O + HCl.

Write the balanced equation for the reaction of benzamide and H_2O + HCl.

Problems and questions

1. Why are drugs that contain the amine functional group most often administered to patients in the form of amine chloride or hydrogen sulfate salts?

2. Draw the structure of the missing substance in the following reaction:

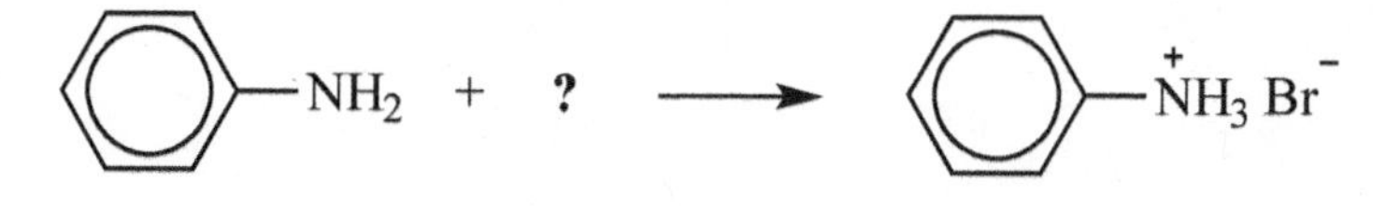

3. Draw the structures of the carboxylic acid and amine from which the following amide could be formed.

$$CH_3-CH_2-\overset{O}{\overset{\|}{C}}-\underset{CH_3}{\underset{|}{N}}-CH_3$$

4. Circle the compounds below whose aqueous solutions are basic.

$$CH_3-CH_2-\overset{O}{\overset{\|}{C}}-NH_2 \qquad CH_3-CH_2-CH_2-NH_2 \qquad NH_2-CH_2-CH_2-\overset{O}{\overset{\|}{C}}-NH_2$$

5. Draw the structure of the missing substance in the following reaction:

$$CH_3-CH_2-\overset{O}{\overset{\|}{C}}-\underset{CH_3}{\underset{|}{N}}-CH_3 + ? \longrightarrow CH_3-CH_2-\overset{O}{\overset{\|}{C}}-O^- K^+ + CH_3-NH-CH_3$$

Goals

- Build molecular models of peptides.
- Isolate casein from milk.
- Calculate the percent of casein in milk.
- Use chemical tests to identify functional groups within a protein.
- Use chemical tests to identify an unknown.
- Use chemical tests to show the effect of denaturing proteins.

Materials

- Gloves
- Molecular model kits
- 150 mL beaker
- Stirring rod
- pH paper
- Red litmus paper
- Hot plate
- Thermometer
- Beaker tongs
- Buchner funnel
- Filter paper
- Suction flask
- Vacuum apparatus
- Watch glass
- Test tubes
- Test tube rack
- Unknown solutions (#1, #2, #3)
- Nonfat milk
- 10% acetic acid in dropper bottles
- 95% ethanol
- 1% glycine
- 1% tyrosine
- 1% proline
- 1% albumin
- 1% gelatin
- 10% NaOH
- 10% HNO_3
- Biuret reagent
- 0.2% ninhydrin reagent
- Concentrated HNO_3
- 1% $Pb(NO_3)_2$
- 1% $AgNO_3$

Discussion

Next to water, proteins are the most abundant substances in almost all cells. They comprise approximately 15% of a cell's overall mass and approximately 50% of a cell's dry mass. All proteins contain the elements carbon, hydrogen, nitrogen, and oxygen; most also contain sulfur. Plants are able to synthesize proteins from CO_2, H_2O, nitrates, sulfates, and other compounds. Animals cannot synthesize all of their required proteins so they must be obtained through regular dietary intake.

Proteins are naturally-occurring polymers in which the monomeric units are amino acids. There are twenty amino acids used in the formation of proteins and each has a carboxylic acid group and an amine group, hence, amino acids. Each amino acid differs in the R group attached to the α-carbon. The α-carbon is the carbon bonded to both the carboxylate group and the amino group. The R group determines the name and properties of the amino acid. The standard twenty amino acids are shown in their zwitterion forms in Table 1. A *zwitterion* is a dipolar ion that carries both a positive and a negative charge as a result of an internal acid–base reaction in an amino acid. Some textbooks show the amino acids in their unionized forms. The unionized form doesn't really exist, it is just an alternate way of drawing an amino acid.

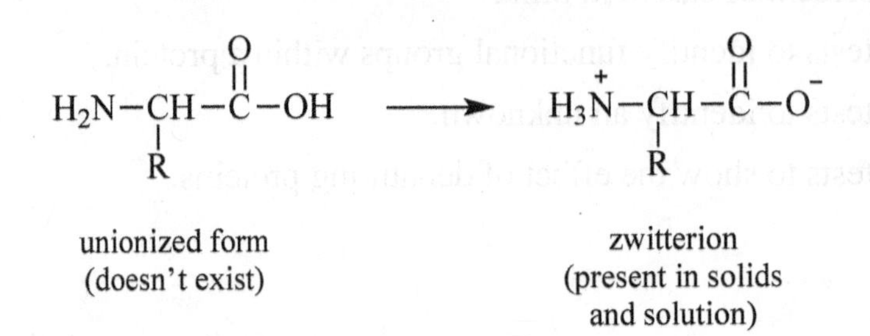

Two or more amino acids react to form *peptides.* These molecules are formed with peptide (amide) bonds connecting the carboxylic acid group of one amino acid and the amine group on the next amino acid. This dehydration reaction also results in the formation of a water molecule. *Dipeptides* contain two amino acid monomers and *tripeptides* contain three amino acid monomers. *Polypeptides* are polymers of amino acids and are commonly referred to as *proteins* when there are at least fifty amino acids in the polymer chain. The formation of a peptide (amide) bond is shown in the following reaction.

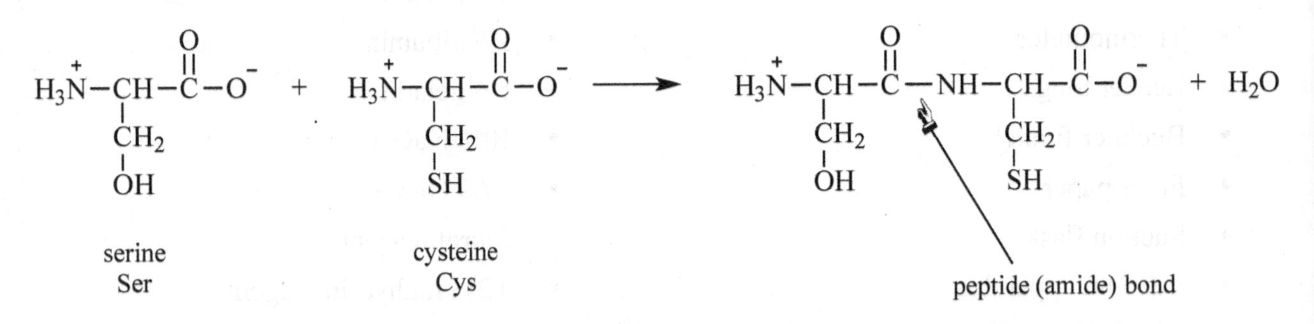

The three letter abbreviations for the amino acid names are generally used to name polypeptides. The order of amino acids is written from left to right and order is important. The dipeptide above is Ser–Cys and this is not to be confused with Cys–Ser, which is a different dipeptide.

Protein structure

There are four levels of protein structure. In order of increasing complexity, they are the primary structure, secondary structure, tertiary structure, and quaternary structure.

The *primary structure* is the order in which the amino acids are linked together in a protein. The two tripeptides Gly–Ala–Phe and Phe–Gly–Ala are composed of the same three amino acids; however, they have different primary structures because the order isn't the same. The primary structure conveys no information about the actual three-dimensional shape of the protein. The shape is determined by the secondary and tertiary structures.

The *secondary structure* is the arrangement in space adopted by the backbone of the protein. The three most common secondary structures are a result of hydrogen bonds along the backbone of the protein: alpha helix, beta-pleated sheet, and the triple helix.

The *tertiary structure* is the overall three-dimensional shape of a protein that results from interactions (attractions and repulsions) between amino acid side chains (R groups). As interactions occur between widely-separated amino acids within the polypepetide chain, segments of the chain twist and bend until the protein acquires a specific three-dimensional shape.

Finally, some proteins have structures involving two or more polypeptide chains that are independent of one another; they are not covalently bonded together. The *quaternary structure* is the organization of the various polypeptide chains. An example of a protein with a quaternary structure is hemoglobin. It consists of two identical α chains and two identical β chains. The interactions of the four chains are responsible for the quaternary structure.

Isoelectric point

Each protein has a specific pH called its *isoelectric point.* This is the pH at which a protein may become insoluble in water and precipitate out of solution. By adjusting the pH of a protein solution, the bonds in the tertiary structure of the protein are disrupted. An example of this reaction can be demonstrated by adding lemon juice or vinegar to milk to form curds.

The most important protein in milk is casein, which reaches its isoelectric point at pH = 4.6. By adding an acid to a neutral milk solution to lower the pH, the casein will separate out as an insoluble precipitate. Any fat that precipitates out with the casein can be removed by rinsing the product with alcohol.

Table 1 ***The 20 standard amino acids, grouped according to side-chain polarity***

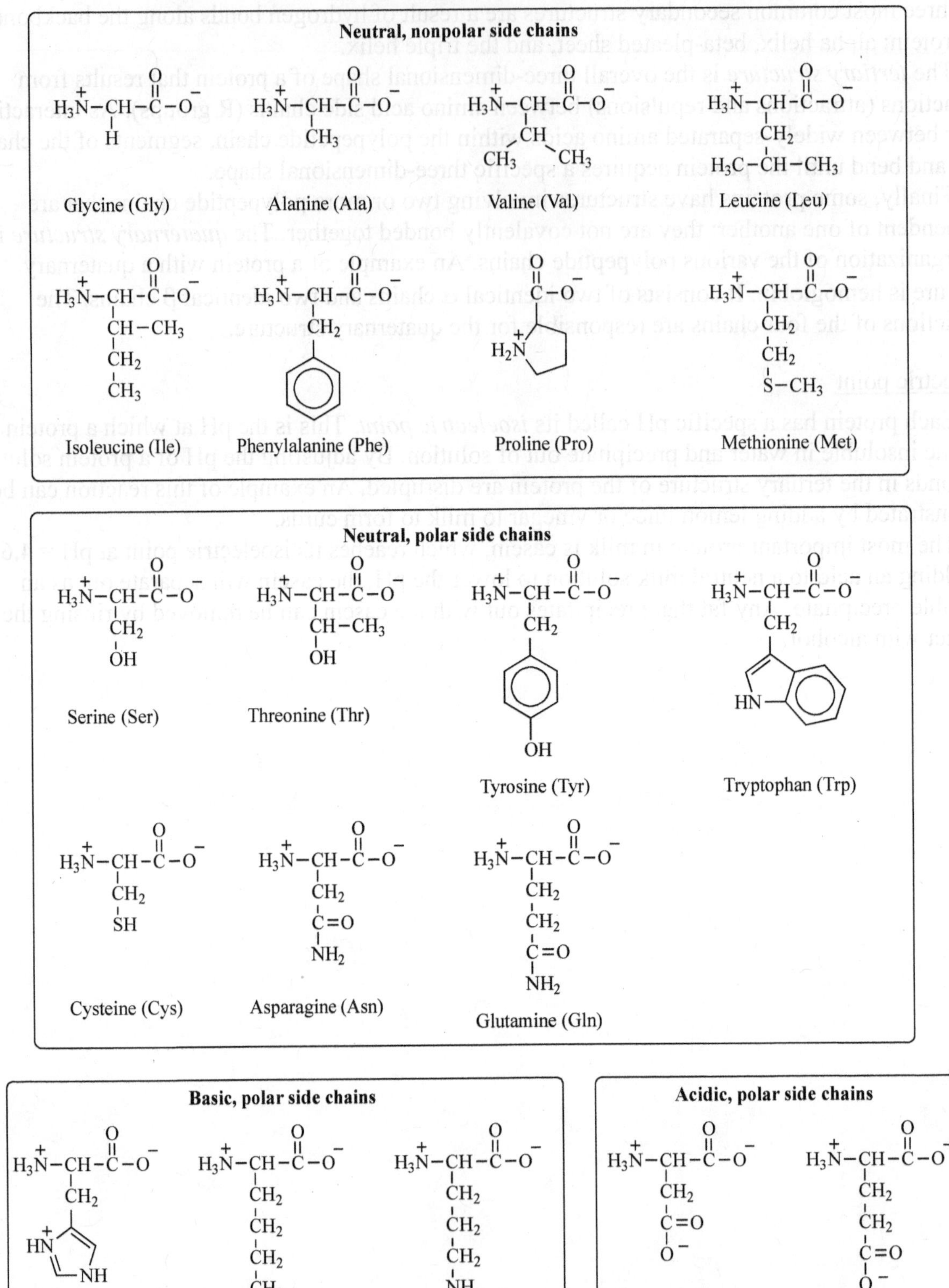

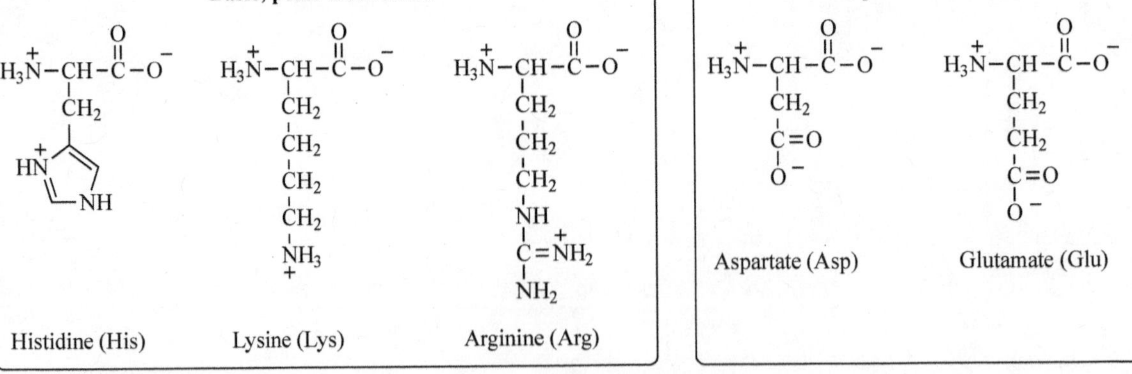

Chemical tests for amino acid R groups

A variety of chemical tests give color changes with the functional groups present in amino acids, peptides, and/or proteins. It is important to note the original color of the reagent and compare it to the product color.

The *biuret test* reagent is a 0.1% $CuSO_4$/NaOH solution. The pale blue color of the reagent turns to a violet color in the presence of at least two peptide bonds. Therefore a positive test result will be observed with a tripeptide or larger polypeptide. The blue color may darken but does not turn violet in the presence of single amino acids or dipeptides.

The *ninhydrin test* reagent is a colorless solution with the molecular structure shown in the following reaction. Notice that two molecules of the reagent are required for the balanced reaction. Also notice that the amino acid is broken into several colorless products, with the nitrogen atom only appearing in the blue-violet product.

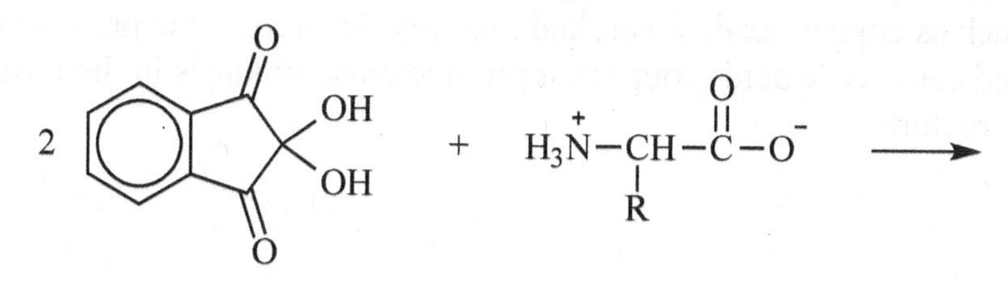

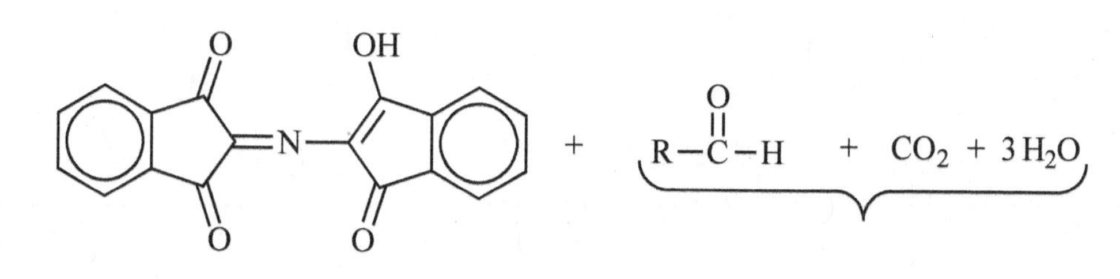

violet-colored molecule
"Ruhemann's purple"

colorless products

Amino acids are colorless and the formation of a blue-violet product indicates the presence of an amino acid and most proteins. Proline and hydroxyproline contain another amino group in their R groups and also react with the ninhydrin solution, but the resulting product is yellow.

The *xanthoproteic test* uses concentrated HNO_3 as the reagent. The nitric acid reacts with tyrosine and tryptophan to give yellow compounds. Making the test solution alkaline can intensify the color to an orange-brown. These two amino acids have R groups containing aromatic rings with amino or hydroxyl substituents. The aromatic ring of phenylalanine has no substituents on it and does not give a positive test.

Denaturation of proteins

Protein molecules have very complex structures that assume precise three-dimensional shapes. The shape of a protein is often key to its function. *Denaturation* is the process in which the bonds that hold together the secondary, tertiary, or quaternary structure of the molecule are disrupted. The denaturated protein will frequently coagulate or form a precipitate. Denaturation may be reversible or irreversible, depending upon the type of protein and the severity of the denaturation. Denaturation may be caused by heat and ultraviolet light, mechanical agitation, organic solvents, acids, bases, detergents, or heavy metal ions.

Heat increases the kinetic energy of a protein solution and may break the weak hydrogen bonds folding the protein into its three-dimensional shape. An example is the frying of an egg. Ultraviolet light and microwave radiation operate similarly to the action of heat.

Organic solvents such as ethanol, 2-propanol, and acetone also destroy hydrogen bonds. A 70% ethanol solution is a disinfectant because it denatures the proteins in bacteria, thus killing them.

Violent whipping or shaking the protein causes the molecules to elongate and then entangle. This is what happens when egg whites are beaten into meringue.

A change in pH caused by the addition of an acid or base may cause the protein to reach its isoelectric point and precipitate from the solution. The acids and bases disrupt the ionic bonds between amino acids containing acidic and basic R groups.

Heavy metal ions such as copper, lead, silver, and mercury denature some proteins by reacting with the sulfur groups and carboxylic acid groups that provide cross-linkages in the proteins' tertiary and quaternary structures.

Experimental Procedures

Eye protection and appropriate clothing must be worn at all times.

Perform the tests slowly and carefully to avoid splashing. Some tests are exothermic and will cause the test tube to become hot to the touch.

Strong acids and bases are used in this experiment. Wear gloves to avoid skin contact. Wash any contacted skin immediately with running water and alert the instructor.

Some reagents will react with the proteins in skin and leave a stain. Wash thoroughly. Be aware that there may be some residual discoloration. The colors will fade over time as the skin sloughs off.

Discard all wastes properly as directed by the instructor.

A. Structures and models

1. Make models of glycine and serine. Choose one other amino acid and make a model. Use Table 1 for the structures of the amino acids. Use the first two models to demonstrate the formation of a peptide (amide) bond between the amino acids to form a dipeptide. Add the third model to make a tripeptide. Complete each model and have it checked by the instructor.
2. Name the dipeptide and tripeptide using the three letter abbreviations. Use the condensed structural formulas and write the balanced equation for the formation of the dipeptide.

B. Isolation of casein

1. Weigh a clean, dry 150 mL beaker. Record the mass on the report sheet.
2. Add approximately 20 mL of nonfat milk to the beaker. Reweigh the beaker with the milk and record the mass.
3. Using a clean, dry stirring rod, place a drop of the milk sample on a strip of pH paper. Compare the color on the strip to the color chart on the container. Record the pH.
4. Use a hot plate to gently warm the beaker and milk until the temperature reaches 50°C. Use a low heat setting and stir constantly with a stirring rod.
5. Use beaker tongs to remove the beaker and milk from the hot plate. Stirring continuously, add 10% acetic acid drop by drop until the casein precipitates from the solution. Continue adding acid until no further precipitation occurs. The total amount of acid added may be 2–3 mL.
6. Use the stirring rod to place a drop of the clear solution on a strip of pH paper. Compare the color on the strip to the color chart on the container. Record the pH. This pH is the isoelectric point for casein.

7. Set up a Buchner funnel apparatus as directed by the instructor. Place a sheet of filter paper in the funnel. Pour the contents of the beaker into the funnel and collect the solid casein. Wash with two 10 mL portions of water. If "non-fat" milk was used, an alcohol rinse is not necessary. If 2% or whole milk was used, rinse the solid casein again, using 10 mL of ethanol.
8. Weigh a clean, dry watch glass. Record the mass.
9. Carefully remove the filter paper from the Buchner funnel. Scrape the casein onto the watch glass. Gently press a dry paper towel onto the casein to help remove any moisture.
10. Reweigh the watch glass with the casein. Record the mass.
11. Calculate the percentage of casein in the milk:

$$\% \text{ casein} = \frac{\text{mass of casein}}{\text{mass of milk}} \times 100\%$$

12. **Save the casein for Part C.** Separate into three match tip size pieces.

C. Chemical tests

Biuret test

1. Label six clean, dry test tubes and place them in a test tube rack.
2. Place 2 mL of liquid or a match tip size sample of solid into the test tubes:
    - Test tube #1.......1% glycine
    - Test tube #2.......1% tyrosine (be sure to shake the bottle before obtaining the sample)
    - Test tube #3.......1% proline
    - Test tube #4.......1% gelatin (be sure to shake the bottle before obtaining the sample)
    - Test tube #5.......casein from Part B (use one match tip size piece)
    - Test tube #6.......unknown
3. Add 2 mL of 10% NaOH to each test tube. Using a clean stirring rod, carefully stir the contents of each test tube.
4. Add 5 drops of biuret reagent (5% $CuSO_4$) to each test tube. Stir. Record the color of each sample.

Ninhydrin test

1. Set up a boiling water bath using a hot plate and a 600 mL beaker. Fill the beaker with approximately 200 mL of tap water.
2. Prepare a new set of samples following steps 1 and 2 from the Biuret test.
3. Add 1 mL of 0.2% ninhydrin reagent to each test tube.
4. Place the test tubes in the boiling water bath for 5 minutes.
5. Remove the test tubes and place in a test tube rack. Record the color of each sample.

Xanthoproteic test

1. Prepare a new set of samples following steps 1 and 2 from the Biuret test.
2. Cautiously add 10 drops of concentrated HNO_3 to each test tube.
3. Place the test tubes in the boiling water bath for 3 minutes.
4. Remove the test tubes and place them in a cold-water bath for 10 minutes.
5. Carefully add 10% NaOH drop by drop until the solution just turns red litmus paper blue. Use a stirring rod to stir after adding each drop and touch the tip of the rod to the litmus paper. The total amount of the base added may be 2–3 mL. **This reaction generates heat. Exercise caution.**
6. Record your observations.

Denaturation of proteins

1. Label seven test tubes and place in a test tube rack.
2. Put 2 mL of 1% albumin egg solution into each test tube.
3. Set aside test tube #7. Use this sample as a comparison when recording your observations for each of the following tests.
4. Place test tube #1 in a boiling water bath for 5 minutes. Record your observations.
5. Add 2 mL of 10% NaOH to test tube #2. Stir. Record your observations.
6. Add 2 mL of 10% HNO_3 to test tube #3. Stir. Record your observations.
7. Add 4 mL of 95% ethanol to test tube #4. Stir. Record your observations.
8. Add 10 drops of 1% $AgNO_3$ to test tube #5. Stir. Record your observations.
9. Add 10 drops of 1% $Pb(NO_3)_2$ to test tube #6. Stir. Record your observations.

Name ______________________ Date ______________________

Partner(s) ______________________ Section ______________________

______________________ Instructor ______________________

1. Complete the following table:

Test	Positive test observation for example: purple solution ⟶ blue solution	What does a positive test indicate? (e.g., aromatic amino acid)
Biuret		
Ninhydrin		
Xanthoproteic		

2. Draw the condensed structural formula for the tripeptide Val–Lys–Trp.

3. What is the primary structure of a protein?

[*Reminder: there are more problems on the next page*]

4. Mercurochrome is a complex organic derivative of mercury and was formerly used as an antiseptic. Why does mercurochrome kill bacteria?

5. Explain why the type of milk used in the isolation of casein determines whether ethanol needs to be used as a rinse solution.

Name ______________________ Date ______________________

Partner(s) ______________________ Section ______________________

______________________ Instructor ______________________

A. Structures and models

	Name using three letter abbreviation(s)	Condensed structural formula	Instructor's approval for models
glycine			
serine			
third amino acid chosen			
dipeptide			
tripeptide			

Write the balanced equation for the formation of the dipeptide.

B. Isolation of casein

Mass of empty beaker ______________

Mass of beaker and milk ______________

pH of milk sample ______________

pH of clear solution after precipitation of casein ______________
(isoelectric point for casein)

Mass of watch glass ______________

Mass of watch glass and casein ______________

Calculations

Mass of milk ______________

Mass of casein ______________

Percent of casein in milk ______________
(show calculations)

C. Chemical tests

Test tube contents	Biuret test observations	Ninhydrin test observations	Xanthroproteic test observations	Test results indicate?
#1 glycine				
#2 tyrosine				
#3 proline				
#4 gelatin				
#5 casein				
#6 Unknown				

Unknown #__________________ is ____________________.

How did you determine the results of your unknown?

Denaturation of proteins

Test tube # Denaturation technique	Observations of denaturation technique on albumin
#1 Heat	
#2 NaOH	
#3 HNO_3	
#4 Ethanol	
#5 $AgNO_3$	
#6 $Pb(NO_3)_2$	

#7 No denaturation	Describe the original albumin appearance.

Did all of the denaturation techniques give the same results? If not, what differences did you observe?

Problems and questions

1. How does a change in pH affect the structural levels of a protein?

2. After working with nitric acid (HNO_3), a student noticed a yellow spot on her hand. Explain why this might have occurred. Be specific.

3. What test might be used to determine whether an unknown compound is an amino acid or a protein?

4. Why is milk or egg whites often given to someone who accidentally ingests a heavy metal ion such as silver or mercury?

5. An appetizer known as ceviche is prepared without heat by placing slices of raw fish in a solution of lemon or lime juice. After 3 or 4 hours, the fish appears to be "cooked." Explain the chemistry of this process.

[*Reminder: there is one more problem on the next page*]

6. *"Little Miss Muffet sat on a tuffet,
   Eating her curds and whey..."*

   Curd is the dairy product obtained by curdling (coagulating) the milk protein casein. It is used in the production of cheese and yogurt. In milk, casein molecules have a net electrical charge. The charged casein molecules are stabilized by the polar water molecules.

   Curds can be separated from the whey by neutralizing the casein. For most cheeses, this is accomplished with a bacterial culture and the enzyme rennet. But a bacterial culture need not be used. What commonly found kitchen ingredient(s) could be added to milk to precipitate the curds? Explain. What is the correct chemical term used to identify the pH at which precipitation of casein occurs?

Goals

- Use nutritional labels to identify vitamins contained in foods.
- Use physical tests to identify vitamins as water soluble or fat-soluble.
- Determine the vitamin C content in a commercial tablet.
- Determine the vitamin C content in various foods.
- Build molecular models showing the reaction of vitamin C with iodine.

Materials

- Gloves
- Molecular model kits
- Test tube rack
- Test tubes
- Push pins to open vitamin capsules
- Stirring rod
- Ring stand
- Buret clamp
- Buret
- 100 mL beaker
- Funnel
- Mortar and pestle
- 250 mL Erlenmeyer flasks
- 10.0 mL or 20.0 mL pipet
- Pipet bulb or pipetter
- Vitamin samples (A, B_2, B_6, C, D, E, and K)
- Nutritional labels from four different food types (cereal, bread, juice, etc.)
- Iodine solution
- Vitamin C tablets (100 mg)
- 0.1 M acetic acid
- 1% starch solution
- At least three different juice samples (light colored, no pulp)

Discussion

Vitamins are organic compounds that are essential in small amounts for the proper functioning of the human body. Vitamins are needed in the microgram to milligram quantities per day. The recommended daily allowance (RDA) of vitamin B_{12} is 2.4 µg per day for an adult. Just 1.0 gram of vitamin B_{12} could theoretically supply the daily needs of approximately 500,000 people. Vitamins are not synthesized in humans and must be obtained from the diet. A well-balanced diet usually meets all the body's vitamin requirements.

There are thirteen known vitamins and they can be divided into two major classes: the water-soluble vitamins and the fat (lipid)-soluble vitamins. There are nine water-soluble vitamins and four fat-soluble vitamins. The water-soluble vitamins must be constantly replenished in the body because they are rapidly eliminated from the body in the urine.

In terms of function, an important difference exists between the water-soluble vitamins and the fat-soluble vitamins. Water-soluble vitamins are required as coenzymes for certain enzyme-catalyzed reactions. Because the coenzymes are not consumed in the reactions and can be reused as part of the catalyst enzyme, only small amounts of vitamins are required in the cells. Fat-soluble vitamins generally do not function as coenzymes in humans and animals, but they are important in processes such as vision, formation of bone, protection from oxidation, and proper blood clotting. This experiment will focus on vitamin C, which has a simple molecular structure compared to most other vitamins. The table below summarizes the vitamins, some food sources, their approximate RDA published in *Merck Manual of Diagnosis and Therapy*, and some symptoms associated with their deficiencies.

Table 1 ***Vitamin examples***

Vitamin	Food sources	RDA (adult male)	Examples of deficiency symptoms
Water-soluble vitamins			
B_1 (thiamin)	watermelon, whole grains, pork, legumes, sunflower seeds	1.2 mg	Beriberi (peripheral neuropathy, heart failure)
B_2 (riboflavin)	mushrooms, asparagus, broccoli, leafy greens, whole grains, milk, cheeses, liver, red meat, legumes, eggs	1.3 mg	dermatitis, cataracts
B_3 (niacin)	mushrooms, asparagus, potatoes, whole grains, tuna, chicken, beef, turkey, legumes, peanuts, sunflower seeds	16 mg	Pellagra (dermatitis, glossitis, gastrointestinal and central nervous system dysfunction)
B_5 (pantothenic acid)	mushrooms, broccoli, avocados, whole grains, meat, legumes, egg yolk	5 mg	muscle cramps, numbness, irritability, fatigue, diarrhea, vomiting, and water retention
B_6 (pyridoxine, pyridoxal, and pyridoxamine)	broccoli, spinach, potatoes, squash, bananas, watermelon, whole wheat, brown rice, chicken, fish, pork, soybeans, sunflower seeds	1.3 mg	Seizures, anemia, neuropathies, seborrheic dermatitis
B_7 (biotin)	fortified cereals, yogurt, liver, soybeans, egg yolk, nuts	30 mg	hair loss, fatigue, depression, nausea, muscle pains, anemia
B_9 (folate)	mushrooms, leafy green, broccoli, asparagus, corn, oranges, fortified grains, organ meats, legumes, sunflower seeds, nuts	400 μg	Megaloblastic anemia, neural tube birth defects, mental confusion
B_{12} (cobalamin)	milk products, beef, poultry, fish, shellfish, egg yolk	2.4 μg	Megaloblastic anemia, neurologic deficits (confusion, paresthesias, ataxia)
C	variety of fruits and vegetables	90 mg	Scurvy (hemorrhages, loose teeth, gingivitis, bone defects)
Fat-soluble vitamins			
A	egg yolks, dairy products	900 μg	night blindness, dry skin
D	liver, fatty fish, egg yolks	15 μg	rickets, weak bones
E	plant oils, leafy greens, whole grains	15 mg	anemia
K	leafy greens, spinach, peas, tomatoes	120 μg	bruising, slow blood clotting

Vitamin C

Vitamin C is a cyclic ester and it exists in two active forms in the body: an oxidized form and a reduced form. Its chemical name, L-ascorbic acid indicates that vitamin C is a weak acid. Although a carboxylic acid functional group is not present, one of the –OH groups exhibits acidic behavior.

The relationship between vitamin deficiencies and diseases has been known for many years. Scurvy is a disease that leads to open sores and loss of movement because of a deficiency of vitamin C. Vitamin C, which acts as a strong antioxidant and is a reducing agent, is an essential enzyme for the formation of collagen and a deficiency of it prevents the enzyme from functioning at its maximum activity level. Collagen abnormalities explain common symptoms associated with scurvy such as skin lesions and broken blood vessels.

James Lind, a ship's surgeon in the British Royal Navy, showed that citrus fruits prevented the disease. He published his work in 1753 but the British navy did not adopt lemons or limes as standard issue for the sailors until 1795. Limes were more popular because they could be found in the British West Indian Colonies, unlike lemons, which were more expensive. This lead to the American use of the nickname "limey" to refer to British sailors. Eventually, the naval connection was lost and limey was used to denote the British people in general.

Food products may have "vitamin C added" prominently displayed on their labels. The current recommended daily allowance (RDA) of vitamin C is 90 mg. That amount is provided by one stalk of broccoli, one medium orange, or eight medium strawberries. Higher dosages (250–10,000 mg) have been suggested to help prevent or ease the symptoms of the common cold. Further medical research is needed to support this claim. The human body cannot store large doses of water-soluble vitamins such as vitamin C and excess amounts are quickly excreted.

Vitamins that are insoluble in water are soluble in non-polar substances such as fat. Excess amounts of fat-soluble vitamins are stored in the body and high levels of accumulation may be toxic.

Analysis of vitamin C

The redox (oxidation-reduction) reaction in this experiment uses an iodine solution to titrate the vitamin C sample. The I_2 molecule is reduced to form the I^- ion and the vitamin C is oxidized. The following equation shows the changes in the vitamin C molecule's structure. Notice that the hydroxyl functional groups oxidize to form carbonyl groups.

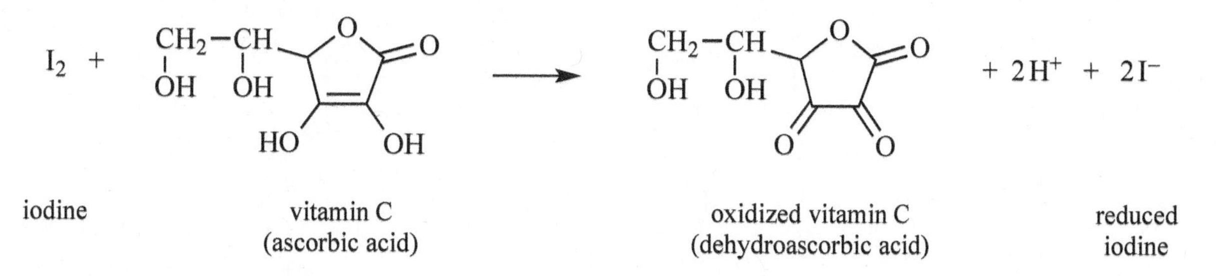

Starch is used as the indicator for the reaction. The reddish-orange iodine solution turns a blue-black color in the presence of the starch polysaccharide.

$$\text{starch polysaccharide} + I_2 \longrightarrow \text{"complex molecule with iodine"}$$

red-orange dark blue-black color

But since the vitamin C reacts with iodine to form a colorless solution, there is no available I_2 to form the dark blue-black color of the starch-iodine complex until all of the vitamin C has been oxidized. When the entire vitamin C sample has been oxidized, any excess iodine solution cannot be reduced and the starch indicator will turn blue-black with the addition of one more drop of iodine. This is known as the *endpoint*. The color change indicates the volume of iodine needed to oxidize the vitamin C sample.

This method is analogous to the titration of vinegar with NaOH, using a phenolphthalein indicator. When the last of the vinegar (acetic acid) had been neutralized, the next drop of NaOH caused the phenolphthalein indicator to change color, signaling that the titration should be stopped.

Vitamin C also is oxidized easily in the presence of oxygen in the air. Vitamin C tablets should be kept dry and in tightly closed containers. Juices and other aqueous solutions should be kept in closed containers. An increase in temperature will increase the rate of oxidation so tablets and solutions should be kept cool.

An initial titration using the iodine solution and a known mass of vitamin C will be used to determine the ratio of mg of vitamin C to mL of iodine solution. That conversion will then be used to determine the amount of vitamin C in juice samples.

Experimental Procedures

Eye protection and appropriate clothing must be worn at all times.

Gloves should be worn since iodine reagent will stain skin, clothing, and lab reports. Wash any contacted skin immediately with running water and alert the instructor.

Discard all wastes properly as directed by the instructor.

A. Solubility of vitamins

1. Label six test tubes place and them in a test tube rack.
2. Place a small amount of a vitamin in the first test tube. If the vitamin is a powder, use a small amount approximately the size of a pinhead. If the vitamin is a capsule, poke a hole in the capsule with a pin and squeeze a match head size amount of the liquid into a test tube. Share any remaining vitamin liquid with other students. Discard the empty capsule into the trash.
3. Add 2 mL of water to the test tube. Use a clean, dry stirring rod to mix the vitamin and water. Carefully watch for the formation of any layers of liquids or dissolution of solids. If a vitamin sample is too large, some of the vitamin may dissolve while the excess remains visible. Record your observations.
4. Repeat the procedure for each vitamin provided.
5. For any vitamins that were **not** water soluble, carry out the next step.
6. Obtain clean, dry test tubes and new samples of the **water insoluble** vitamins. Repeat the procedure in step 3 using methylene chloride (a nonpolar solvent) instead of water. Record your observations.
7. Read the nutritional labels on the food products provided. Record the name of the food and the vitamins listed on the label.

B. Standardization of vitamin C

1. Set up a ring stand, buret clamp, and buret as directed by the lab instructor and shown in Figure 1.
2. Place a small beaker under the buret tip. Rinse the buret with a few milliliters of iodine solution. Run the solution through the tip into the beaker. Discard this rinse solution.
3. Close the stopcock. Place a funnel in the top of the buret. **Slowly** fill the buret with iodine solution. Drain a small amount out to fill the buret tip. Record the initial reading of the iodine solution in the buret. All volume readings should be recorded to the correct significant figure.

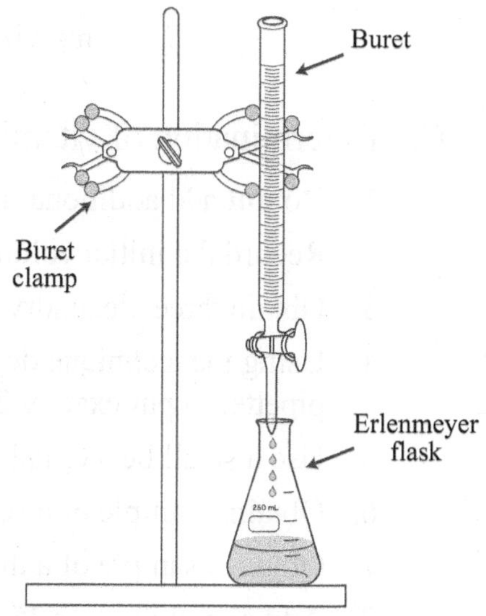

Figure 1 ***Titration apparatus***

4. Obtain a vitamin C tablet. Record the active amount of vitamin C as stated on the label. Do not "weigh" the tablet. Other inactive ingredients are also in the tablet.
5. Use a clean, dry mortar and pestle to crush the tablet. Carefully transfer the entire crushed tablet to a 250 mL Erlenmeyer flask.
6. Add 50 mL of deionized water, 2 mL of 0.1 M acetic acid and 10 drops of 1% starch solution to the flask. Swirl to mix.
7. Place the flask under the buret. The tip of the buret should be just inside the flask.
8. Open the stopcock and add iodine solution to the flask while carefully swirling the flask to mix. The blue-black color should disappear quickly. Placing a piece of white paper under the flask will make it easier to observe the color change.
9. When the blue-black color starts to take longer to disappear, slow down the flow of iodine and swirl the flask after the addition of each drop. The *endpoint* is reached when a medium blue color remains for at least 20 seconds after swirling the flask. Record the final reading of the iodine solution in the buret. The *endpoint* is extremely sensitive. One drop of iodine solution can make the difference between the desired medium blue color and a deep blue-black color.
10. Calculate the volume of iodine solution used in the titration by subtracting the initial volume reading from the final volume reading.
11. Calculate the ratio of the milligrams of vitamin C to the milliliters of iodine solution. This is the conversion factor that will be used for the titration of the juice samples.

$$\frac{\text{mass (mg) of vitamin C}}{\text{volume (mL) of iodine solution used}} = \frac{\text{? mg vitamin C}}{\text{1 mL iodine}}$$

This conversion factor may also be written:

_____?_____mg vitamin C = 1 mL iodine

C. Determination of vitamin C content of liquid foods

1. Do **not** add additional iodine solution to the buret.
2. Record the initial volume of the iodine solution (remaining from Part B) in the buret.
3. Obtain three clean, dry 250 mL Erlenmeyer flasks.
4. Using the technique demonstrated by the lab instructor, use a pipet and pipet bulb or pipetter to put exactly 20.0 mL of a juice sample into the first flask.
5. Use a small beaker half-filled with deionized water to rinse out the pipet.
6. Obtain a sample of a second type of juice for the second flask. Rinse out the pipet.
7. Obtain a sample of a third type of juice for the third flask.
8. Add approximately 25 mL of deionized water to each flask.
9. Add 1 mL of 0.1 M acetic acid to each flask.

10. Add 5 drops of starch indicator to each flask. Swirl each flask.
11. Notice that the juice samples may have a color while the vitamin C tablet sample was colorless. The presence of the juice color may make the endpoint color transition more difficult to see. For example, if the juice has a yellow color, the combination of yellow and the desired blue endpoint will appear green.
12. Titrate the first juice sample using the procedures in step B.7 through B.10. Be sure to record the final volume of the iodine solution.
13. Repeat for the remaining two juice samples. Be sure to have enough iodine solution in the buret and to **record the initial and final volume readings** for each sample.
14. Calculate the milligrams of vitamin C in each juice sample. Use the conversion factor obtained in step B.11.

D. Molecular model of vitamin C reaction

1. Make molecular models of vitamin C and an iodine molecule. Use "flexible bonds" to form and attach the hydroxyl groups to the ring. Show the models to the instructor and then demonstrate the reaction that occurs when the vitamin C is oxidized and the iodine molecule is reduced. Have the instructor approve the models.
2. Use condensed structural formulas to write the balanced equation for the reaction of vitamin C and iodine.

Pre-Lab Questions: Vitamins

Name ______________________ Date ______________________

Partner(s) ______________________ Section ______________________

______________________ Instructor ______________________

1. Draw the structure of vitamin C. This compound is also called ascorbic acid even though it is shown as a cyclic ester. Draw a circle around the ester functional group. This compound contains two hydroxyl groups that oxidize. Draw squares around those hydroxyl groups.

2. Why is it possible to overdose with fat-soluble vitamins?

3. What types of foods do you eat to obtain vitamin C?

4. Would purple grape juice be an appropriate juice to use in this experiment? Why or why not?

Name ______________________ Date ______________________

Partner(s) ______________________ Section ______________________

______________________ Instructor ______________________

A. Solubility of vitamins

Vitamin	Solubility in water	Solubility in methylene chloride
A		
B_2		
B_6		
C		
D		
E		
K		

Record the name of the food product and the vitamins listed on the label. Write the vitamins under the correct heading.

Food product	Water-soluble vitamins	Fat-soluble vitamins

B. Standardization of vitamin C

Mass of active vitamin C (from label) ____________

Final iodine buret reading ____________

Initial iodine buret reading ____________

Volume of iodine used ____________

Calculated conversion factor ____________ $\frac{\text{mg vitamin C}}{\text{1 mL iodine}}$

(show calculations)

C. Vitamin C in liquid foods

	Sample #1	Sample #2	Sample #3
Identity of juice	________	________	________
Volume of juice used	________	________	________
Final iodine buret reading	________	________	________
Initial iodine buret reading	________	________	________
Volume of iodine used	________	________	________
Amount of vitamin C in juice sample (show calculations)	________	________	________

Sample #1:

Sample #2:

Sample #3:

D. Molecular models

Instructor's approval of models and reaction _______________

Use condensed structural formulas to draw the balanced equation for the redox reaction of vitamin C and iodine.

Problems and questions

1. Which disease is associated with a diet low in vitamin C?

2. Why is there a daily requirement for vitamins, especially water-soluble vitamins?

3. Why should vitamin C tablets be stored in dry, closed containers?

4. The RDA for vitamin C is 90 mg. How many milliliters of each of the juice samples titrated in this experiment would be needed to meet the minimum daily requirement? Show your calculations.

   Juice sample #1:

   Juice sample #2:

   Juice sample #3:

Goals

- Use nutritional labels to identify fats, carbohydrates and proteins contained in food products.
- Use hydrolysis reactions to simulate the digestion of foods.
- Use chemical tests to identify the products of the digestion of foods.

Materials

- Gloves
- Hot plate
- Droppers
- 400 mL beaker
- 250 mL beaker
- 100 mL beakers
- 250 mL Erlenmeyer flask
- Rubber stopper for flask
- Rubber stoppers for test tubes or parafilm
- Stirring rod
- Watch glasses
- Test tubes (two sizes: ~15 × 125 mm and ~12 × 75 mm)
- Test tube rack
- Spot plate
- Phenolphthalein
- Nutritional labels from three different foods containing proteins, carbohydrates, and fats
- 1% starch solution
- Amylase solution (fresh)
- Iodine solution
- Benedict's reagent
- Fibrin (raw, **not frozen**, chicken—match-head size samples)
- Cooked egg white pieces (~1 cm^2 paper-thin slices)
- HCl solution (pH = 2.0)
- Pancreatin powder
- Pancreatin solution
- Cooking oil (vegetable oil)
- 0.5% NaOH solution
- Bile salts (e.g., choleate)

Discussion

Digestion is the process that converts large food molecules into smaller molecules that can be absorbed and used by cells. The digestive reactions are enzyme-catalyzed hydrolysis reactions. Since the enzyme proteins are temperature and pH sensitive, it is necessary to replicate their optimum conditions when simulating the digestive process in the laboratory. In order to follow the progress of digestion reactions, chemical tests will be used to identify the products.

Carbohydrates

Monosaccharides, also known as simple sugars, are absorbed into the bloodstream through the small intestine. But larger carbohydrate molecules such as starch (amylose) must be broken down to form monosaccharides. The hydrolysis of amylose begins in the mouth in the presence of the enzyme *salivary amylase.* Since the starch is only in the mouth for a short time, the hydrolysis is not complete. The products of the partial hydrolysis include *dextrins*, which are smaller polysaccharide segments, as well as maltose (disaccharide) and glucose (monosaccharide) molecules. The hydrolysis continues in the small intestine in the presence of the enzyme *pancreatic amylase*. Hydrolysis is completed in the intestine by the enzyme *maltase*. The other form of starch (amylopectin) undergoes a similar digestive process using different enzymes. The cellulose polysaccharide is not digested because humans lack the enzyme *cellulase*, which is capable of breaking the β-1,4 glycosidic linkages. Cellulose is a *dietary fiber* and passes through the digestive system unchanged.

The digestion of starch (amylose) is summarized in the following table.

Reactant(s)	Site	Enzyme	Product(s)
starch (amylose)	mouth	salivary amylase	dextrins, maltose, glucose
dextrins, maltose, glucose	small intestine	pancreatic amylase	maltose
maltose	small intestine	maltase	glucose

The hydrolysis of disaccharides is also enzyme-catalyzed. The disaccharides and their corresponding enzymes are shown in the following table.

Reactant(s)	Site	Enzyme	Product(s)
lactose (milk products)	small intestine	lactase	galactose, glucose
maltose (grains)	small intestine	maltase	glucose
sucrose (table sugar)	small intestine	sucrase	glucose, fructose

Two tests will be used to identify the products of carbohydrate digestion:

<u>Iodine solution</u>

starch polysaccharide + I_2 ⟶ iodine-starch complex

colorless — red-orange — dark blue-black color

<u>Benedict's test</u>

glucose + $2\,Cu^{2+}$ ⟶ oxidized sugar + $Cu_2O(s)$

colorless — blue acidic solution — colored precipitate (red, brown, green, or yellow colors are typically seen)

Proteins

The digestion of proteins begins in the acidic (pH ~ 2) environment of the stomach. The HCl present in the stomach denatures the proteins' tertiary structures. The unfolded proteins expose several peptide bonds that are hydrolyzed in the presence of the enzyme *pepsin*.

The digestion of proteins continues in the small intestine where the smaller, more soluble peptide fragments are further broken down in the alkaline (pH ~ 8) environment. The presence of the enzymes *trypsin, chymotrypsin, carboxypeptidase,* and *pancreatic peptidase* catalyze the hydrolysis of these smaller peptide segments. The products are water-soluble amino acids that are absorbed through the intestinal wall into the bloodstream.

Cooking foods that contain protein also aids digestion because the heat denatures the protein structures and the peptidases can more effectively catalyze the hydrolysis reactions. Some foods such as soybeans contain proteins that are competitive inhibitors of digestive peptidases and cooking destroys these competitive inhibitors. The digestion of proteins is summarized below.

Reactant(s)	Site	Enzyme	Product(s)
proteins (animal, vegetable)	cooking	(heat)	denatured protein structure
proteins	stomach (acidic environment)	pepsin	smaller peptides
smaller peptides	small intestine (basic environment)	trypsin, chymotrypsin, carboxypeptidase, pancreatic peptidase	amino acids

Lipids

Lipids include a variety of water-insoluble compounds. The digestion of lipids occurs entirely in the small intestine. Cholesterol is a lipid that requires no digestion and is absorbed directly into the lymphatic system when it reaches the small intestine. Large fat globules formed from solid fats and vegetable oils (triglycerides) are emulsified by bile salts, resulting in the formation of smaller droplets. The emulsified triglycerides then undergo hydrolysis catalyzed by the enzymes known as *lipases*. The products of the hydrolysis of the triglycerides are glycerol and fatty acids. However, not all of the triglycerides are completely hydrolyzed and may only be reduced to diglycerides or monoglycerides. The process is shown schematically below.

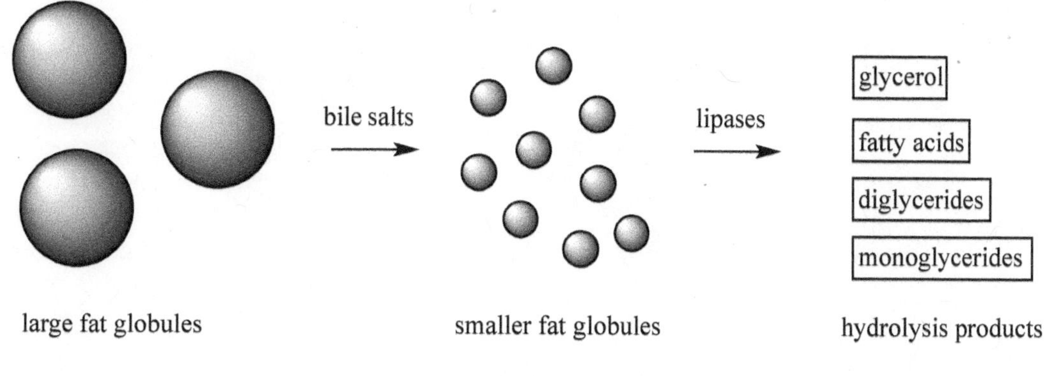

The digestion of lipids is summarized in the following table.

Reactant(s)	Site	Enzyme	Product(s)
cholesterol	small intestine	none	none
triglycerides (large globules)	small intestine	non-enzyme bile salts	smaller globules
emulsified smaller globules	small intestine	lipases	glycerol, fatty acids, diglycerides, monoglycerides

<u>Emulsification</u>

Emulsification is a process that causes two immiscible liquids to mix together even though their polarities are different. In the case of oil (nonpolar) and water (polar), it would be expected that no mixing would occur. However, if a mixture of water and oil is shaken vigorously, the oil droplets break up into smaller droplets and a colloid is formed. A *colloid* is a homogeneous mixture that contains dispersed particles that are intermediate in size between those of a true solution and those of a heterogeneous mixture. The water-oil colloid is unstable and eventually the two substances will separate into two layers.

An *emulsifier* is a substance that can disperse and stabilize the water-insoluble substances as colloidal particles in an aqueous solution. For example, soaps and detergents emulsify the surrounding grease particles in micelles.

In the body, fats tend to form large globules in the digestive system because they are insoluble in water. The number of enzyme molecules that can surround the fat globules and begin digestion is limited. Although bile salts do not catalyze any digestion reaction, they break up the large globules into smaller ones, thereby increasing the surface area dramatically. Thus, more enzymes can surround the fat globules and digestion proceeds quicker and more thoroughly.

Experimental Procedures

Eye protection and appropriate clothing must be worn at all times. Students must wear gloves.

Wash any contacted skin immediately with running water and alert the instructor.

Discard all wastes properly as directed by the instructor. Take care not to put any solids down the drain.

A. Food nutritional labels

1. Record the dietary information found on the food nutritional labels provided.
2. Fill out the table on the report sheet.

B. Carbohydrates

1. Set up a boiling water bath using a hot plate and a 600 mL beaker. Fill the beaker with approximately 200 mL of tap water.
2. Label two test tubes and pour 4 mL of 1% starch solution into each one.
3. In the first test tube add 10 drops of water. Stir with a clean stirring rod.
4. In the second test tube add 10 drops of amylase solution. Stir with a clean stirring rod.
5. Set aside in a test tube rack. **Record the time.**
6. Obtain a spot plate.
7. **After five minutes**, use a clean dropper to transfer 4 drops of the solution in the first test tube to the spot plate.
8. Use another clean dropper and transfer 4 drops of the solution in the second test tube to a different well in the spot plate.
9. Add 1 drop of iodine solution to each sample on the spot plate. Record your observations.
10. **After five more minutes (total of 10 minutes elapsed time):**
    - Use a clean dropper to transfer 4 drops of the solution in the first test tube to a clean well in the spot plate.
    - Use a clean dropper to transfer 4 drops of the solution in the second test tube to a clean well in the spot plate.
11. Add 1 drop of iodine solution to each new sample on the spot plate. Record your observations.
12. To identify the hydrolysis products of the carbohydrate digestion, add 3 mL of Benedict's reagent to each of the solutions remaining in the two test tubes. Stir.
13. Place the test tubes in a boiling water bath for 5 minutes. Record your observations.

C. Proteins

1. Set up a 37°C warm water bath using a hot plate and a 600 mL beaker. Fill the beaker with approximately 200 mL of tap water. Use this water bath for Part C and Part D to simulate the temperature of the human body.
2. Label two test tubes. Prepare each test using a match-head sized sample of the protein listed as follows:
   - Test tube #1.......................... fibrin
   - Test tube #2.......................... egg white (sliced paper thin)
3. Add 2 mL of HCl solution (pH = 2.0) to each test tube.
4. Warm the test tubes in the warm water bath for 5 minutes. Record the initial appearance of each sample.
5. Add 5 mL of pancreatin solution to each test tube. Use a clean stirring rod to stir the contents of each test tube. Place the test tubes back into the warm water bath for 10 minutes. Look for signs of increased solubility of the protein sample. Changes may be difficult to see. Look closely for cloudiness in the solution or fuzziness on the egg white sample. Record your observations.

D. Lipids

A schematic of the digestion procedure is shown below:

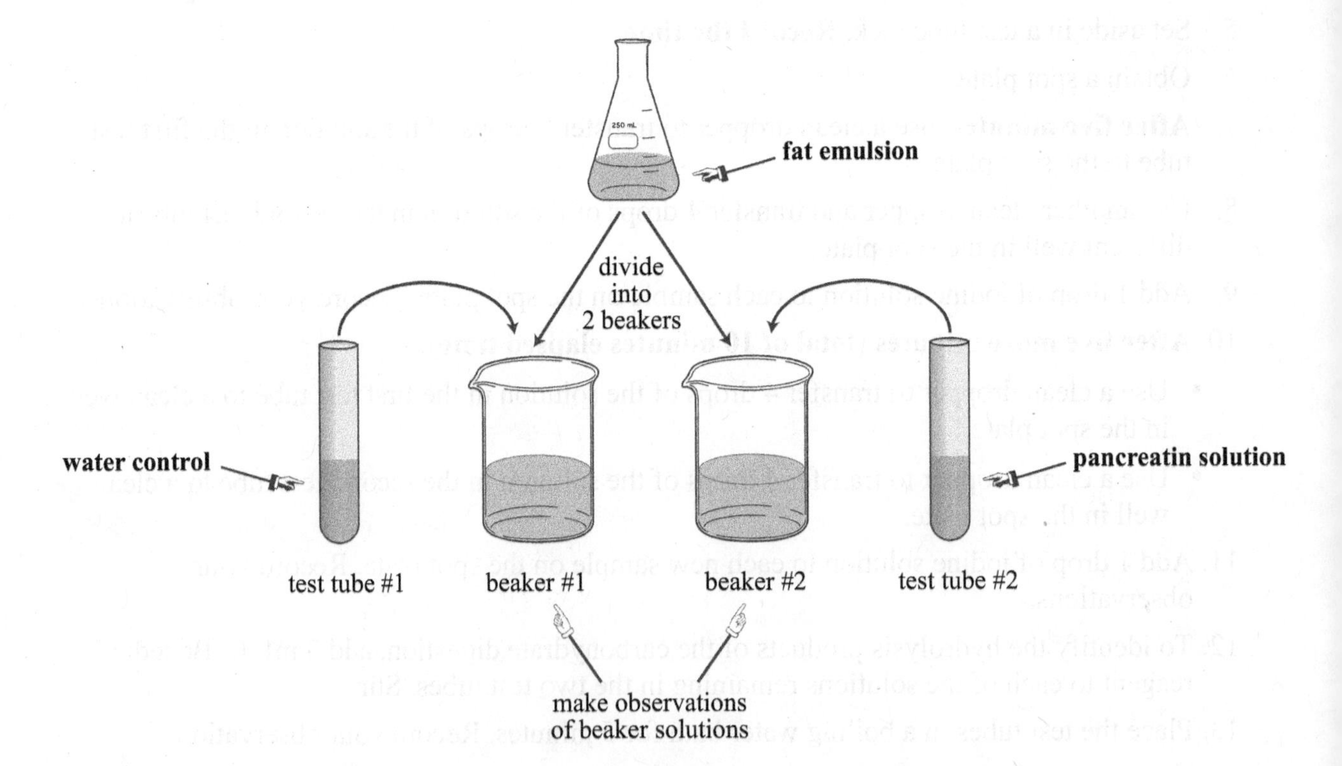

Prepare the fat emulsion

1. Obtain a 250 mL Erlenmeyer flask. Pour approximately 100 mL of tap water and approximately 1 mL of cooking oil in the flask.
2. Close the flask with a rubber stopper and shake it vigorously until a milky liquid forms. Record the appearance of the flask's contents. Although using emulsifiers like bile salts results in a better emulsion, they are not necessary for this experiment.
3. While swirling the flask, add 3 drops of 0.5% NaOH and 3 drops of phenolphthalein solution to this emulsion. The solution should turn pink. Record the appearance of the fat emulsion in the Erlenmeyer flask.

Prepare the pancreatin solution

4. Obtain two of the larger (15 × 125 mm) test tubes and label them #1 and #2.
5. Pour approximately 5 mL of water in each test tube.
6. Test tube #1 will be the control solution. Do not add anything else into this test tube.
7. Add approximately 50 mg (a small spatula tip-full, about one-half of a match head) of pancreatin into test tube #2.
8. Use a rubber stopper or parafilm to seal test tube #2. Shake vigorously until a suspension forms.

Digestion procedure

9. Obtain two 100 mL beakers. Divide the pink fat emulsion solution from step D.3 between these two beakers.
10. Pour the contents of test tube #1 (water control) into the first beaker.
11. Pour the contents of test tube #2 (water and pancreatin) into the second beaker.
12. Stir each beaker with a clean stirring rod and compare the colors of the two beakers. Remember that phenolphthalein is pink in a basic solution and colorless in an acidic solution. Record and explain your observations.

Comparison of emulsification techniques

13. Obtain two of the smaller (12 × 75 mm) test tubes. Fill each test tube half full with water and add 3 drops of cooking oil to each one.
14. Add approximately 50 mg (a small spatula tip-full, about one-half of a match head) of bile salts to the test tube #2.
15. Use rubber stoppers or parafilm to seal the test tubes. Shake both test tubes vigorously.
16. Obtain two watch glasses and pour the contents of each test tube onto a watch glass.
17. Compare the size of the globules in each watch glass. Record and explain your observations.

Name ______________________ Date ______________________

Partner(s) ______________________ Section ______________________

______________________ Instructor ______________________

1. What is the name of the substrate and the type of bond hydrolyzed by the following enzymes?

   a) lactase

   b) lipase

   c) peptidase

2. Name the products formed when the following compounds are completely hydrolyzed.

   a) maltose

   b) a triglyceride

   c) a protein

3. What is the optimum temperature for most enzymes?

4. What is the optimum pH for an enzyme found in the stomach?

5. What is the optimum pH for an enzyme found in the small intestine?

6. What is the purpose of emulsifying fats in the digestive system?

Name ______________________ Date ______________________

Partner(s) ______________________ Section ______________________

______________________ Instructor ______________________

A. Nutritional labels

Food	Serving size	Amount of carbohydrates per serving	Amount of proteins per serving	Amount of fats per serving

Are all food items that are fat-free (0 grams fat per serving) healthy food in terms of nutrition? Explain.

Besides macronutrients (carbohydrates, proteins, and fats), list three other nutritional information items that appear on the nutritional label.

B. Carbohydrates

Contents of test tube	Total ELAPSED time	Iodine tests: Observations on spot plate	What do the observations indicate?
#1 starch + water	5 minutes		
	10 minutes		
#2 starch + amylase	5 minutes		
	10 minutes		

Contents of test tube	Benedict's tests: Observations	What do the observations indicate?
#1 starch + water + Benedict's		
#2 starch + amylase + Benedict's		

What is the monosaccharide produced by the complete hydrolysis of starch?

Which test tube solution (#1 or #2) gave the best results for the hydrolysis of starch? Explain.

C. Proteins

Contents of test tube	Initial observations	Observations after 20 minutes
#1		
#2		

Which of the test tubes showed the greatest amount of protein digestion? What observations support your answer?

Fibrin is a protein that aids in blood clot formation. Egg whites contain about 40 different proteins. Did you notice any difference in the initial appearances of the samples? Did they hydrolyze in a similar manner? Give a possible explanation for your observations.

D. Lipids—Digestion

Record the appearance of the fat emulsion in the Erlenmeyer flask.

Contents of flask	Observations

Record the appearance of the beaker mixtures.

Contents of beaker	Observations
#1 fat emulsion, NaOH, phenolphthalein, water	
#2 fat emulsion, NaOH, phenolphthalein, pancreatin	

In which beaker did digestion of the oil occur? What observations support your answer?

Why did the color of the solution in beaker #2 change from pink to colorless? (Hint: think about the products of the digestion reaction.)

D. Lipids—Comparison of emulsification techniques

Contents of test tube	Observations
#1	
#2	

Which test tube showed the smaller fat globules? Explain.

How do bile salts aid in digestion?

Problems and questions

1. Where does carbohydrate digestion begin in the body? What is the name of the enzyme involved in this initial digestive process? Based on the name of the enzyme, on what form of starch is it most effective?

2. Only a small fraction of starch in food is digested before it enters the small intestine. Why?

3. Where does protein digestion begin in the body? What substance was used in this experiment for the digestion of proteins? Based on the name of the enzyme, what organ in the body produces it?

4. What is the function of HCl in the stomach?

5. Based on the observations of the emulsification comparison tests, would you expect better digestion of fats with the globules in the test tube that was shaken, or the test tube that was shaken and also had the bile salts added? Explain how you arrived at your answer.

Goals

- Build molecular models of salicylic acid and acetic anhydride and use them to demonstrate the formation of acetylsalicylic acid.
- Synthesize and purify acetylsalicylic acid (aspirin).
- Use chemical tests to identify impurities in the prepared aspirin.
- Use chemical tests to identify impurities and starch in commercial aspirin.
- Build molecular models of salicylic acid and methanol and use them to demonstrate the formation of methyl salicylate.
- Synthesize methyl salicylate (oil of wintergreen).

Materials

- Gloves
- Molecular model kits
- Test tube rack
- Test tubes
- Stirring rod
- pH paper and color charts
- Hot plate
- Thermometer
- 400 mL beaker
- 125 mL Erlenmeyer flask
- Buchner funnel
- Filter paper
- Filter flask
- Vacuum trap
- Ice bath
- Metal spatula
- 20 cm test tube
- Evaporating dish
- Salicylic acid
- Acetic anhydride
- 85% H_3PO_4
- Ice-cold deionized water
- Powdered commercial aspirin
- 1% $FeCl_3$
- Iodine solution
- Methanol
- Concentrated H_2SO_4

Discussion

Esters are formed by the esterification reaction between a carboxylic acid functional group and an alcohol functional group. Since salicylic acid is a molecule that contains both of these functional groups, it can react in two different ways to give different products. The reactions and their products are dependent upon which of the functional groups is involved in the esterification reaction.

The following equation shows the reaction of salicylic acid with methanol to form the ester methyl salicylate, commonly known as oil of wintergreen. As with most esterification reactions, a molecule of water is also formed as a product. Methyl salicylate is used as a topical analgesic in creams for muscle aches, a mild antiseptic in throat lozenges and mouthwashes, and a distinctive food flavoring.

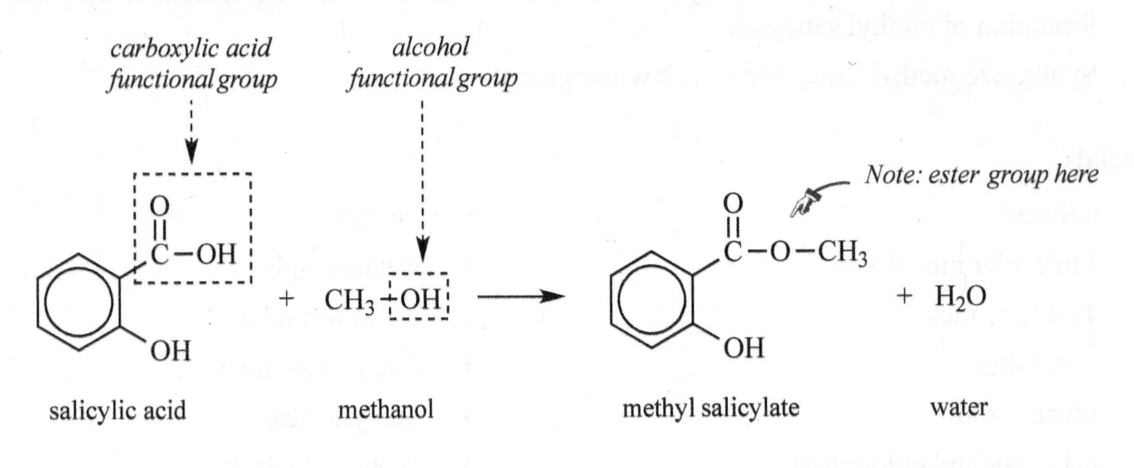

The following equation shows the phenol group on the salicylic acid reacting with acetic acid to form acetylsalicylic acid (ASA), commonly known as aspirin. Notice the water molecule formed as a product.

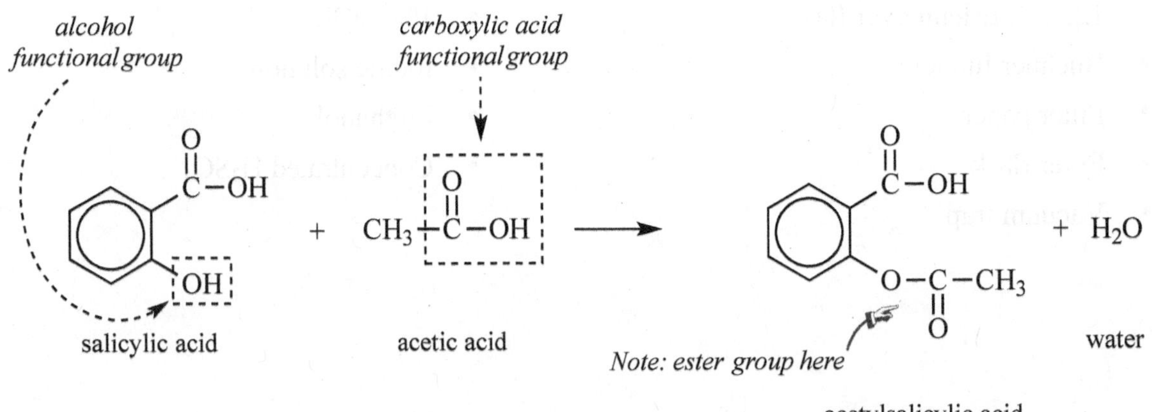

Because the reaction using acetic acid is slow, acetic anhydride will be used instead of acetic acid for this experiment. The following equation shows the formation of the aspirin molecule, but the second product is acetic acid instead of water.

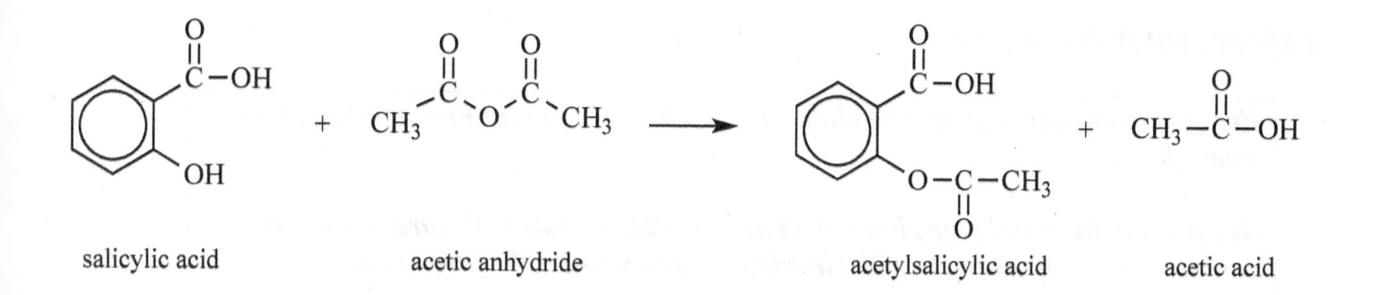

The aspirin requires purification because not all of the original salicylic acid is converted to aspirin. Also, in addition to aspirin, acetic acid is a product. The $FeCl_3$ test is specific for the presence of a phenol group and may be used to test for impurities. The formation of a violet/purple color easily identifies the presence of unreacted salicylic acid. Commercially prepared aspirin may also give a positive phenol test since the tablets tend to break down over time to produce salicylic acid and acetic acid.

A brief history of aspirin

Aspirin is an analgesic (a substance that provides relief from inflammation) and an anti-pyretic (providing relief for fevers). The medicinal properties of salicylic acid have been known for thousands of years. The word "aspirin" comes from *Spiraea*, a biological genus of shrubs that contains salicylic acid. Salicylic acid (from the Latin *salix*, willow tree) can also be found in jasmine, beans, peas, clover, and certain grasses and trees.

The ancient Egyptians used willow bark as a remedy for aches and pains, but they did not have the chemical knowledge to understand why it reduced fevers and inflammation. Hippocrates (c. 460 – c. 370 BC), the father of western medicine (think of the Hippocratic Oath), advised patients to chew on willow leaves and bark to reduce pain and fevers.

In the 1800s, researchers across Europe studied salicylic acid. French pharmacist Henri Leroux isolated it in 1829. Hermann Kolbe (the most eminent organic chemist of his time) and Rudolf Schmitt synthesized salicylic acid in the laboratory in 1874, but when administered in large doses, the acidity irritated the mucous membranes in the mouth and the digestive tract of patients. Patients also experienced nausea and vomiting, and some even went into a coma.

In the late 1890s, Felix Hoffmann, a chemist working at Bayer AG in Germany, used acetylsalicylic acid to alleviate his father's rheumatism. The ester produced fewer side effects than the acid. Beginning in 1899, Bayer distributed a powder with this ingredient to physicians to give to patients. The drug became a hit and, in 1915, it was sold as over-the-counter tablets.

The Nobel Prize in Physiology or Medicine in 1982 was awarded to Sune K. Bergström, Bengt I. Samuelsson, and John R. Vane who discovered why an aspirin a day reduces heart attack risk. It inhibits production of hormones called prostaglandins. Prostaglandins are responsible for the formation of clots that lead to heart attacks and strokes, and aspirin prevents that clotting.

Today's aspirin tablets contain more than just acetylsalicylic acid. Other ingredients such as caffeine and buffers are often added to aspirin to enhance its effectiveness. Since the total mass of all of the active ingredients in a dose is less than a half of a gram, starch is added as a binder and filler. The starch allows aspirin to be formed into small but manageable sized tablets. The starch is easily detected by performing the iodine test, which produces a blue-black color in the presence of starch.

Experimental Procedures

Eye protection and appropriate clothing must be worn at all times. Students must wear gloves.

Acetic anhydride and phosphoric acid are caustic. Handle with care. Wash any contacted skin immediately with running water and alert the instructor.

The acetic acid vapors given off in the preparation of aspirin are irritating. Stay a safe distance away from the top of the flask.

The aspirin prepared in the experiment is impure.
DO NOT TAKE INTERNALLY!

Discard all wastes properly as directed by the instructor.

A. Molecular models

1. Make models of salicylic acid, acetic anhydride, and methanol. Have the instructor approve the models.
2. Use the models to demonstrate the esterification reaction to form aspirin and acetic acid.
3. Reassemble the salicylic acid model and demonstrate the esterification reaction to form oil of wintergreen and water. Have the instructor approve the reactions.

B. Preparation of aspirin

1. Set up a boiling water bath using a hot plate and a 600 mL beaker. Fill the beaker with approximately 200 mL of tap water.
2. Obtain a 125 mL Erlenmeyer flask.
3. Weigh out 2.00 g salicylic acid in a weigh boat. Transfer the salicylic acid into the Erlenmeyer flask.
4. **Working under the hood,** add 5 mL of acetic anhydride to the flask.
5. Carefully add 10 drops of 85% H_3PO_4 to the flask. Stir with a stirring rod.
6. Place the flask into the boiling water bath. Continue to stir with the stirring rod until all of the solid dissolves.
7. Carefully remove the flask from the boiling water bath. Set aside until it cools to room temperature. Save the water bath for Part D. Let the water bath cool down to 70°C.
8. While waiting for the flask to cool, set up a Buchner funnel and suction flask to the vacuum as demonstrated by the instructor and shown in Figure 1. Place a piece of filter paper in the funnel. It should cover the flat surface of the funnel. Moisten the filter paper with distilled water.

9. **Perform this step under the hood.** After the flask has cooled, very slowly add 20 drops of deionized water to the mixture. Any unreacted acetic anhydride will react vigorously with the water and splatter.
10. After the reaction has subsided, add 50 mL of **ice-cold deionized water**.
11. Cool the mixture by placing the flask in an ice bath for 10 minutes.
12. Use a stirring rod to stir the mixture and gently scratch the inside bottom of the flask. Crystals of aspirin should form.
13. Turn on the vacuum for the Buchner funnel apparatus.
14. Pour the cooled aspirin mixture onto the filter paper in the Buchner funnel.
15. Pour 10 mL of **ice-cold deionized water** into the flask to help rinse out and transfer any remaining aspirin from the flask and onto the filter paper. Repeat two more times to ensure all of the aspirin crystals are transferred and rinsed.
16. Use a stirring rod to carefully spread the crystals onto the filter paper. Take care not to tear the paper.
17. Continue to run the vacuum for a few minutes to help the crystals dry. Turn off the vacuum.
18. Carefully use a spatula to lift the filter paper and crystals out of the funnel. Place the paper containing the crystals onto a dry paper towel and let the crystals dry.

Figure 1 ***Buchner funnel setup***

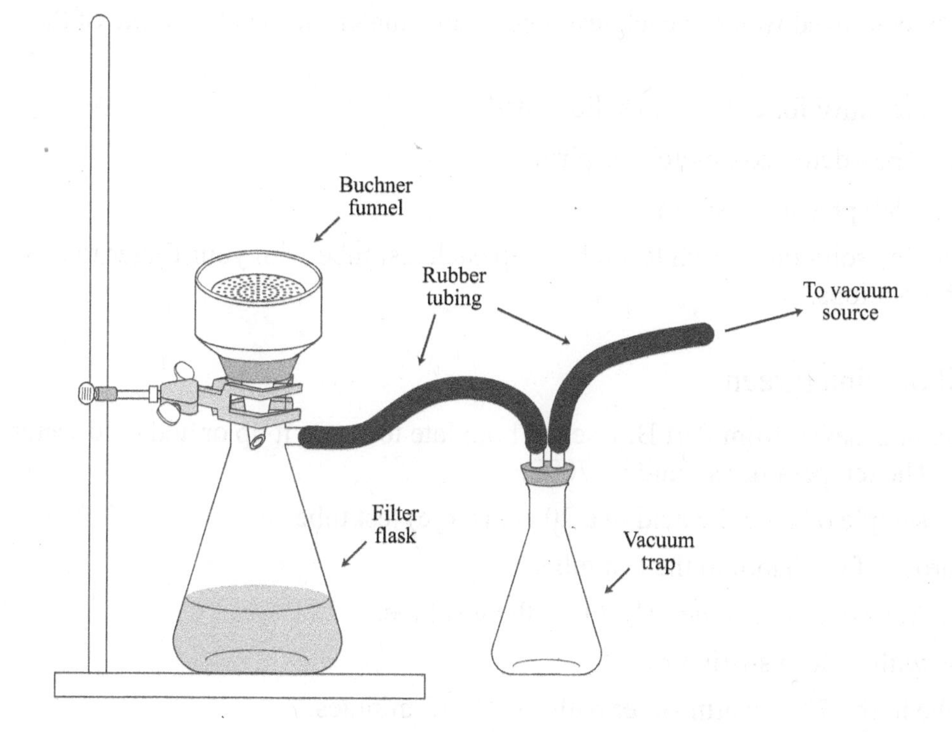

C. Chemical tests on aspirin

pH test

1. Label four test tubes.
2. Place 20 drops of deionized water in each test tube. Add a match-head size sample of the following:
   - Test tube #1.......control for color—no solid added
   - Test tube #2.......powdered commercial aspirin
   - Test tube #3.......lab prepared aspirin
   - Test tube #4.......salicylic acid

   Record the contents of each test tube in Table C of the report sheet.
3. Use a clean, dry stirring rod to test the pH of the solution in the first test tube. Dip the stirring rod into the test tube and then touch the tip to a strip of pH paper. Compare the color of the paper with the color chart provided. Record the pH. Repeat for each test tube. Save the test tubes for the phenol test.

Phenol test

4. Add 1 drop of 1% $FeCl_3$ solution to each test tube saved from part C.3. Tap each test tube with your finger to mix. Record your observations.

Starch test

5. Label three test tubes.
6. Place 20 drops of deionized water in each test tube. Add a match-head size sample of the following:
   - Test tube #1.......control for color—no solid added
   - Test tube #2.......powdered commercial aspirin
   - Test tube #3.......lab prepared aspirin
7. Add 1 drop of iodine solution to each test tube. Tap each test tube with your finger to mix. Record your observations.

D. Preparation of oil of wintergreen

1. Adjust the water bath saved from Part B. Use the hot plate to warm it up or add cold water to cool it down. The temperature should be 70°C.
2. Place a pea-size sample of salicylic acid in a 20 cm (large) test tube.
3. Add about 20 drops of methanol to the test tube.
4. Carefully add 3 drops of concentrated H_2SO_4 to the test tube.
5. Stir the solution with a clean stirring rod.
6. Heat the test tube in the 70°C warm water bath for 10–15 minutes.
7. Pour a small amount of the contents of the test tube into an evaporating dish. **Carefully** waft the fumes and record your observations.

Name ______________________ Date ______________________

Partner(s) ______________________ Section ______________________

______________________ Instructor ______________________

1. Draw the structure of salicylic acid. Circle the carboxylic acid group. Draw a square around the phenol group.

2. What functional group does a positive $FeCl_3$ test indicate? What is the appearance of a positive test?

3. What does a positive iodine test indicate? What is the appearance of a positive test?

4. Why is ice-cold water needed to rinse the aspirin product? (Hint: what would happen if warm water was used?)

5. Why should you **not** ingest the aspirin you prepare in the lab?

REPORT SHEET: ASPIRIN

Name ______________________ Date ______________________

Partner(s) ______________________ Section ______________________

______________________ Instructor ______________________

A. Molecular models

Use condensed structural formulas to write the balanced equation for the reaction of salicylic acid with acetic anhydride. Use the models to demonstrate the reaction to the instructor.

Instructor's approval of models and reaction: ______________________

Use condensed structural formulas to write the balanced equation for the reaction of salicylic acid with methanol. Use the models to demonstrate the reaction to the instructor.

Instructor's approval of models and reaction: ______________________

B. Preparation of aspirin

Describe the appearance of your aspirin product.

C. Chemical tests on aspirin

Contents of test tube	pH	Phenol test observations	Starch test observations	What do these observations indicate?
#1				
#2				
#3				
#4			X	

Explain the differences you observed regarding the pH of each test tube.

Did any of the test tubes show an insoluble substance in water? What was the substance? Support your answer.

D. Preparation of oil of wintergreen

Describe the appearance and other properties of your oil of wintergreen product.

Problems and questions

1. Irritation of the stomach is caused by what functional group on the aspirin molecule?

2. If a bottle of aspirin tablets has the aroma of vinegar, it is time to discard those tablets. Explain why, and include the chemical equation for the hydrolysis of aspirin in the explanation.

3. Why is starch added to aspirin?

4. In step B.9 of this experiment, excess acetic anhydride is destroyed by the addition of water, which converts the anhydride into acetic acid. Write the balanced equation for this reaction.

5. $FeCl_3$ is used to test for the presence of unreacted salicylic acid in this experiment. A chemistry journal article from 1941 described using salicylic acid to test for the presence of Fe^{3+}. Instead of using salicylic acid, its sodium salt (sodium salicylate) was used because the salt is more soluble in water than the carboxylic acid. Sodium salicylate is prepared by reacting salicylic acid with sodium hydroxide. Write the neutralization reaction of salicylic acid with NaOH to produce sodium salicylate. Remember, the other product is water.